Engineering the Environment

ENGINEERING THE ENVIRONMENT

PHYTOTRONS *and the* QUEST *for* CLIMATE CONTROL *in the* COLD WAR

DAVID P. D. MUNNS

UNIVERSITY *of* PITTSBURGH PRESS

The following website was created to offer a richer overview of the various permutations of trons in modern history. For a chronological diagram of the history of trons, or to post references or materials to other tron projects, please visit:

www.worldoftrons.com

Published by the University of Pittsburgh Press, Pittsburgh, Pa., 15260

Manufactured in the United States of America
Printed on acid-free paper
10 9 8 7 6 5 4 3 2 1

ISBN 13: 978-0-8229-4474-4
ISBN 10: 0-8229-4474-X

Cataloging-in-Publication data is available from the Library of Congress

Jacket art: Climatron exterior at night, as reflected in tropical lilly pools. © Missouri Botanical Garden Archives, http://www.mobot.org
Jacket design by Alex Wolfe

"Ass!" said the Director. "Hasn't it occurred to you that an Epsilon embryo must have an Epsilon environment as well as an Epsilon heredity?"

— Aldous Huxley, *Brave New World*

Smith is not a man. He is an intelligent creature with the genes and ancestry of a man, but he is not a man. . . . He's been brought up by a race which has nothing in common with us. . . . He's a man by ancestry, a Martian by environment.

— Robert A. Heinlein, *Stranger in a Strange Land*

Observe her, comrades! This is a Bene Gesserit Reverend Mother, patient in a patient cause. She could wait with her sisters ninety generations for the proper combination of genes and environment to produce the one person their schemes required.

— Frank Herbert, *Dune*

CONTENTS

ACKNOWLEDGMENTS

On Regent Street in London is the latest form of ecotourism, The National Geographic Store. Deftly combining high-quality materials with local manufacturing and a global vision through human and natural photojournalism, the store is abuzz. I needed a coat for New York. The National Geographic Store has an extensive selection of all degrees of winter coats, both fashionable and able to ward off varying types of arctic winter. The real test of a National Geographic coat, however, comes through identifying precisely what extremes of temperature and wind you are going to encounter and matching them with your jacket. In order to properly make that assessment, the National Geographic Store has installed a climate-controlled room. Three sides are Perspex, which allows all the other consumers to consume the spectacle of the person being subjected to well-below-freezing temperatures and windchills, while the fourth wall supports the refrigeration unit, a wind tunnel, and the infrared sensors that measure the temperature differential all over your body's surface. Dressed in your coat, you can judge how it performs, how comfortable you are, and whether you need to ratchet up a notch in order to defend your body's core temperature against the elements out in the wide world. In a very real sense you are participating in your own controlled-environment experiment.

I had known about the strange controlled-environment laboratories for biology called *phytotrons* for a number of years, but my time in London invigorated my search for them in a variety of ways. Imperial College's academic community of scholars, graduate students, and especially earnest master's candidates once more fueled intellectual fires. Moreover, I finally had time to finish the manuscript on my history of the radio astronomy community, published as *A Single Sky: How an International Community Forged the Science of Radio Astronomy* (MIT Press, 2013). I also turned to publishing the long-delayed case of the Australian phytotron. The central issue for the new radio astronomers was in deciding whether their new science was really an "astronomy" or a "physics" or perhaps both, and what that said for the nature of science neatly divided into discrete disciplines. The creation of phytotrons, coincidentally at the same immediate postwar moment, saw botanists and plant physiologists confront the same problem. The consistent staggering claim was that phytotrons were the cyclotrons of biology. Even more intriguing was a claim from the well-known British cotton breeder S. C. Harland in the *New Scientist* in 1958: "The phytotron is to botany and agriculture what the radio telescope is to astronomy."[1] The radio telescope gave the astronomers a new vision that has uncovered an incredible universe that we can only listen to. Likewise, the phytotron offered a new vision of life and of biology as the study of life.

The story of phytotrons says that the study of biology became an exercise in technological control after the Second World War. This book describes how groups of technologist biologists understood that their new facilities called *phytotrons* effectively made the plant sciences analogous to the physical sciences through control over the physical environment and pursuit of basic science. In so doing they specified what the "environment" meant in the life sciences, a definition that by the end of the century had largely been erased by another new science of the twentieth century, namely, genetics and molecular biology. In part, the history of phytotrons is especially valuable not only because it is largely absent from the history of science but also because it complements the well-studied story of the discovery of the gene. While a biology of the molecular has successfully confronted the scourge of cancer and other diseases that terrify so many, a biology of the environment can contribute toward the threat of climate change that threatens everyone. My hope is that by bringing to light a forgotten part of modern biology, the now recent incarnation of phytotrons, called *Ecotrons*, can establish a biological science of climate change through the experimental study of the whole and not just the parts.

All that began in London where I had the great fortune to meet Hannah Gay, who had just completed her monumental history of Imperial College and who told me of their "Ecotron." I now had a beginning and an end—the first phytotron in Caltech and the Ecotron at Imperial College. In the middle went the various cases the chapter titles outline. I knew about most and needed to research and visit them all. A survey of the notes will show that the various personal papers of the phytotronists examined during that period have been crucial, as well as the institutional settings that have helped preserve the records even while memories fade. Deserving special mention though is the kind donation of Frits Went's papers to the Missouri Botanical Garden archives by his son, who invaluably saved the lifework of one of the most significant plant scientists of the twentieth century and the founder of phytotrons.

On moving to John Jay College of the City University of New York, I was generously given the opportunity to take a sabbatical term and plow through the research in Australia, California, Saint Louis, Madison, Paris, London, Philadelphia, and Cambridge. For that invaluable opportunity I thank my chair, Allison Kavey, and our provost, Jane Bowers. Among the visits were opportunities to view the continuing work of controlled-growth chambers: my thanks to Jim Klug for a wonderful tour of the growth chambers at Michigan State, and Peter Volk for sharing some grand memories. My appreciation too to William and Melissa Laing in New Zealand for their wonderful and thoughtful correspondence and to the previews of their documentary on the New Zealand Climate Laboratory. Over the years I have been variously and generously supported in my efforts to recover the people of the phytotron: historians, like armies, march on their stomachs.

My appreciation goes to the Maurice Biot Fund supporting archival research at the California Institute of Technology. An early grant from the Rockefeller Archives Center, North Tarrytown, New York, formed an important foundation for my research. I thank the Friends of the University of Wisconsin-Madison Library for their grant to visit the Biotron papers, especially Tom Garver for his friendly welcome. This work was also supported in part by a grant from the City University of New York PSC-CUNY Research Award Program, as well as a grant from the Office for the Advancement of Research at John Jay College.

Parts of this book have previously appeared in "The Phytotronist and the Phenotype: Plant Physiology, Big Science, and a Cold War Biology of the Whole Plant," *Studies in the History and Philosophy of Biological and Biomedical Sciences Part C* 50 (2015), 29–40; "'The Awe in Which Biologists Hold Physicists': Frits Went's First Phytotron at Caltech, and an Experimental Definition of the Biological Environment," *History and Philosophy of the Life Sciences* 36, no. 2 (2014), 209–31; and "Controlling the Environment: The Australian Phytotron and Postcolonial Science," *British Scholar* 2, no. 2 (2010), 197–226. I thank the publishers for permission to reproduce them. Likewise, I thank the many institutions that permitted me to reproduce the wonderful illustrations that help make this story.

Few projects can succeed without the detailed knowledge and diligence of the librarians and archivists on whom the historian is grateful to rely. To get inside multiple controlled environments, I would like to thank the Caltech Archives staff, Shelly Irwin, Mariella Sopano Pelligrino, and Loma Karklins for a wonderful time in Southern California; Andrew Colligan at the Missouri Botanical Garden archives and library; Rosanne Walker at the Adolph Basser library of the Australian Academy of Science; Thomas Harkins at the Duke University Archives; Lajos Bordas of the Dentistry Library at Sydney University; David Null at the University of Wisconsin-Madison archives; Stephen Simon at the LeEster T. Mertz library at the New York Botanical Garden; Karen Stewart at the Desert Research Institute; Isabelle Dujonc au Dépôt des archives du CNRS (Gif-sur-Yvette), and Etienne Wintenberger au Dépôt des archives du CNRS (Paris). Likewise, I have had the able assistance of two students over the years who have sped the process along with their research skills: my thanks go to Lucas Riley and Anjelica Camacho. Furthermore, I thank the legions of unnamed secretaries, typists, and file clerks of the Cold War era for the bountiful copies of immediately legible resources through which the past comes alive.

Then there is the long labor of turning a morass of paper, quotes, diagrams, recording, inscriptions, and other assorted stuff into a work that explains who some people thought they were when they lived. Only through the patient exhumation of others' understandings can we achieve the most significant work of the historian, knowing ourselves through knowing others; history is not written for the past (they're all dead, my old social history professor said) but for the present. For helping me realize that ambition, I owe a deep debt to Andrew Warwick, who took a young man and told him of the world. Likewise, my profound thanks to Allison Kavey for her support and guidance—borrowing that pencil all those years ago was the best move I ever made. Lord Robert Winston of Imperial College, London, was the source of many excellent conversations and an inspirational enthusiast of science and science studies. Likewise, Graham Hollister-Short's conversations about technology just kept on thrilling. As the project developed, Kärin Nickelsen heroically read the entire manuscript and her direct Germanic comments recast several chapters in new and richer ways. Angela Creager's valuable reading of an initial chapter has also meant that many subsequent pages benefited from her project-shaping comments. Stalwartly, Bruce Hunt, Luis Campos, Colin Milburn, Jim Endersby, Catherine Jackson, Karen Rader, Nicolas Rasmussen, Susan Lindee, Betty Smocovitis, Rachel Ankeny, Gail and Mark Schmitt (and the fidotrons), Matt Wisnioski, Frank Bongiorno, Abigail Woods, Serafina Cuomo, Greg Raddick, Christian Joas, Lucie Gerber, Peter Redfield, Caterina Schürch, Bruno Strasser, Helen Anne Curry, and Sharon Kingsland have all listened patiently to my various ravings about trons and gently prodded me back in better directions. Jim Collins gave a splendid commentary on an early paper, while Kim Kleinman lent me early aid with materials about the Climatron. I remain tremendously grateful to the extensive, insightful, and often painfully true comments of my anonymous reviewers. They performed a Herculean task of commenting and editing, and this book would only be a shadow without them. Likewise, to Abby Collier and Alex Wolfe at the University of Pittsburgh Press, who took on this unwieldy project and shaped it into something worthwhile.

There are also the silent partners in one's work without whom little would taste as sweet: Diane Kagoyire, J. J. Shirley, Walter Fralix, the "Yes" Appersons, the Wisnioskis, the Windeyers, the Borises, the Griffins, the Worrells, the DeLeons, the Hungarians, A. J. Benitez and Brad Oister, Kenneth Moore and Derek Bishop, Eric Kolb, Scott Knowles, Dara Byrne, Vivian Ewalefo, and Ezine Okpo, all helped in more ways than they know. Joseph DeLeon quite simply completes my world. Thanks to my parents, Peter G. and Susan Munns, and to Lillian, Max, Trudi, and David MacKay for their patience with their son/brother/uncle's continuing wanderings.

Finally, to Paris, where life becomes art.

ABBREVIATIONS

The Cold War era is almost known by its myriad acronyms. Wherever possible, I have kept their usage to a minimum, but an inevitable list is necessary.

AA	Australian Archives
AAS	Australian Academy of Science
AEC	Atomic Energy Commission (United States)
ASPP	American Society of Plant Physiologists
BSA	Botanical Society of America
CIEP	Committee on International Exchange of Persons (United States)
CIT	California Institute of Technology (Archives)
CNRS	Centre National de la Recherche Scientifique (France)
CSIRO	Commonwealth Scientific and Industrial Research Organisation (Australia)
CSR	Commonwealth (Colonial) Sugar Refining
DSIR	Department of Scientific and Industrial Research (Britain)
MBG	Missouri Botanical Garden
MSU	Michigan State University
NAA	National Archives of Australia
NAS	National Academy of Sciences (United States)
NASA	National Aeronautics and Space Administration
NCSU	North Carolina State University
NIH	National Institutes of Health (United States)
NLA	National Library of Australia
NSF	National Science Foundation (United States)
ONR	Office of Naval Research (United States)
RAC	Rockefeller Archives Center Tarrytown, New York
USDA	United States Department of Agriculture

PRELUDE

THE WORLD OF TRONS

Tron. What have you become?
—TRON: Legacy

THIS BOOK concerns the rise and importance of a *tron* in the life sciences, the evocatively named *phytotron*. Phytotrons were, and still are, computer-controlled environmental laboratories consisting of any number of rooms or smaller cabinets, all able to produce any set of climatic conditions. Because the growth and development of any organism depends on its genes and its environment, plant scientists required the ability to create reproducible climates in order to conduct experiments that tested plants' (and some animals') responses to various environmental conditions. Moreover, as we shall see, phytotrons were only the first of an entire family of *trons* for biology. Following the first phytotron came the Climatron, Biotron, and Ecotron, all increasingly elaborate facilities to control climate. There were also a number of smaller associated biological technologies like the assimitron, which measured the CO_2 uptake of a canopy, the dasotron, which studied small ecologies, and the rhizotron, which is a viewing chamber where one can view tree roots and various arthropods that live underground.[1]

Our modern world of science and technology sees trons everywhere. According to the *Oxford English Dictionary* (*OED*), *tron* derives from "a weighing machine," or "the place where the tron was set up." One can still visit Trongate in Glasgow and the Tron Kirk in Edinburgh. In the past century, trons became a ubiquitous part of people's new modern lives, initially through radio: the first real vacuum tubes, Irving Langmuir's "kenotron" and "pliotron" date from around 1915. The name of the kenotron was explicitly drawn from the Greek roots of *keno* for "empty" and *tron* for "tool." Subsequently, the klystron and the rhumbatron became vital components of the radio industry in the 1930s. Trons helped win the Second World War. Heralded as the most important invention of the war, the resonant cavity magnetron—no, not the atomic bomb—developed at the University of Manchester was the heart of every radar set. Later, Radiation Laboratory engineers at the Massachusetts Institute of Technology (MIT) designed the hydrogen thyrotron modulator for Project Cindy—the name of a high-resolution radar set (at about 1 cm) for smaller ships, like PT-boats, for ship search work.[2] In short, trons starred in the Battle of Britain and the war in the Pacific, and assisted in the rescue of a young JFK.

Postwar, a creation of the 1930s, the cyclotron, a particle accelerator and one of the most famous instruments in the history of science, begat another tron lineage that grew to dominate nuclear physics. As cyclotrons proliferated, newer and larger accelerators like the synchrotron and then the Cosmotron (with its twenty-four ignitron rectifiers[3]), Bevatron, and Tevatron offered Cold War era physicists the possibility of creating new elements and peering inside the atom. Moreover, as much in the physical as in the life sciences, trons were not just devices, they were an entire class of cultural objects. It was not just a particle accelerator, it was a Cosmotron! And, as this book describes, it was not just a plant research laboratory, it was a phytotron!

To understand the phytotron and the worldview of those living in the Cold War era, I follow the suffix *-tron*. I take up Robert Proctor's challenge to grapple with the "pragmatics of language," though with technological and scientific instruments and facilities rather than disciplinary regimes. A suffix like *-tron* is, in Proctor's terms, an "embodied symbol."[4] When scientists built and then named their new device a *tron,* whether it was a cyclotron or a phytotron, they inscribed a set of meanings for the world to see, much as ancient knights displayed heraldic shields. The history of any one of those biological and physical instruments is important in its own right, but following the lineages of the trons of physics or biology offers insights, as we shall see, into how scientists, governments, industries, and the public understood that strange period of peace lined by imminent nuclear annihilation called the Cold War. Above all, the suffix *-tron* signals the centrality of modernism to postwar science, namely, the idea that technology would solve social problems and scientists would be the technologists to master both nature and society. Consequently, I argue in this book that in the life sciences, modernist trons speak of an era that demanded control, whether control over nature, control over populations, or ultimately control over minds and thoughts, and put its hope for that control in technology. Trons evince a people that sought security and salvation in machines and systems.

In the spirit of the *OED*'s meanings derived from use, the unexpected example of the "Eggatron" serves as an archetypical tron developed for a life science, and illustrates much of the worldview of biologists in the Cold War. In 1962, a scientific journal announced that "an electronic device, inevitably called the 'Eggatron,' records . . . data in such a way that [it] can be fed directly into an electronic computer."[5] In essence the Eggatron was a digital counter that recorded when an egg was laid—the result being recorded on paper tape readable by early generation computers—in an effort to produce hens that laid more than a single egg per day, as nature, both genes and environment, dictated.[6] The journal credited the conception of the Eggatron to Dr. P. J. Claringbold of Sydney University's Veterinary Physiology Department, while its actual design was the labor of Dr. Rathgeber of the Physics Department. I draw my reader's attention to the following facts about the case of the Eggatron. First, it was a physiologist who dreamed up the Eggatron; plant physiologists will be prominent characters in the development of phytotrons. Second, the device linked biological data to computation; large, centralized computer control systems were the heart of all phytotronic facilities. Third, the development of a tron required cooperation with another scientist, significantly a physicist; phytotrons required expansive networks of scientists and engineers. And fourth—the most damning fact of all—the declared *inevitability* of a technoscientific object named with the suffix *-tron*.

Just in case the reader suspects that the Eggatron was not a sufficient exemplar, please also consider the Algatron, which was an audacious attempt at a closed ecological system of living and growing algae to provide for oxygen generation/carbon dioxide absorption as well as "microbiological waste conversion" for "humans sealed within an isolated capsule," on its way to the Moon, Mars, or even "indefinitely long periods of time" on their way to the stars.[7] Built by a pair of sanitary engineers from the University of California at Berkeley, William Oswald and Clarence Golueke, the Algatron was an effort to replicate and control in a space ecosystem "the mutual interdependence of organisms within an isolated environment" as a way of modeling waste management on earth, itself an isolated environment or biosphere.[8] Their tron system of waste extraction and management formed part of a modern cybernetic imaginary focused around the idea that people are part of, and not just autonomous within, the planetary biosphere. Rather infamously, the technology of the "fecal bag" was employed throughout the American space program of the 1960s over the Algatron on the recommendation of doctors and National Aeronautics and Space Administration (NASA) engineers that treated human waste as a diseased product to be isolated and contained. However, so odious was the smell, feel, look, process, and psychology of fecal bags for early astronauts that some preferred starving rather than eating and subsequently having to defecate into the bag. Frank Borman, accompanying Jim Lovell in Gemini VII managed to go nine days without having to use the "fecal bags," a new record.[9] Even though fecal bags have now been used for over fifty years, no astronaut has suggested that they are the "best" solution to the problem at hand, and thus they provide an evocative example of one key lesson from the history of technology: "technologies . . . may be best because they have triumphed, rather than triumphed because they are best," in the succinct phrasing of the historian of technology, Donald MacKenzie.[10] Incidentally, since the end of the Cold War, we have learned that the Soviet Union had also developed a similar bioregenerative system much like the Algatron, and continued to develop the system into a fully functioning and tested closed ecological system called the BIOS-3, which completed a successful test run with human occupants eating algae and recycling their air, water, and urine in 1965.[11]

Another exemplar is the pyrotron, built by Australian bushfire researchers to model the spread of fire. To control the uncontrollable and to explain the complex interaction between fire, fuel, and forest bushfire researchers deployed tron technology. Notably, while bushfire scientists "used to conduct their research in the field" such "tests were at the mercy of the wind and weather and often failed to give good results." Instead, a BBC journalist reported, "the pyrotron allows for small-scale but physically accurate, very controlled repeatable tests."[12] Like the Eggatron and, as this books details, phytotrons, the pyrotron readily displays the embodied values of scientists, namely, their emphasis on repeatability and the desire for control even as they sought security in machines and systems from the threat of food scarcity or fire.

To call a plant laboratory a phytotron, an egg counter an eggatron, or a fire laboratory a pyrotron was not really inevitable but was rather a clear and conscious choice. It means something when people give objects nicknames and cognomens. In the best spirit of Umberto Eco, we must follow such semiotic signifiers wherever they take us.[13] We need to keep in mind, as Brother William of Baskerville learns in Eco's novel *The Name of the Rose*, that just because people recognize and act on a pattern they see it does not necessarily mean the pattern is, in fact, true. We only know that, regardless, nicknames signify the patterns that govern people's beliefs and actions. Of course, no good mystery would be complete without a red herring; in this case, the red herring is what most readers will be familiar with, namely, the "tron" particles, the electron, neutron, mesotron, and positron. G. J. Stoney coined the word *electron* in 1891, and it made its way into wider use through his nephew G. F. FitzGerald, who in 1894 convinced Joseph Larmor to adopt the word for what Larmor had been calling just "ions." However, as the eminent historian of science Bruce Hunt notes, unlike Langmuir's kenotron Stoney was not really using a "-tron" suffix but rather an "-on" one; it just happened that he was adding it to a root, "electr-," that ended in "tr." The same thing happened a little later with "neutron," a word that was clearly an analogy to "electron" but actually coined long before the particle was discovered experimentally in 1932. The positron followed in 1933.[14] Lastly, just before the Cold War, a small decision in April 1939 finally corrected the terminology for the elementary particle, the mesotron, to be properly renamed the muon. At the time it was a small moment of no special import, shortly to be overshadowed by war and the atomic bomb. It is a historical curiosity that the letter writer was C. G. Darwin, the grandson of Charles, who argued that while the electronic uses of the suffix "-tron" were already too common to be altered, the word *mesotron* was known to hardly anybody and could be changed into a standardized "-on" nomenclature without "widespread trouble."[15]

The younger Darwin notwithstanding, in fact trons-as-devices have formed the very bedrock of culture, a semiotic pattern, over the past sixty years. Trons have littered popular culture. Imagined through comic books and B-grade science fiction, the latter half of the twentieth century was lived via prospecting with a "Detectron" metal detector after 1949, or grooving on a "Mellotron" electromechanical keyboard in England in the 1960s, perhaps attempting to replicate the new Stevie Wonder song "Higher Ground," recorded through his "Mu-Tron"; it was seen with "Unitron" reflecting telescopes in the backyards of new suburbs free of city lights by young amateur astronomers, while their fathers wore "Accutron" electronic wristwatches to work—"it's not a time piece; it's a conversation piece"; it was witnessed by crowds of tens of thousands gaining better views of questionable plays on the Jumbotron, powerfully combining the most American of devices with the most American of sports.[16] Trons form numerous cultural touchstones, prominently the Disney film *Tron,* which enthralled audiences in 1982 and spawned a sequel in 2010, *Tron: Legacy,* as well as a string of computer games. Speaking like the Metatron, Optimus Prime has battled Megatron; five robot lions came together to form the defender *Voltron* on the hugely popular 1980s TV show. In the 1970s, Woody Allen emerged from the Orgasmatron in his film *Sleeper,* Scantron-style exams began their reign of both terrorizing and shoddily educating children the world over, even as the Gravitron thrilled them at amusement parks.[17] Lastly, in order to interview significant historical personages including Robert McNamara, the documentary director Errol Morris has forged historical memory itself through his Interrotron, a name that "reminds him of alien devices in '50s science-fiction movies."[18]

At the same time, the appeal of popular science proliferated new trons. One example, the Phototron 2™, was available from the late 1980s until just recently. It is particularly apropos because the Phototron 2™ was a small version of a phytotron, the life science technology I will be exploring in much more detail in the pages to come. According to its publicity material, the Phototron 2™ "allows uniform control of up to 23 physical/chemical environmental factors" in a one-meter tall, hexagonal design that uses vertical fluorescent lights combined with a "Base Nutrient Formula™" "calibrated based on known factors including light spectrum, intensity and output, wattage as a measurement of heat, air exchange, [and] calculated water evaporation/transpiration rates."[19] To support your controlled indoor growing Phototron 2™, the corporation further offers the "Feed-A-Tron™" patented watering system.[20] While such units now seem adept at supplying the growing personal marijuana market, in the 1990s the Phototron's designers and promoters proudly reported it "in use" at the NASA-Marshall Space and Flight Center as part of the study of reclamation and recycling technologies and systems for the then proposed international space station and "virtually all subsequent, future, long-duration, manned space exploration missions."[21] The Phototron 2™ even starred on an episode of Martha Stewart's television show in May 2011.

All this suggests a far broader history of trons. The major theme concerns the cultural imagination of calculability and the engineering ideal to negotiate a period of both fear and modernist technological optimism. To begin that history, I have outlined the history of many other trons online at www.worldoftrons.com. I encourage anyone with information about other tron projects to post references or materials to this Web site, which aims to offer a richer overview of the various permutations of trons in modern history. Quite simply, from the algatron to the zootron, the history of science is a world of trons.

Engineering the Environment

INTRODUCTION

THE AGE OF BIOLOGY

> An organism is the product of its genetic constitution and its environment . . . no matter how uniform plants are genotypically, they cannot be phenotypically uniform or reproducible, unless they have developed under strictly uniform conditions.
>
> **— Frits Went, 1957**

A LITERARY and cinematic sensation, Andy Weir's *The Martian* is engineering erotica. The novel thrills with minute technical details of communications, rocket fuel, transplanetary orbital calculations, and botany. The action concerns a lone astronaut left on Mars struggling to survive for 1,425 days using only the materials that equipped a 6-person, 30-day mission. Food is an early crisis: the astronaut has only 400 days of meals plus 12 whole potatoes. Combining his expertise in botany and engineering, the astronaut first works to create in his Mars habitat the perfect Earth conditions for his particular potatoes, namely, a temperature of 25.5°C, plenty of light, and 250 liters of water. Consequently, his potatoes grow at a predicted rate to maturity in 40 days, thus successfully conjuring sufficient food to last until his ultimate rescue at the end of the novel. Unlike so many of the technical details deployed throughout the novel, the ideal conditions for growing potatoes are just a factoid. Whereas readers of the novel get to discover how to make water in a process occupying twenty pages, the discovery of the ideal growing conditions of the particular potatoes brought to Mars is given one line.[1] Undoubtedly, making water from rocket fuel is tough, but getting a potato's maximum growth in minimum time is also tough. Back on Earth, current consumers wandering supermarkets full of fruit and vegetables making decisions about a potato's or tomato's look and texture and guessing about taste perhaps barely appreciate that the discoveries of the incredibly complex processes of growing plants have constituted some of the most important knowledge of all time. For although the sciences and technologies of plants have not yet saved a single astronaut on Mars, they have helped feed the multiplying people of the Earth.

Starting around the eighteenth century, European empires went to great lengths to collect and cultivate new plants. In the nineteenth century, the science of agriculture emerged as a proper function of many states to produce new breeds of crops and livestock and to make productivity gains through the development of new farming practices.[2] As many sciences moved into laboratories, the study of plants moved into greenhouses. Under glass, experimenters sought to reveal how the environment regulates and controls elements of plant growth, flowering, and development; notably, Charles Darwin had his greenhouse heated. Subsequently, in the late nineteenth century, genetics and plant physiology emerged as the two great new experimental sciences for understanding plants. Although the story of the geneticists' discoveries of genes and their wondrous promise is widespread, the corresponding story of knowledge about the plant physiologists' technologies of plants' environments is far less well known. Yet today, the wealth, variety, and sheer uniformity of everything people eat from apples to zucchini owes much to both the pioneering efforts of commercial facilities that fixated on a few systems and variables of climatic control as well as those scientific institutions that experimented with plant varieties and variable environments. Quite simply, the sciences of genes and environments have underpinned the new agricultural revolutions through the Green Revolution to modern hydroponics.

Engineering the Environment tells the history of one class of laboratories that created artificial climates and helped make those discoveries possible. They were called *phytotrons,* a name that resounded with all the promise of the dawning atomic age. For plant scientists, especially botanists and plant physiologists, phytotrons offered to "make it possible to study plant behaviour in its broadest sense under a diversity of climatic conditions where it is possible to vary each factor without appreciably altering the others."[3] A phytotron was a facility consisting of any number of rooms or smaller cabinets, in each of which any desired set of environmental conditions could be produced and monitored by new computers. Plant scientists used the ability to produce and then reproduce any climate to conduct experiments on the environmental responses of plants. And for over sixty years now, phytotrons have continued to be part of the global experimental study of the effect of environments on growth and development. They now serve on the front lines to attack the growing threat of climate change and uncertainty about its effects on the planetary food supply and biosphere. In the near fictional future, Andy Weir's astronaut builds a phytotron on Mars to survive—as his potato crop nears maturity, Weir's astronaut thanks "the billions of dollars' worth of life support equipment" in his habitat, which "maintains perfect growing temperatures and moisture at all time."[4]

When it opened in 1949, the first phytotron at the California Institute of Technology (Caltech) was a wonder of environmental systems engineering. It possessed new fluorescent tube lighting that controlled light, new air-conditioning systems and thermostats that controlled temperature, new devices of humidity regulation and nutrient standardization. Postwar, the study of plants also required a radioactivity room and a wind tunnel for early experiments in airflow across single leaves, whole plants, and rooms of plants. In a second-generation phytotron like the one in Stockholm any temperature between +5°C and +40°C could be maintained to an accuracy of ±0.2°C, or 0.5 percent; a fivefold improvement over the original phytotron in just twenty years.[5] Subsequently, the third-generation phytotron, named the Biotron at the University of Wisconsin-Madison, went even farther building soundproof rooms, dark rooms, and below-freezing rooms, and extended controlled environment experimentation to animals as well as plants. In all, like the more familiar story of the cyberneticans of the Cold War era, plant scientists in phytotrons obsessed about control over everything from their experimental black boxes, to their professional lives, and the wider geopolitical struggle of the era.[6] To establish the biological response to the environment required control: "What is important in a phytotron," the deputy director of France's national phytotron, Jean Paul Nitsch, told an audience in 1969, "is *the degree of control* over the various environmental factors."[7] Importantly, early phytotrons sought not only to control the technologies that made environments but also to govern the scientist users themselves.

Centrally, new computer systems at the heart of every phytotron gave control of control. In recurrent images of the era, computer panels occupied prominent and visible spaces in the first phytotron at Caltech, the Climatron, and the Biotron.[8] Those computers were not the desktops and laptops of today, though; they were the room-sized mechanisms of electronic and social control.[9] Opening in 1965 at the Royal College of Forestry in Stockholm, the "control room" in the Swedish phytotron, for instance, centralized the "timers regulating the photo- and the thermoperiods in the individual climate rooms." At the same time, housed in the control room was the "control system using thermocouples and multipoint recorders [*sic*] the temperature, the humidity, and the light conditions at certain points in all climate rooms." Overseeing regulation and monitoring was a third control system, "an elaborate alarm system to warn of malfunction"; on nights and weekends, the alarm system could "by a telephone robot" alert "any desired home number."[10] Computerized, phytotrons realized one vision of high modernism where every season would be created, charted, and overseen by the central regulating equipment of the control room. Consequently, as this book shows, learning about plants meant learning about the technology to replicate any biological environment. Plant science in the phytotron was timed and recorded, monitored and warned, called and regulated—a science governed by machine.

New assemblages of technologies to produce and control artificial climates reshaped the very boundaries of being human and offered ever-greater control, notably as a few went into space, some went deep under the sea in atomic submarines for months on end, and most went to their new middle-class jobs high above the street in clean and modern air-conditioned high-rise office buildings.[11] Like spaceships, skyscrapers, and airports, phytotrons sat squarely within the architectural, artistic, and scientific movement known as modernism, which, as Peder Anker traced, saw technology as the key to just social and natural organizations.[12] It was some of the grandest thinking of the era. For a technocrat modernist such as the Lloyd Berkner, a "growing technological capability" led straightforwardly to "knowledge of nature" by which "man acquires greater control."[13] Le Corbusier's vision of his "plan" for a new architecture and a new city included notable new forms that would best encompass the totality of needs and wants, from his famed "City of Towers" to any single house that would, of course, have a controlled environment with "baths, sun, hot-water, cold-water, warmth at will, conservation of food, hygiene, beauty in the sense of good proportion."[14] Purposively designed to create a new experimental and ordered plant science, the designs of phytotrons resonate with such high modernist visionaries. Indeed, while Le Corbusier encapsulated the spirit of modernism in the bon mot, "a house is a machine for living in," for over thirty years it seemed to some scientists that a phytotron was a laboratory for doing plant science in.

Little wonder then that as ever grander facilities took shape around the world through the 1950s and 1960s, Pierre Chouard, the director of *le grand phytotron* outside Paris, upon his retirement in the 1970s announced that biology was "entering . . . a Phytotronic era."[15]

BEFORE THE PHYTOTRONIC ERA

A variety of efforts to control one or more elements of the environment arose as part of the broad turn toward experimentation across the biological sciences.[16] Alongside open-air field trials, those cheap and popular mainstays of agriculture and horticulture then and now, greenhouses could hold a climate approximately steady for the benefit of a whole range of plant species. Greenhouses and fields served as places of agricultural experimentation on new breeds as well as new techniques of farming.[17] Technology was celebrated as much as botany and agriculture in the grand Victorian palm house at Kew Gardens and the grander art-deco-styled greenhouses of the Jardin des Plantes in Paris of the 1920s. Prior to the Second World War and the widespread availability of air-conditioning, the "control of air temperature by heating" was the achievement that elevated the glasshouse above the field for commercial growers as well as for botanists and physiologists.[18] Greenhouse technology saw exotic plants grown en masse in unnatural locales, such as the tropical palms grown in London and Paris, or the roses grown by one Illinois producer who possessed nearly a million square feet under glass by the 1920s.[19] But by the middle of the twentieth century, according to one experimenter, greenhouse conditions might suffice for agricultural production but experimental science demanded repeatability and control: the "chief physical characteristic of the average glasshouse environment," he complained, "is its great variability." In ten minutes, light intensity could change by 50 percent, air temperature by 10 percent, and the air itself by

30 percent. Such environments, the experimenter denounced, were "not quite haphazard but prehistoric, or rather pre-scientific!"[20]

Since the scientific revolution, scientists have sought to control the experimental environment of their instruments, laboratories, and objects. One well-known example saw the mere body temperature of more than one experimenter in the room with the apparatus at one time undermined James Joule's measurement of specific heat.[21] The emergence of experimental biology in general, and the discipline of plant physiology in particular, gave rise to one of the first attempts to claim mastery over the biological environment. Called the Vivarium, the facility opened in Vienna in 1903, and offered innovative technologies and systems brought to bear on zoological and plant physiological problems.[22] Later in the 1920s, scientists at the Boyce Thompson Institute in New York built "two constant-condition rooms" to address emerging experimental work on environments after the landmark studies of W. W. Garner and H. A. Allard indicated that day length governed flowering.[23] Later still in the 1930s, German biochemists could lay claim to running the "best-equipped biochemical research facilities in Germany and the world," the director of the Kaiser Wilhelm Institute for Biochemistry in Berlin advertised, because they had built adjustable controlled chambers that stabilized the environments for their new ultracentrifuges and electrophoresis apparatuses.[24] By the mid-1950s controlled environment facilities had become plainly ubiquitous: as the leader of Australia's major plant research group, Otto Frankel, reported after a tour through the United States, "Controlled environment facilities are now, at least to some degree, part and parcel of every botanical institution."[25]

Phytotrons unified and extended earlier piecemeal efforts to claim total control of the whole environment. In both walk-in rooms and smaller reach-in cabinets, phytotrons produced and reproduced whole complex climates of many variables. In the first phytotrons each individual room was held at a constant unique temperature. As figure I.1 shows, the Australian phytotron, for example, had rooms maintaining 9°C, 12°C, 16°C, 20°C, 23°C, 26°C, 30°C, and 34°C. Because some of the earliest controlled-environment experiments showed that plants reacted differently in daytime temperatures and nighttime temperatures, the first experiments to observe the effect(s) of varying the daytime versus the nighttime temperature saw experimenters move their plants from higher to lower temperatures over the course of a daily, or any other variable or

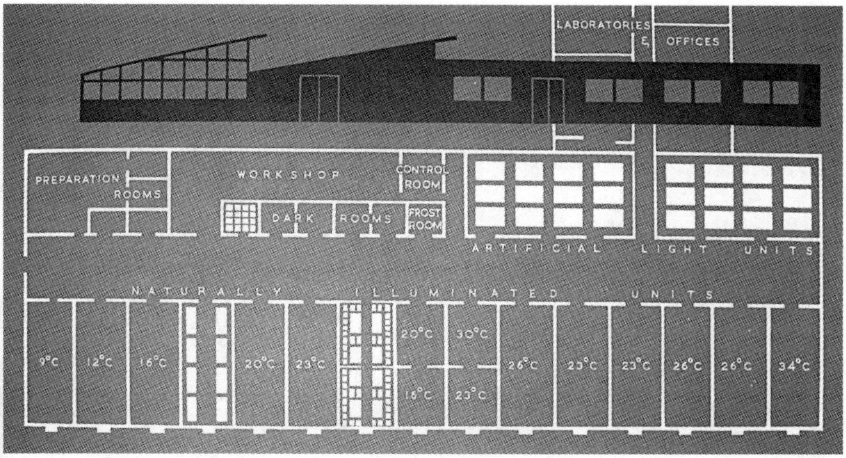

Figure I.1. "Plan of the Proposed Australian Phytotron." From "What Is Needed for an Australian Phytotron?" March 28, 1958. NAA Series A4940, file C2060. Image courtesy of the National Archives of Australia.

constant, routine. This rendered the variable "temperature" experimentally controllable. Even a brute force approach that tested each successive environmental variable and every variety of plant would serve to pinpoint specific environmental conditions to maximize growth. Expecting that more knowledge would surely come from greater technology, the next generation of phytotrons expanded in technological reach, in their ranges of environmental variables, and also in the degree of control over each variable. The phytotron in Stockholm offered a humidity-controlled room and a custom built computer, as well as a low-temperature room that extended the temperature range down to –25°C for the study of Nordic forests.[26] After that, phytotron technology compressed whole environments into smaller cabinets able to be set to any desired combination of environmental conditions, which are still in use today.

By the middle of the twentieth century, plenty of plant scientists, broadly including botanists, foresters, horticulturists, plant pathologists, and plant physiologists, used controlled and monitored environments to establish connections between specific environmental conditions and the mechanisms of flowering, trace elements and plant nutrition, photosynthesis, and plant heredity.[27] With control over the entire interrelated complex of the environment it seemed to many plant scientists that they had at last cracked the great puzzle, namely, the study of plant behavior

Figure I.2. "Glasshouse of Phytotron CSIRO, Canberra." 1967. NAA Series A1200, L66896. Courtesy of the National Archives of Australia.

from their genes as well as their environments. The plant breeder, phytotronists' advertised, already provided control over plants' "genetic constitution."[28] Phytotrons offered similar mastery over the environment through technology. While the control of temperature, humidity, airflow, and day length was achieved by the 1950s, the control and study of light has preoccupied the builders of phytotrons since the 1960s (and proved, as we shall see, to be a more complex technological and biological problem). Work in phytotrons helped botanists and plant physiologists better understand all the "hottest topics" of plant physiology of the 1920s and 1930s—phenomena such as photoperiodism (the response of plants to day length) and vernalization (the response of plants to temperature), as

Figure I.3. Kennebec potatoes grown . . . From Went, *The Experimental Control of Plant Growth*, plate XII. Reproduced with permission from John Wiley & Sons, Ltd.

well as the actions of auxins, plant hormones, and chemical herbicides.[29] Later, the name of phenomena of *phytochrome*, a photoreversible pigment came, the story went, from a "combination of *phytotron* and *Kodachrome*."[30]

One notable success of plant science in phytotrons was the ability "to determine the precise limits of productivity of plants."[31] The first phytotron, for instance, hosted two years of experiments on Kennebec potatoes and pinpointed that the greatest weight of tubers came from a combination of 20°C day temperatures with 14°C night temperatures (fig. I.3), in contrast to Andy Weir's astronaut's Idaho potatoes, which required an optimal temperature of 25.5°C. However, experiments in the phytotron also discovered that growth cannot be reduced to one environmental condition: the yield of Kennebec potatoes decreased by a fifth as the length of daylight went from eight to sixteen hours, and then decreased by four-fifths at twenty-four hours.[32] Remarkably, plants need nighttime, and thus, although unmentioned in the novel, Andy Weir's astronaut in fact helped his potatoes grow each time he turned out the lights.

Another celebrated "great discovery" by none other than the founder of the first phytotron, Frits Went, was that tomato fruit only set "over a limited and experimentally determinable range of night temperatures."[33] Went was a central figure in twentieth-century plant science.[34] Having achieved controlled conditions, Went spoke widely about how "with 3 parameters, . . . day temperature, night temperature and length of day, we can describe any climate at any particular time of the year in terms which are important for plant growth and plant distribution." To visualize the optimal climates for particular varieties, Went had three-dimensional models built, consisting of wire-frames forming lines that marked the growth of plants across several environmental measures, which were reproduced in a variety of publications (fig. I.4). What his model showed was revolutionary: in his new environmentally controlled laboratory, which had just gained its cognomen *phytotron*, plant science could now experiment on the "environment." Went spoke dramatically to the assembled audience of the International Botanical Congress in Stockholm in 1950, and illustrated his talk with results from several plants, but particularly highlighted the case of *Saintpaulia*, or the African violet. African violets required 25°C day temperatures and 22°C night temperatures, Went told his listeners, thus its optimal point existed outside the ellipse that described the climate in Pasadena, California, where he lived. However, the question of whether African violets could be grown inside their houses "sharply divided" the gardening public of Pasadena. Many swore they could be grown. Just as many dismissed even the possibility. The explanation, Went happily claimed, stemmed from the dependence of the plant's growth and development being intimately linked to three variables of climate, phototemperature (day temperature), nyctotemperature (night temperature), and photoperiod (length of light). People who left their windows open during the night could not grow African violets, but those who closed their windows could. And so the reveal: "you tell me which plants you grow in your house, [I will] tell you how you live."[35]

Such models dramatically illustrated the power of controlled-environment plant science everywhere from headline articles in *Science* magazine to popular picturesque 1960s coffee-table books.[36] Readers of *The World of Plants* (volume 3 of Doubleday's *Encyclopedia of the Life Sciences*)

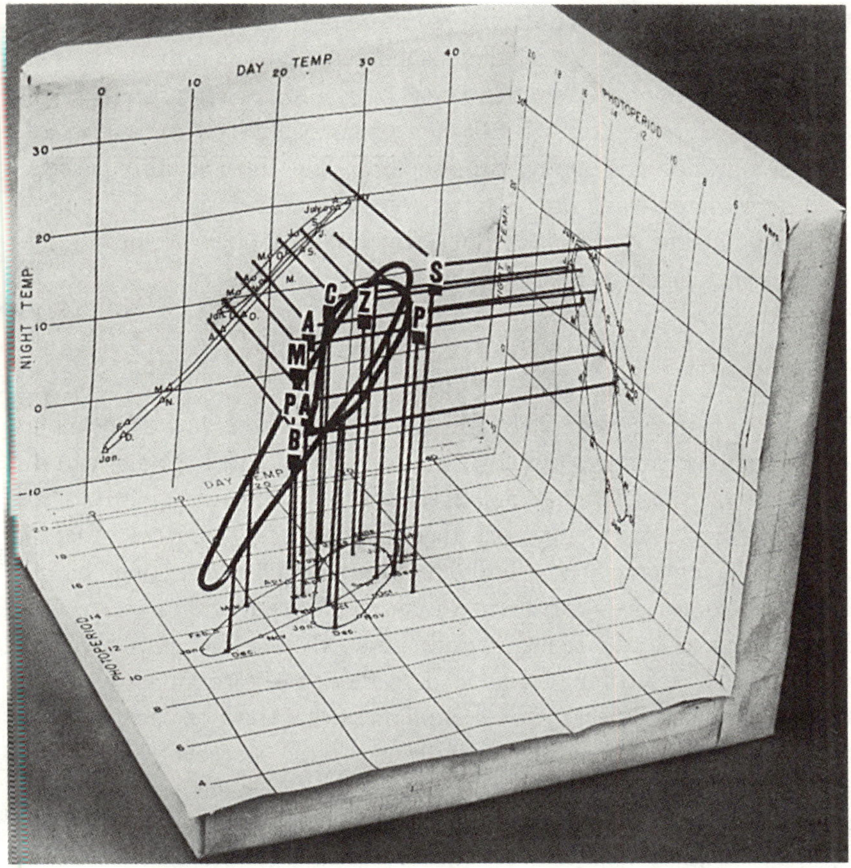

Figure I.4. Three-dimensional axis graph showing the relationship between phototemperature, nyctotemperature, and photoperiod for two locations, Pasadena, California, and Denver, Colorado for a number of garden flowers over an approximate 12 month period. The optimal growing conditions for various plants are indicated by the letters: S=Saintpaulia; Z=Zinnia; B-Bellis perennis; M=Mattiola; PA=Papaver rudicaule; A=Ageratum; C=Callistephus. From Went, *The Experimental Control of Plant Growth*, plate XXII. Reproduced with permission from John Wiley & Sons, Ltd.

were treated to color images of the newest facilities such as the Missouri Botanical Garden's Climatron and France's phytotron at Gif-sur-Yvette. They read how a phytotron's "reproducible . . . experimental conditions" gave the "basic laws of the physiology of plants." Readers noted how scientific methods of control were at work in agriculture and botany to render the world regular, stable, and wonderful. They were shown multidimensional graphs that displayed the point of maximum growth and photographs showing development across a range of environmental variables. They were offered startling facts of nature, for example, that "cold conditions are necessary to break the dormancy of seeds" in peaches and apples, and they saw photographs showing that apple seeds exposed to cold germinated while ones kept at constant temperature did not.[37] As Went said just after his phytotron opened, "modern research cannot do without such laboratories any more."[38]

THE PHYTOTRONIC ERA

The pursuit of technological control over organisms and experiments has been and remains a fundamental agent of change for biology in the twentieth century. To explain how scientists think about the world and how they create knowledge, historians have long followed and observed what technologies they have built and used, notably those famed biological technologies like electron microscopes, ultracentrifuges, electrophoresis apparatus, and radioisotopes that have shaped biologists' ability to see and trace molecular processes.[39] Running parallel to the technologies that have helped reveal genes, technologies that have revealed the biological environment have been equally important. Moreover, just like the instruments of the physicists "fix what it is to *be* an experimenter," so too have the use and embrace of molecular or environmental technologies defined what it meant to be a geneticist or molecular biologist as much as what it meant to be a botanist, plant physiologist, or ecologist.[40]

This book argues that the construction of technologies to control the biological environment had three immense consequences. First, control enabled the "environment" to be defined as a part of an experimental science of life. Second, phytotrons saw some biologists become technologists in their pursuit of biological knowledge. Third, the construction of new laboratories with elaborate technological systems to control and regulate elements of any climate saw feedback emerge as a powerful challenge to reductionism, not only because the technological control of one climatic variable destabilized another but also because it revealed organisms as complex products of genes and environments. In sum, the study of life became an exercise in technological control over both genes and environments and so the knowledge of the machine equaled knowledge of the plant.

Like so much in the early Cold War, phytotrons were built with fresh memories of depression, global war, and then widespread Malthusian predictions of population explosions, with "algae burgers" proposed to head off the world's "greatest single cause of unrest," namely, hunger.[41] Believing that a revolution in the scientific attack on the global issue of food was necessary, governments as well as the sugar, tobacco, forest, rice, and tomato industries all supported new phytotrons. Went's "great discovery," for example, underpinned the Campbell Soup Company's large research project in the late 1950s, including the building of new rooms in the Caltech phytotron to develop and test varieties of tomato to find those that would set fruit in the hot conditions of the southwest United States.

Phytotrons were the practical application of science to increase productivity. Testing progenies for potentially successful adaptations to particular climates, often far removed from their local climate, occupied much of the ordinary work in phytotrons after 1949. Caltech's physiologists lauded their facility as saving valuable time and money for breeders because far fewer plants up to the F_4 generation would need to be tested for far less time in the controlled conditions of a phytotron. Even better, breeders need no longer take the risk, Caltech biology division chairman James Bonner quipped, of not the right "kind of summers" ruining everything.[42] Likewise, several Australian plant physiologists maintained years afterward that phytotrons had made it possible "to accelerate and make more reproducible many kinds of research on plants at all levels of organization from the sub-cellular to the community."[43] Similarly, the Swedish Royal College of Forestry declared their phytotron to be a boon to the Nordic forest industry barely a year after the facility opened because it "made it possible to determine the various photo- and thermoperiodic systems controlling the growth of different provenances of European conifers."[44] Finally, by rationalizing the identification of new useful plants for particular environments, phytotrons played a small role in the now famous Green Revolution. The identification of best-correlated varieties and environments was considered so important that the Australian government donated a phytotron to the International Rice Research Institute in 1974 to study the most significant staple crop grown under the most diverse climatic conditions, namely, rice.[45] Whether for the forests of Sweden or for the agriculture of Australia, California, or France, something like an "engineering science" style of biology established control over genes and environments and promised a new biological world of economic and social benefits.[46]

At the same time, those scientists who built phytotrons believed that biological science ought to be ultimately directed toward gaining basic knowledge, not just increasing portion size. For many plant scientists the real revolution ignited by phytotrons was that the basic science study of living organisms under controlled environments might reveal biological laws. Historians of science have long noted the commonplace cultivation of an image as a basic science in many biological and physical sciences in the Cold War era, in part, because through the pursuit of "basic science" one might achieve an elevated status within the scientific community.[47] In the moral economy of most sciences, the pursuit of mere applications remains distinctly second-class, no matter how useful they might be, unless they are directed toward basic knowledge. For the plant physiologist Lloyd Evans, once a postdoctoral student of Went, and later the designer and director of the Australian phytotron, and later still the president of the Australian Academy of Science, the choice for a young scientist between "pure or applied" always remained "that old intellectual class distinction."[48]

The distinction sat at the heart of the major changes sweeping over science during the Cold War. Something like half of the era's scientists and engineers worked secretively, albeit dutifully, in a variety of applied-science projects connected to the variegated goals of the military-industrial complex from building ICBMs (intercontinental ballistic missiles), to radar dishes to listen to Soviet radio signals bounced off the moon, to using atomic bombs to build harbors, or cloud-seeding experiments to create or guide tropical cyclones over Vietnam,[49] not least, said one, because such "research is the last and only defense against communism."[50] New technology both enmeshed scientists in much sought-after applications and permitted grander experiments.[51] Plenty of biologists worked with, and gained much from, the military-industrial complex, notably new technologies like the sudden and widespread availability of radioisotopes as trace elements as well as mutation agents.[52] To many, the expansion of medical and biological science was understood in no small part as a salve for American science over the wound of their development of atomic weapons.[53]

At the same time, the pursuit of idealized "pure science" (also termed basic or fundamental science) highlighted the gulf between scientists' own values concerning knowledge for its own sake and the demands of their patrons for useful science and applications in exchange for public support. Caltech's president Lee DuBridge succinctly grasped the paradox: "How 'pure' can the research program as a whole be and still command community and public support, or how 'practical' can it be allowed to become without losing the essential spirit of true scholarship—the search for new knowledge?"[54] Counterintuitively, in the early Cold War an apolitical stance often served an overt political purpose for science's patrons, and consequently, although the military-industrial complex garnered substantial and growing criticism in the 1970s stating that it denied openness, stifled cooperation, and hindered international and interdisciplinary exchange, it actually also broadly supported pure science for decades.[55] The reason, as the historian of science Nicolas Rasmussen argued, that molecular biology initially flourished was because it embraced "the mantle of the apolitical scientists' scientists" to ensure government support by explicitly rejecting calls to political action and engaging only in the search for "truth."[56] In other words, molecular biology first prospered in part because of its "political significance," Rasmussen said, as an "offshoot of genetics," the science "notoriously subject to suppression by Stalin," as well as its promise of wondrous medical cures.[57] Went perceptively, albeit privately, noted in his diary that government support for science in the Cold War world stemmed essentially from "the competition with Russia," a cause, he considered upon reflection, that was "hardly mentioned" at the major scientific symposium on the problem of basic research featuring Robert Oppenheimer and even President Eisenhower and at which DuBridge had spoken.[58] Consequently, for two decades after 1945, funding bodies such as the National Science Foundation (NSF) in the United States, the Commonwealth Scientific and Industrial Research Organisation (CSIRO) in Australia, and the Centre national de la recherche scientifique (CNRS) in France possessed political support to fund basic science largely because a much-touted science independent of politics could be used as a cudgel against Soviet science and the "monster" of politicized science embodied by the Lysenko affair.[59]

For any postwar plant scientist, no charge was more damaging or inflammatory than that of being labeled a Lysenkoist, a follower of the Soviet agronomist Trofim Lysenko, who ruined Soviet agricultural productivity and encouraged the purge of Russian geneticists in the late 1930s.[60] In effect, genetics became synonymous with anticommunism while plant physiology became suspicious in the McCarthy era, and rumors circulated darkly about plant physiologists and Lysenkoism. As writers, directors, and actors painfully experienced during the House Un-American Activities Committee (HUAC) hearings of the late 1940s, even a loose connection to communists ruined reputations and damned careers. Seeing Reds under every bed, any connection reinforced paranoid suspicions. It may well have been enough, for instance, that a 1948 symposium on vernalization and photoperiodism that contained a frontispiece of a woodcut of Lysenko and a brief description of that "excellent prophet's" prewar work by Eric Ashby, starred many future users of phytotrons including Went, Sterling Hendricks, and Anton Lang, and a foreword by Kenneth Thimann. While neither Went, Hendricks, Lang, nor Thimann went beyond merely mentioning Lysenko's "controversial hypotheses," it seems likely that even such an innocent association cast a long shadow over the reputation of the entire subject of plant physiology.[61] There were whispers, even years later, that some people believed Went to be a Lysenkoist, as the biographers of George Beadle suggest without attribution.[62] Suspicions lingered for decades, and contributed to the lack of recognition for the achievements of plant physiologists.[63]

Was it any wonder then that Went went to such extremes to divorce his controlled-environment laboratories from industrial or political applications? The epitome of effort to label research in phytotrons "basic science" came in 1957 when Went prophesied that the facilities gave no less than a "Theoretical Botany" comparable to a generally accepted "Theoretical Physics."[64] Went advocated that phytotrons aimed to reveal the "universal" factors of growth and flowering and argued that the "general understanding" of the development of a plant had been hindered simply by "inadequate experimental techniques."[65] The message that flowed out to plant scientists the world over was that the experimental control available in phytotrons at last permitted botany and plant physiology to become a basic science akin to physics and free of any association with Lysenko. John Holloway, a forest ecologist in New Zealand, for example, spurred his country's investment in a phytotron because scientists possessed "no real knowledge of the physiology of any New Zealand forest species. All we have are a few deductions based on primitive autecological observations."[66] Only with a phytotron, as Holloway succeeded in arguing to his fellow scientists and his national Department of Scientific and Industrial Research in New Zealand, could biological science claim to know the causes of phenomena. From his ever-expanding phytotron outside Paris, Chouard said, "Phytotronics is the methodological key in plant research, to which phytotrons . . . are the necessary logistics."[67]

This mantle lasted into the 1970s. After that, plant physiologists saw interest in basic science wane, along with their fortunes. By the 1980s, as the president and later historian of the American Society for Plant Physiologists J. B. Hanson noted, support for "fundamental biology [was] a poorer third" behind "medicine, which received the bulk of the funding" and "agriculture a poor second."[68] Eschewing practical applications for dreams of large theoretical breakthroughs, the plant physiologists in their increasingly costly phytotrons struggled, as their best patrons, the NSF and the National Institutes of Health (NIH), shifted from idealistic supporters of basic science to be compelled politically to stipulate practical goals in the 1970s.[69] In just the past two decades, however, support for phytotrons has modestly increased once more because of the urgent commercial and governmental need to understand the biological effects of climate change. At the same time, thanks to scholars like Kärin Nickelsen it is also only now becoming apparent that the history of science in the twentieth century is woefully incomplete without the story of the plant physiologists.[70] Though plants underpin life on this planet my hope is that this book might offer some insight into the continued lowly status of the study of plants among scientists, their historians, and the wider public.

AN IMPORTANT DEVICE NO ONE HAS HEARD OF

The story of phytotrons is little told, and the word itself exotic and unfamiliar.[71] Yet, across at least two dozen institutions in the middle of the twentieth century, a new community of scientists built and used phytotrons. As we shall see, the history of phytotrons replicates many features of the early story of computers, notably about creating "agents of control" as much as couriers of information, as the historian of computing Paul Ceruzzi has argued.[72] Readers will particularly note just how far removed our present conception of computing and biology is from the past: to look at modern computing in the present is to see an information age of personal computing as much as to look at modern biology is to see a genomic age of personal health and wonder how it could ever have been otherwise. These views are now so persuasive that they, in fact, quite readily hide their own early histories, both of biology and computing, and contribute to the general marginalization of plants for historians of science and the public.[73] In other words, the story of phytotrons is little known because the dominant narratives about the discovery of genes are so inescapable they in effect erase the fact that the study of life was also the discovery of controlled environments.

Now forgotten, a scientific community took shape united by a desire to experiment on organisms' environments. The community drew people from a huge range of fields including botany, forestry, horticulture, plant pathology, agronomy, genetics, entomology, and agriculture, but especially plant physiology, brought together often for single research projects though occasionally for whole careers. They sometimes called themselves phytotronists.[74] A comparative history is necessary to tell the story of phytotrons and the phytotronists because science after 1945 was built between disciplines, by multiple instruments, and above all internationally. Globally, a host of phytotrons occupied large portions of research budgets variously in Sweden, New Zealand, Canada, Hungary, Germany, the Netherlands, India, and Japan, while smaller units appeared in Austria, Israel, China, South Africa, Great Britain, and Taiwan.[75] Across all were continual efforts to create a biological science of the whole plant via the construction of increasingly elaborate and expensive technological systems. Many countries agreed with Went in California, Mitchell in New Zealand, and Chouard in Paris, who said "one *big phytotron* at least is necessary for a large country with welcome facilities for those who need such sophisticated equipment."[76] I devote a chapter to one of the largest phytotrons in Australia (chapter 4), but the great phytotron of the Soviet Union and the later Biotron Institute in Japan are the two most significant institutions not addressed in this book. Constrained by language and other barriers, I am in great sympathy with Paul Edwards in dreaming of fully international histories of sciences that work on the planetary scale—I look forward to studies on each of these in the near future.[77] In the meantime, this book dwells primarily on the American experience, not least because Americans built the first and the greatest number of phytotrons, nearly a dozen, variously at Caltech, Duke, Yale, North Carolina State, and Michigan State Universities, along with the related Climatron in Saint Louis and finally the Biotron at the University of Wisconsin-Madison.

The larger arc of the book argues that the story of phytotrons is the complementary half of the story of genetics, namely, the discovery of the biological environment alongside genes. Within that larger narrative are two story arcs. The first, broadly the story of the creation and work of the first phytotron and then the Climatron unites chapters 1 to 3 and ends with a Coda that wraps up the life and career of Frits Went. The second arc describes the creation and work of two, second-generation phytotrons, first in Australia (chapter 4) and then in North Carolina (chapter 5) as comparative examples, and then crescendos with the case of what was supposed to be the apex of phytotronics, the American national Biotron (chapter 6). A second coda briefly discusses the decline of phytotrons in the 1970s and 1980s. My Conclusion offers some thoughts about how the history of phytotrons might aid recent efforts to determine the biological effects of climate change. It highlights that in the Ecotron (1989–2010), a few biologists have constructed whole controlled environments and ecosystems, while even more recent incarnations have been equipped for carbon dioxide (CO_2) measurement, such as those at Michigan State University's Plant Research Laboratory, the Biotron Institute at Kyushu University in Japan, and at the world's newest phytotron at the University of Saskatchewan since 2011.

Throughout, one clue helps reveal the story of phytotrons and phytotronists, namely, the suffix *-tron* itself. Coming after the "physicists' war," plant scientists explicitly appropriated the embodied symbol of the tron from the famous devices of modern physics like cyclotrons and synchrotrons.[78] Of course, many life sciences appropriated metaphors and practices of the physical sciences in the twentieth century.[79] Even so, the phytotronists' usage seems extreme; both Went and the director of *le grand phytotron* outside Paris, Pierre Chouard, swore that "the cyclotron . . . fulfills about the same function in physics as the phytotron does in plant science."[80]

To explain what it meant for a scientist to liken a phytotron to a cyclotron, I follow above all the lead of the historian of science Evelyn Fox Keller, who powerfully noted how "the ways in which [scientists] talk about scientific objects . . . actively influence the kind of evidence [they] seek."[81] The explanation is that in their facilities of environmental control, those biologists became technologists. Moreover, by equating knowledge of the machine with knowledge of the plant, the study of life became an exercise in the technological control over both genes and environments. Phytotrons, then, sit at the intersection of biology and technology, as do many parts of modern life science.[82] Critically, an extensive literature in the history of technology has demonstrated that social processes shape the adoption, understanding, and use of technological systems as much as scientific ideas.[83] Phytotrons, like the computers that regulated them, embodied scientists' and governments' modernist convictions that even the largest social problems could and would be solved by new sciences, new technologies, and new technoscientific infrastructures.[84] In perhaps the most dramatic example, modernism came to Missouri, when the Missouri Botanical Garden not only had the garden's old Palm House demolished but also had the palms themselves left out to die in order to build the Climatron: "The immediate present—and the palms—were sacrificed to the future," declared the garden's *Bulletin*. Trons reflected the optimistic future of modernism where the past needed to be swept aside, echoing the famed exhortation of Ezra Pound—"Make it New!"[85]

PHENOTYPE = GENOTYPE + ENVIRONMENT

The history of biology has been broadly focused on biologists' struggles to specify and then measure the phenomenon in question. Across many approaches to the life sciences, in the twentieth century alone, Linus Pauling notably pursued chemical molecules, others viruses, some cells, still others various animals, and a few ponds.[86] For the plant scientists in phytotrons (among many others), the axiom "GENOTYPE + ENVIRONMENT = PHENOTYPE" spoke to what an organism was.[87] Everyone agreed: from professional ecologists—"an organism without environment is inconceivable"[88]—to gardeners—"plants are the result of their environment," as one indoor gardening book stated, referring its readers back to "the principles of botany."[89] The director of the Duke University phytotron, Paul Kramer, traced this concept back to the German physiologist Georg Klebs who suggested just before the First World War that "hereditary potentials" and "environmental factors" combined to produce a plant's "processes and conditions" that dictated the "quality and quantity of growth."[90] Went championed time and time again that "the ultimate shape and size of a plant depends both on its genetic constitution and on the environmental conditions under which it grew up."[91] Established by the doyens of plant science, the principle flowed down to undergraduates. In one textbook on plants by Went's Caltech colleague Arthur Galston, for instance, students read that "with any given genotype, tremendous control over growth may be exerted by obvious influences in the environment" such as light and temperature.[92] Students read in another undergraduate textbook from 1964 titled *Physics in Botany* that "it is now known that genetical factors are responsible for the time of appearance of flowers, but also, if the environmental conditions are unfavourable, the passage from vegetative to reproductive growth may be retarded or even stopped altogether."[93]

Moreover, the equation suggested a path of research, namely, that the process of measuring the actual characteristics of any whole organism, or "phenotype," required the genotype as well as the environment to be made experimental. Consequently, in the same era that many worked to specify and measure genes, in phytotrons botanists and plant physiologists worked simultaneously to specify and measure environments. Thus to their builders, the phytotron's creation was really the endpoint of a long struggle to control the environment, at least since the famed nineteenth-century physiologist Jacques Loeb, who saw, as the historian of science Philip Pauly noted, "the main prerequisite for success in biological manipulation was command of a wide range of forces active in the organism's environment."[94] Even as innovative breeding techniques for plants and animals had generated a great variety of new crosses and hybrids, early twentieth-century work with early controlled environmental experimentation struggled with and finally conceded that wide deviations in environmental conditions like temperature undermined any conclusions about even basic relationships such as how the length of day affects flowering.[95] Common solutions included agricultural sciences' crop testing which employed active strategies to minimize the variation around the mean yield such as planting trials in several locations across several years, while ecological studies generated intense inquiries into the nature, methods, and successes of statistical sampling of areas and species.[96]

The plant physiologists scoffed at such rustic and inexact measures: as late as 1969, one French plant physiologist noted during a conference of the International Biological Program how "plant physiologists have always had a justified skepticism about field research, particularly on natural ecosystems. Experimental difficulties are severe."[97] Such attitudes underpinned the technological drive to fully replicate and control the environment. Quite simply, the phytotron technologically solved the scientific dilemma presented by the field, namely, being able to exactly repeat climatic conditions, and consequently, when both halves of the

Figure I.5. Twenty-seven-month-old Bourbon trees grown in constant day/night temperatures of, from left to right, 17/12°C, 20/14°C, 30/23°C, 26/20°C, and 23/17°C. From Went, *The Experimental Control of Plant Growth*, plate XIX. Reproduced with permission from John Wiley & Sons, Ltd.

equation were fully reproducible biology would be reproducible. As Went preached in his magnum opus, once biologists accepted that "an organism is the product of its genetic constitution and its environment" it necessarily followed that "no matter how uniform plants are genotypically, they cannot be phenotypically uniform or reproducible, unless they have developed under strictly uniform conditions."[98] Went offered visual evidence that genetically identical trees grown at different temperatures appeared radically different (fig. I.5). For nearly three decades, with the combination of simultaneous advances in both genetics and environments, plant scientists savored their ability to generate reproducible experimental objects for biological study. They called it the Age of Biology.

THE AGE OF BIOLOGY

In 1966 the American National Academy of Sciences (NAS) received a comprehensive report compiled from over a thousand questionnaires sent to a representative third of the estimated number of plant scientists active in teaching and research in the United States. The report, *The Plant Sciences Now and in the Coming Decade,* was a wide-angle snapshot of the biological community concerned with plants. Chaired by Kenneth Thimann who was then at Harvard but had stood by Went's side as the first phytotron opened at Caltech, the panel declared recent discoveries so revolutionary that "the second half of our century" might be called "the beginning of the 'Age of Biology.'"[99] As the panel tellingly concluded, the reality of most plant scientists' working lives was that the advent of "new concepts in biochemistry and genetics" was as important as "the availability of new technological tools such as computers [and] controlled environments."[100] Of course, plant biology was a science of genes and environments, of new concepts and new tools, but the leading figures of American biology saw a far grander vision of their science. By adding a technological mastery of controlled environments to breakthroughs in genetics, they lived and worked at a time when DNA + phytotrons = Age of Biology.

Importantly, the NAS panel's conclusion challenges historians of science to appreciate how the new concepts and the new tools appeared equally significant to, and seemed equally necessary to, the future of the plant sciences. Geneticists, of course, had made great strides in understanding the genotype, including finding new ways to create huge new numbers of crosses, hybrids, and mutants, while plant geneticists soon turned toward evolutionary biology via work on the phenomena of polyploidy, hybridization, and apomixes.[101] Before the Second World War, there was Thomas Hunt Morgan's sweeping genetics work on the fruit fly, and the discovery of molecules that promised to be "magic bullets" such as plant hormones for agriculture, chemicals like DDT, and above all medical cures for the pharmaceutical industry.[102] Postwar, the pursuit of the gene drove the molecularization of the study of heredity, ultimately spurred gene technology and genomics by century's end, and created a culture of heredity. The celebrated moments for the culture of heredity remain the discovery of the structure of DNA in 1953, followed by the technique of recombinant DNA (1972), which permitted manipulation and thus the construction of the Human Genome Project (1991). By being variously informational, traceable, and reductionist, the historian of science Hans-Jörg Rheinberger argued, "the gene" came to be considered "the representative unit of the genotype and the ultimate determiner of the phenotype and, with that, executor of life" over the course of the twentieth century.[103] Indeed, the historian of science Angela Creager described how the conceptualization and manipulation of genes via new techniques, notably radioisotopes, served as "key ingredients of a postwar episteme of understanding life in molecular terms."[104]

The gene's once fellow traveler, however, has been erased in historical memory. Historians of science have noted that it remains one of the great unspoken assumptions of modern biology that many biologists of all stripes considered experimental organisms *identical enough*—not actually identical just sufficiently similar, regardless of the environmental conditions of their development.[105] In addition, it has long been a comfortable and convenient way to simplify the analyses of environment historians, scientists, policymakers, and even social and political theorists. One of the earliest environmental historians, H. H. Lamb, for example, stipulated his "assumption that the climate, the opportunities which it offers and the constraints it places upon man and the environment are effectively constant" for the nineteenth and twentieth centuries.[106] There were also immediately practical reasons for the erasure: in the genetics research program, mutations were the objects sought after by Mendelian genetics, not adaptive changes from environmental conditions,[107] while early molecular biology was undemanding of variable environmental conditions: William Laing from New Zealand's Climate Laboratory remembered that "growth conditions were simple (37°C with shaking)."[108] *E. coli* featured as an ideal reductionist model organism for early molecular biology because, as Evelyn Fox Keller explained, the environment plays no role in the development of the bacterium.[109] In contrast, plant physiologists railed against the "view of an organism as solely active and the environment as solely passive" as "a one-sided picture" as early as the 1920s.[110] All scientists exercised a choice, as the future directors of the Duke phytotron and the Wisconsin Biotron, Paul Kramer and Theodore Kozlowski, respectively, noted in the introduction to their textbook on the physiology of trees in 1960, stressing how "emphasis is placed on the effects of environmental conditions on physiological processes of the organism as a whole, rather than a wholly biochemical one in which emphasis is placed on the details of the processes themselves."[111]

The larger issue is that the erasure of a biological science of the experimental environment has appeared, falsely, as a natural consequence of the triumphant molecular view of life. As the biologist Richard Lewontin and Richard Levins explained, as DNA became a fetish, organisms in modern biology were active and richly described, but the environments in which they grew and developed were considered passive agents, minimally understood, and in any event largely outside biological disciplines.[112] But historians have also helped erase the study of the biological environment: when the historian of science Lily Kay concluded that the discovery of the structure of DNA resolved "what had been defined for decades as the central problem in the life sciences,"[113] she reinforced the erasure of the environment at no less than the very institute (Caltech) and over the same period when the plant physiologists sought to make a biological science of the "environment" alongside the science of the gene in the first phytotron.

Thankfully, Evelyn Fox Keller first stressed the now common view that the history of genetics has overshadowed a larger history of experimental life science.[114] *Engineering the Environment* offers part of what has been overshadowed, namely, the story of a global science of the biological environment at work alongside the science of the biological gene. As every chapter in this book illustrates, the act of both specifying the components of the "environment" and defining the proper measurement of each component preoccupied the plant scientists in phytotrons. Those chapters will serve, as the historian of science Peter Bowler once cautioned, to "try to demythologize the past"[115] by recovering the exotic world of phytotrons. Moreover, they collectively offer some reflection on the topics of what history remembers, what parts of lives and works get retold, what become the famed experiments, and what gets cast into the dustbin of history.[116]

"PAUSE TO THINK WHAT WE ACTUALLY MEAN BY CLIMATE"

Frits Went asked scientists to "pause for a moment to think what we actually mean by climate."[117] Plant physiologists knew that plants grew and developed in complex whole environments from the late nineteenth century onward, and had demanded the ability to claim "with confidence

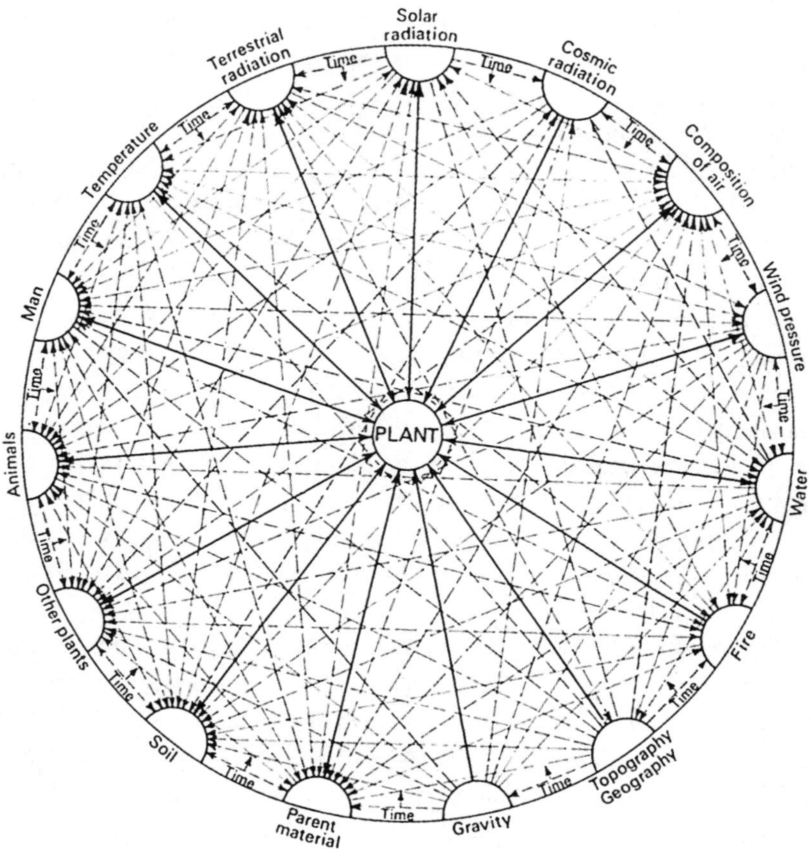

Figure I.6. Dwight Billings's "Holocoenotic" diagram. From Billings, "The Environmental Complex," 256.

that a certain change takes place in the plant when, and only when, accompanied by a *single* change in the environment." Lacking any mechanism for control, Ludwig Jost's 1903 *Lectures on Plant Physiology* resigned plant physiologists to accept the rarity of any "physiological observation" under nature's complex conditions that might confirm "that the special alteration in the surroundings is the cause of the special phenomenon in the plant."[118] But what did they mean by a plant's "environment?" Some plant physiology texts were incredibly broad: one pointed out how, "in the small zone inhabited by living creatures, there is an infinite variety of different types of environment." "The land" possesses soil, which in turn possesses minerals and other organisms, moisture, and atmosphere, a pH, and a temperature; the "atmosphere" possesses the characteristics of humidity, nutrients (like carbon dioxide), respiration, temperature, pressure, wind, and the atmosphere's "optical properties."[119] The text told students that in order to grasp an understanding of the processes governing growth and development required appreciating how all those characteristics exerted feedback effects on "all activities of living things [which] are the expressions of literally thousands of processes being carried out at the same time within the cells."[120]

Others stressed interrelated and interlocking complexity: notably, the ecologist and plant physiologist Dwight Billings at Duke University (later a cosponsor of the Duke phytotron) defined "the environment of a plant as sum of all external forces and substances affecting the growth, structure, and reproduction of that plant," including "heat, light, water, [and] elements." He displayed this complex array of interacting variables in his "holocoenotic environmental complex" diagram (fig. I.6). Importantly, even for Billings as an ecologist, "other plants" were one of only fifteen distinguishable factors defining the environment of a "plant."[121] Phytotronists made each variable into a discrete technology to study the equation of the environment. Their conception of the study of life saw the climate broken into discrete variables, each encased in technological innovations, especially air-conditioning, new forms of lighting, cheap electricity, herbicides, and pesticides. In the creation of both systems of control and controlling systems, plant scientists in phytotrons established a biological measure and meaning for the "environment" that centered on mutually dependent variables. The evocative climax came in the early 1970s from a French phytotronist, who gave the "environment"—*climat*—a mathematical expression:

> Even the notion of climate encompasses a complex of many variables: temperature, light, rain, humidity, winds; which vary according to intensity, length, periodicity, quality, and orientation. If it were possible to write them as an equation, one obtains: Climate = (T + E + P + H + V) × (i + d + p + q + o) or 25 principle components but with a lot of secondary variations. For example, [the product] Eq represents the diverse radiation of the solar spectrum, or of a multiplication of factors: Eq_1 = red, Eq_2 = orange, Eq_3 = yellow, etc.[122]

Scientists' categories reveal much about their intellectual process. "Environment," read the epigraph to one of the last major summaries of phytotrons in 1980, "seems to be the key word in an amazing number of unsolved or partially solved problems in biology." The summary's author, Robert Downs, director of the North Carolina State University phytotron and one of the most significant phytotronists in the United States, spoke for many in the by-then declining community as he surveyed not only the international array of phytotrons but their ongoing scientific work. Phytotrons, Downs said, had worked to define what "scientists" considered the great unknown of the science of biology, namely, the proper "measurement of environmental parameters [which] has long been a problem in biology."[123] In other words, Downs revealed that the biological environment was what some biologists had struggled to specify and then measure. To those ends, "phytotrons provide the means of dissecting the environment," said Henry Hellmers at the Duke University phytotron, the twin of Downs's North Carolina State facility.[124]

In contrast, other biologists have described quite different conceptions of what the biological environment is. Evolutionary biologist G. Ledyard Stebbins, for example, highlighted that for many biologists the "environment" was essentially a disciplinary problem. Stebbins rhetorically asked the meaning of the "concept of the environment" in his 1982 book, and he answered that different disciplines viewed the category differently: "naturalists and ecologists" regard the "other organisms" in an ecosystem as the "most significant factors in their environment," whereas "physicists, chemists, geologists and biologists . . . think of the environment chiefly in terms of its physical features—climate, temperature, moisture, soil, and atmosphere."[125] In contrast, thirty years earlier Billings's diagram defined "other organisms" as an equivalent component to any particular climatic feature.

To understand those divergent views, I build on the historian of biology Garland Allen's classic narrative of twentieth-century biology that saw "naturalist" biologists emphasize the phenotype while "experimentalist" biologists focused on the genotype.[126] The emergence of an experimental science of the environment suggests that there was a third group, the "technologist" biologists who emphasized control over the environment. This third way reshaped the meanings of those core concepts of biology, namely, "phenotype," "genotype," and "environment," because whereas botanists of earlier generations had concluded that a plant existed as only an individual expression of a range of possible expressions for their type, the phytotronists regarded an environment as integral to the plant itself, not just noise in the phenotype.[127]

Not coincidentally, the founder of the first phytotron highlighted these issues. All biologists accepted "phenotypic variability . . . as a basic property of living matter," Went said, in that any single set of genes presented a range of expressions in nature. Plant breeders and horticulturalists had worked to reduce variability, but the assumption that "atomistic reactions" controlled variability had driven them, Went claimed, toward "statistical analysis." To Went, this had been a distraction: instead, "if phenotypic variability had been considered as being largely due to environment, more serious efforts would have been made to control the external environment of growing plants." Writing in no less than the American *Proceedings of the National Academy of Sciences,* Went announced in 1953 that his experience "show[ed] that environment rather than the atomistic nature of biological reactions is responsible for phenotypic variability."[128] Went remained convinced until his death that the underlying problem with the creation of biological knowledge was the absence of controlled environments, and that only through such environments could science study variability and ascertain actual causes. In short, embracing technology and reductionism, as much at the genetic scale as at the environmental, technologist biologists sought the laws of growth via their ability to control the environment added to the geneticists' established control over the genotype. Control over environmental variability offered control over organism variability and thus gave experimental certainty.

Consequently, the story of the creation of phytotrons was at its heart a modernist project in the Age of Biology. The phytotronists built their facilities and began their research convinced that knowledge of the technological systems would bring knowledge of living systems. The first phytotron certainly began as a reductionist, imperial project to establish which climatic condition governed growth, or flowering, or fruit set; at the same moment the geneticists saw great success via reductionism. And like the geneticists, plant scientists had also been trained to consider genes and environments separately: Otto Frankel's concluding remarks at the opening of the Australian phytotron mentioned how "our minds are conditioned to regard 'genetic' and 'environmental' pathways as distinct and, in a mechanistic sense, unrelated."[129] In one's new phytotron, of course, a scientist could keep a set of genes or series of environments constant so that the other could be the subject of experiment. Yet it is also clear from Frankel's remarks that a leader in plant science was already dissatisfied with this simple reductionism. The experimental production of environments was unexpectedly complex.

In fact, this book concludes that what the technologist biologists learned was that their conception of the environment demanded interlocking assemblages of feedback-laden technological systems to replicate and control interrelated variables. Struggling with their array of devices to maintain conditions at preset levels, they came to appreciate how feedback was constitutive of the environment, the plant, and life itself. Environments and organisms, and even nature itself, were differential equations not linear functions. Consequently, toward the end of the century, conceptions of the "environment" became more explicitly cybernetic, notably as ecology and climatology developed an array of computer models.[130] In plant physiology, however, the cybernetic meaning of the "environment" always remained tied to the physiologist's pursuit of experiment control over climates. As recently as 2007, Henry Hellmers explained how the "effects of environmental factors are more often synergistic or antithetic than additive because plant growth is capable of adapting, within limits, to environmental changes."[131] At the same time, the conception of the plant itself became more cybernetic: Chouard's colleague in France, N. de Bilderling, noted, "as a matter of fact, certain environmental factors provoke reactions from the plant which, by interacting with others, can conceal direct actions and thereby complicate our understanding of phenomena and our explanation of the way factors act."[132]

Alongside new cybernetic conceptions of nature and organisms, came a new cybernetic understanding of plant science itself. By the late 1960s, it appeared that disciplinary and epistemological systems of biology were predicated on feedback, in a situation similar to the historian Paul Edwards's analysis of the vast infrastructure machine at work to collect, process, and evaluate climate data.[133] In the most overt exemplar, a trio from the botanical institute of the University of Würzburg at a meeting of physiologists during the International Biological Program (unfortunately timed to take place only a year after the Soviet suppression of Czechoslovakia in 1968) diagrammed the evolving place of phytotrons within experimental biology for their audience (fig. I.7). For O. L. Lange, E. D. Schulze, and W. Koch, experiments in phytotrons

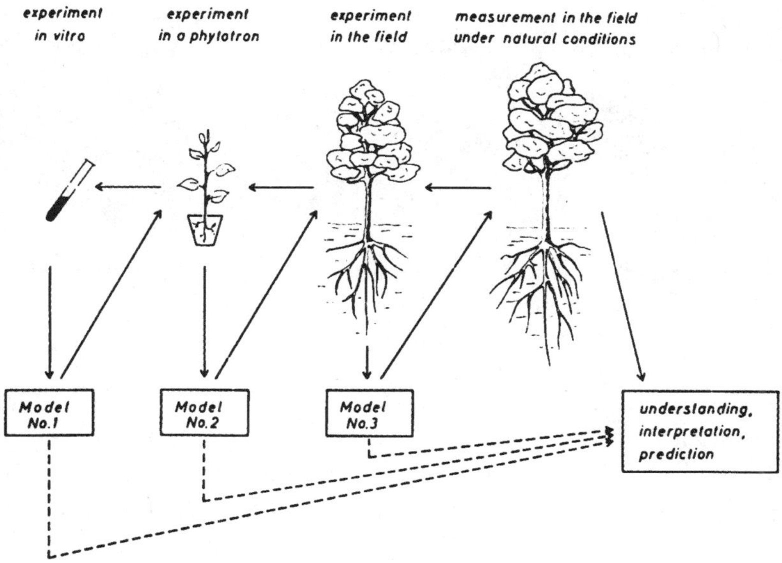

Fig. 1. Diagram for the necessary investigations to explain photosynthetic productivity of plants in the field.

Figure I.7. Diagram for the necessary investigations to explain photosynthetic productivity of plants in the field. From O.L. Lange, E.D. Schulze, and W. Koch, "Evaluation of Photosynthesis Measurements Taken in the Field," in *Prediction and Measurement of Photosynthetic Productivity: Proceedings of the IBP/PP Technical Meeting*, Trebon, 14–21 September 1969 (Wageningen: Centre for Agricultural Publishing and Documentation, 1970), 339–352, 340. Reproduced courtesy of Wageningen UR.

occurred in between in vitro or test-tube experiments and isolated fields trials. Moreover, they specified the relationship between in vitro experimentation, experiments in phytotrons, experiments in the field, and finally measurements under natural conditions as a network of feedback arrows. It was via integrating several levels of control, linked via feedback loops, that "understanding, interpretation, and prediction" would emerge in biological science. Those loops involved experiments in vitro, in phytotron, and in field.[134] Twenty years later, Thomas Yuill offered the University of Wisconsin-Madison Biotron to his "colleagues" in exactly the same way, highlighting that the Biotron provided "facilities that will fit between the laboratory and the field."[135]

The director of the Hungarian phytotron at Martonvásár Sándor Rajki called phytotrons "the grand experiment" of modern biology.[136] In short, plant scientists saw that no one discipline held a monopoly

Figure I.8. "What can a Phytotron do for Australia?" From "What Is Needed for an Australian Phytotron?" March 28, 1958. NAA, Series A4940, file C2060. Reproduced courtesy of the National Archives of Australia.

on life science because the way to grasp an understanding of life's processes was as a series of cybernetic feedback loops between test tubes, phytotrons, and field experiments. Although the first phytotrons were built as universal facilities, what they demonstrated to their creators in practice over the Age of Biology was that no one style of experimentation was independently sufficient, the science of the plant required them all.

AN AIR-CONDITIONED EDEN

In the sunshine of the 1950s, the hand of man finally took hold of capricious nature through controlled environments—at least that was the provocative image in the booklet promoting an Australian phytotron (fig. I.8). The computer-like square product symbolized a conviction that through modernist technological science, nature would be made regular, controlled, and predictable. "What can a Phytotron do for Australia?" the text accompanying the image asked. Building on Steven Shapin and Simon Schaffer's fundamental observation that solutions to problems of knowledge are embedded in practical solution of the problem of social order, the icon conveys that the technological control over the environment assured the social goal of control in the Cold War era.[137] In the case of phytotrons, control meant to grip the randomness of the prescientific glasshouse or field and forge a future through human hands where regular, consistent heads of wheat would stabilize humanity. And that, Andy Weir's astronaut might conclude, makes the story of the experimental control of the biological environment one of truly "extreme botany."

CHAPTER 1

"THE AWE IN WHICH BIOLOGISTS HOLD PHYSICISTS"

Building the First Phytotron at Caltech

> We decided to call it a phytotron—phytos from the Greek word for plant, and tron as in cyclotron, a big complicated machine. [It] is going to do for plant biology what the cyclotron has done for physics.
>
> —**James Bonner, 1980**

THE PHYTOTRON and the cyclotron were both born in California. Grandiose ambitions to control nature from the whole environment down to the atom matched the aspirations of that American western frontier state. Already famous, Ernest Lawrence's cyclotron lay in the north of the state at the public University of California at Berkeley. After 1945, it became godlike with the revelation that fissile material for the atomic bomb was produced via Lawrence-designed Calutrons.[1] In 1949, Frits Went's phytotron opened in the south at the private California Institute of Technology, Caltech, in Los Angeles. The gleaming modernism of vast air-conditioning systems that made the first phytotron possible seems especially Angeleno, that City of Quartz, grown on the consumption of electricity and water all used to produce a cool breeze throughout the year.[2] The hubris of envisioning a botanical science of reproducible plants and climatic variables evokes the creation of entirely new dreamscapes under the white Hollywood sign. Surely, a new American assurance imagined, the twin technologies of trons would reign in the second half

Figure 1.1. Caltech's phytotron, officially "The Earhart Plant Research, Campbell, and Dolk Laboratories." June 1961. Photo 10.23–29. Courtesy of the Archives. California Institute of Technology.

of the twentieth century, for as the cyclotron brought an end to war (for what war could now be fought against the might of the atom?), so the phytotron would bring an end to hunger.

In the sunshine of postwar California the prospect of an eternally full table, that dream of every wish-fulfillment story, seemed certain because plant scientists had cracked how to repeatedly achieve the maximum size of all crops, fruits, and plants.[3] With geneticists at Caltech and elsewhere having solved the problem of genes, the answer, according to a pair of plant physiologists, Frits Went and Henry Eversole, was to control the environment through technology. Already famous for his hormone research, Went was one of the leaders of the botanical and plant physiological communities, disciplines that had long demanded greater experimental control over the environment. Eversole wanted to grow better orchids. To those ends, they began to air-condition a set of greenhouses during the Second World War.[4] Subsequently, Went designed and then gained institutional and private support for an entire laboratory to impose a rigorous control over the total environment for experimental botany and plant physiology.

Officially named the Earhart Plant Research Laboratory, Went's conception was nicknamed the "phytotron" shortly after it opened in 1949. It operated until 1972. The creation of the first phytotron is now a largely forgotten path of modern biology, yet at the time the further technological pursuit of controlled environments appeared to be the next rational step in the development of plant science. As Went repeatedly explained, the first phytotron was a technological solution to a scientific problem, namely, a lack of experimental standards and control that had long held back progress in botany. As he later opined in his seminal book, *The Experimental Control of Plant Growth,* "control of the environment was long recognized as essential in physics and chemistry where temperature, radiation and other factors have been carefully adjusted in all experiments." Unlike those disciplinary hallmarks of modern science, Went lamented how "botanists have had to work in uncontrolled aerial surroundings of plants grown in the open or in conventional greenhouses."[5] His phytotron, Went said, established experimental control over plants.

This chapter explains the demand for an experimental plant science by situating the creation of the first phytotron in the institutional context of Caltech as well as in the disciplinary context of plant physiology. In her history of molecular biology, Lily Kay noted that Caltech's plant physiologists vied to dominate Caltech biology in the immediate postwar years.[6] What she did not mention, however, was what the biology division's catalogue lionized in 1950: "the newly completed Earhart Plant Research Laboratory . . . a unique instrument for the study of plant growth under complete weather control. All the elements of climate, such as light, temperature, humidity, wind, rain, and gas-content of air, can be controlled simultaneously. The old and the new research laboratories offer the opportunity to study plants under different synthetic climatic conditions, yet with complete reproducibility of experimental results."[7]

To understand the creation of the phytotron circa 1945, and, not coincidentally, molecular biology in the same time and place, we must appreciate that plant physiology was actually the better-known part of Caltech biology. As one institute trustee described it, "the exceptional strength of the Biology Department, not only in plant physiology but also in genetics and biochemistry make the California Institute of Technology an exceptionally favorable place for initiating a fundamental and far-reaching program of this kind."[8] The trustee inverts the usual story of the development of twentieth-century biology, namely, that while genetics and biochemistry shone, one should not forget about plant physiology. In fact, the stories of plant physiology and genetics at mid-century are rather equivalent: Caltech hosted these two remarkable new trajectories of biology that grew in a similar institutional context and via a similar process, through the explicit appropriation of metaphors, concepts, and practices from the physical sciences. In other words, while the institute has been renowned as an important site of the development of genetics, biophysics, and the study of master molecules toward a new experimental science of molecular biology,[9] Caltech was also where botany and plant physiology specified a new experimental science of the biological environment and then measured that environment.

Critically, Went's new facility was not just a new greenhouse or laboratory, it was a "phytotron." As Sharon Traweek sagely observed, "the artifacts we make remind us of who we are, how we are expected to act, and what we value."[10] Following that wisdom, this chapter explores the measure and meaning of the cognomen "phytotron" as it tells of the founding of the facility. To explore the facets of that name is to reveal layers of meanings attached to the study of life in the twentieth century. The historian of science Sharon Kingsland has argued that Went's phytotron represented "the laboratory equivalent to the 'hybrid culture' that [Robert] Kohler described as emerging in the field" and "a countercultural movement of resistance against the divisive trends of molecular biology."[11] In contrast, I argue that Went's creation of a laboratory of environmental control was actually deeply conservative, not overturning but rather securing botany and plant physiology through the standards of practice of the physical sciences. The name "phytotron" encapsulated how new technologies of light, air-conditioning, and computing were understood to have finally given botanists and plant physiologists the control over their plants' growing environments they had yearned for since the nineteenth century. As expected, technological control over the environment heralded experimentation for all manner of biological sciences, especially, Went said, ecology. As Went mentioned in Caltech's own popular science magazine, *Engineering and Science*, "Ecology is not an experimental science as yet" because ecology was too descriptive with too little regard for the "reasons" behind the relationships between organisms.[12] And indeed, when he later reflected on the "last 30 years [of] phytotrons," Went believed that they had "helped to make ecology an experimental rather than a descriptive science."[13] In short, the name "phytotron" is our doorway to understanding what the idea of an "experimental science" meant to plant scientists like Went at mid-century.

FRITS WENT

Beginning as a small Pasadena institute, Caltech rose to fame in the 1920s on the back of tremendous successes in astronomy, chemistry, and physics.[14] The institute subsequently became a major American center of biology when, under the direction of its influential, Nobel Prize winning president, Robert A. Millikan, Caltech recruited the "foremost biologist in the United States," Thomas Hunt Morgan, to continue his groundbreaking genetics work using mutations in the fruit fly (*Drosophila melanogaster*).[15] On the West Coast, Morgan and the entire group of fly researchers delivered spectacularly; the historian Robert Kohler assessed the Caltech period as their most productive.[16] Subsequently, Morgan recruited Went in 1932 to lead the plant physiology group with a distinct emphasis on the discovery of the mechanisms of growth and development, investigation of the chemical nature of growth substances, and further integration of biology with the physical sciences.[17]

The plant physiologist James Bonner said understatedly that, postwar, Caltech was clearly a "favorable place." The members of its research staff were among the best in their fields, and were abundantly supported: asked to provide a general statement in 1945, Bonner wrote simply, "our division is, from a material standpoint, one of the best and most luxuriously equipped in the United States."[18] For Bonner, all that Caltech required was a new infusion of "a bright, stimulating, and everchanging younger generation" of graduate students,[19] which it would receive in droves with the GI Bill within a few years. Over sixty research fellows and a host of graduate students regularly augmented the professorial staff. A later postdoc of Went, Lloyd Evans recalled "the galaxy of biological stars in the plant physiology group: Frits Went, James Bonner, Art Galston, George Laties, and occasionally that agent-provocateur from UCLA, Sam Wildman."[20] Went remains perhaps the least well-known of a remarkable groups of scientists at a remarkable institution during a remarkable time (fig. 1.2). It was a campus of scientific "giants," the later biology division chairman Robert Sinsheimer celebrated: "Linus Pauling was chairman of chemistry. Frits Went, who was a leader in plant biology, was there. . . . In genetics, [Sterling] Emerson was still there. [And] then, of course, in

Figure 1.2. Caltech's Biology Division faculty, 1947. From left to right, standing: Keighley, Sturtevant, Went, Haagen-Smit, Wildman, Beadle, Lewis, Wiersma, Mitchell, Van Harreveld, Alles, Anderson; seated back row: Emerson; Dubnoff; seated front: Borsook, Bonner, Tyler, Horowitz. Photo 10.24–162. Courtesy of the Archives California Institute of Technology.

physics, there was [Richard P.] Feynman . . . and [Murray] Gell-Mann. . . . So you could say it was pretty deep company."[21]

Alongside such luminaries, Went led part of the scientific community over a career spanning nearly fifty years, during which he served as president of the American Botanical Society, the American Society of Plant Physiologists (ASPP), and the American Institute of Biological Sciences. According to modern biology textbooks Went rose to fame by completing Julius Sachs's work on geotopical movement, the cause of roots' downward growth, and the discovery of the plant hormone auxin.[22] Went's auxin discovery in the mid-1920s solved a vexing problem of plant growth, the movement of roots downward and the growth of the plant upward. It was a well-known problem, which led the nineteenth century German plant physiologist Sachs to challenge the explanation of the reaction of plant roots to gravity of no less an authority than Charles Darwin.

At the heart of the Sachs/Darwin debate was a dispute over claims about knowledge produced in the laboratory versus that from nature, a debate rekindled in phytotrons a generation later.[23] As the historian Soraya de Chadarevian described, Darwin the naturalist, the country gentleman whose diligence and care outweighed any shortcomings of exacting instrumentation or laboratories, contrasted with Sachs the experimentalist who disciplined both the physical space of his laboratory and his subject. Importantly, it was Sachs's approach to botanical experimentation that spread rapidly, and Sachs himself became a hero of plant physiologists, establishing, according to one early historian, the "standards of biological work by reference to experiment."[24] Consequently, the perceived need to bring greater attention to experimental control underpinned much of the work of the American Society of Plant Physiologists, itself created in 1923 "from a desire to give plant physiology identity and recognition as a distinct branch of plant science."[25] As far as Went was concerned, that identity hinged on a "criteria of high-class experimental work."[26]

While a young graduate student in the mid-1920s in Utrecht in the Netherlands, Went isolated the signaling substance he called *Wuchsstof* (later termed "auxin") from the root tips of seedlings in a singular triumph of experimental plant physiology. Of direct relevance to the creation of the first phytotron a generation later, his isolation of plant hormones necessitated laboriously measuring seedlings' responses under controlled experimental conditions, and validated the laboratory-based modern scientific discipline of plant physiology, and Sachs's experimental principles.[27] Afterward, Went was widely considered Sachs's heir. Scientists marked Went's commitment to Sachs's conviction that only a disciplined, completely controlled experimental space could answer modern physiological questions.[28] The early plant physiologist and historian Theodorus Weevers praised Went for his "careful research" on physiological processes, especially the effects of changing temperature on respiration in plants.[29] The director of the New York Botanical Garden, William Campbell Steere, lauded Went as "an outstanding botanist with an unusually broad understanding of his subject."[30] Moreover, when Went's colleague Kenneth Thimann later advertised the discovery of auxin as the product of "a continuing chain of closely-knit research and deduction," one on which a science like "genetics rests," it supports the view that mid-century genetics and plant physiology shared a goal of creating an experimental biology via control.[31]

For the scientific community, the discovery of auxin was a "momentous" result, serving to "catalyze a revolution in the science of Plant Physiology," according to one distinguished plant physiologist, Samuel Wildman. "Went's discovery would be looked back upon as the starting point of research destined to result in the multi-million dollar industries now engaged in supplying world agriculture with weed-killing, defoliant, and growth-stimulating chemicals."[32] At the same time, Went's discovery of auxin was also a moment that served as the beginning of the search for "master molecules" that might regulate and govern growth. The pursuit of master molecules has been seen as important for the history of molecular biology insofar as the identification of molecules' structural and physiological processes became the core research focus of many researchers working toward the medical search for "magic bullets" like penicillin in the borderlands of biology, chemistry, and physics.[33] On moving to Caltech in the 1930s, Went worked alongside Bonner and then later with Thimann on plant hormones geared toward agriculturally productive science and more engineering aspects of plant growth and development.[34] Bonner, Thimann, and Went became central players in the interwar plant physiologists' utopian goal of nothing less than complete "growth control" particularly "the structure and mechanism of hormones and growth factors" centered on the search for master molecules.[35] In 1938, Went coauthored a seminal book with Thimann, titled *Phytohormones*, before moving toward his initial studies of controlled laboratory design during the war years. Bonner and Thimann continued working on plant physiological research on growth hormones, which paid off spectacularly not in terms of growth hormones but another member of that same class of substances: 2,4-dichlorophenoxyacetic acid, better known as 2,4-D. Locally, it was no wonder that, as Bonner later said, "everybody who worked in modern plant physiology had to come to Caltech."[36]

Plant physiology represented some of the most exciting biological work of the 1930s and 1940s. In the year after the discovery of the structure of DNA, 1953, the long-serving executive secretary of the American Society of Plant Physiologists, J. Fisher Stanfield, considered "plant Physiology [the] major undergraduate subject and the most mature." From his perspective, "zoology had always dominated the biology departments," but physiology would triumph because it embraced an "increasing complexity and the physio-chemical approach born of technological advances."[37] Went saw himself in the vanguard of plant physiology. As he later said, plant physiologists turned the study of plants away from early "display gardens" of "largely horticultural interest" toward modern "experimental science" in "experimental greenhouses" and had consequently revealed "new knowledge about plant nutrition, plant hormones, [and] plant responses to their environment."[38] Exemplary physiological work included Melvin Calvin and Andrew Benson's famed determination of the cyclic path of carbon in photosynthesis between 1946 and 1954, courtesy of radioisotope tracers, particularly carbon-14.[39] In 1956, the American Society of Plant Physiologists awarded Calvin a lifetime membership, the Hales Award: in 1958 Went received the next award.[40]

THE MOST VITAL POSTWAR RESEARCH INVESTMENT

By the 1930s, evidence was mounting that greater control over the experimental environment revealed important facets of plant growth and development. Among the notable moments for Went were the "precise" laboratory experiments of Anton Blaauw, whose interwar work on tulips had established that while the optimal temperature to initiate flowering is 20°C, the optimal temperature for development then abruptly drops to 8°C for three weeks followed by 9°C for a further ten weeks.[41] For plant physiologists, such painstaking and counterintuitive results indicated the general need for facilities that could control the environmental conditions of growing plants for weeks or months at a time. Sam Wildman later recalled that plant physiologists' ongoing search for plant hormones related to the auxins similarly relied on going to "great trouble to insure that temperature and humidity remained constant in the dark-room where . . . assays were performed."[42] Establishing the boundaries of, and control over, their experimental environment was believed vital for plant physiologists, and, consequently, a variety of facilities for environmental control emerged. Herman Spoehr at the Carnegie Institution's Department of Plant Biology, for example, built rudimentary constant-temperature chambers, while others, like the famed American ecologist Victor Shelford, could only dream of the better-controlled laboratory conditions that physiological work required.[43] Moreover, conveying how crucial environment control was to plant physiologists, it can hardly be a coincidence that Spoehr's and Went's tenure as president and vice-president, respectively, of the American Society of Plant Physiologists followed their early pursuit of environmental control.

To gain control, Went and Eversole began tinkering with a smaller greenhouse around 1942 or 1943, testing "different methods of introduction of air into greenhouses."[44] Apparently dissatisfied with the limited control of the environment of his greenhouses, Went and Eversole developed the first complete controlled-environment spaces for botany and plant physiology. Eversole's interest in temperature control came through an interest in growing orchids for Americans who had become obsessed with the tropical plant.[45] Went would later remind his readers that air-conditioning (cooling rather than heating) was the core technology that permitted a truly replicable environmental space. One of the earliest papers brought together an electrical engineer, Lauriston Marshall, an air-conditioning engineer, Arthur Hess, Eversole, and Went. The four authors explained that since 1931, they had conducted laboratory experiments with growing plants under "partial control of temperature and relative humidity." Writing in 1948, they now looked to expand their "new methods" to achieve "greater crop yields" than those of commercial greenhouses. Cooperation between engineers and plant physiologists brought together unusual domains of expertise, and results that were disseminated in unusual places; publishing in the major journal of the American Society of Refrigeration Engineers, *Refrigerating Engineering*, may have been a first for a plant physiologist. Yet it seems clear that theirs was a long-standing productive interdisciplinary effort to evaluate the effects of increased control of temperature and humidity on plant growth, and to obtain "knowledge of the mechanisms of plant growth."[46]

In their early experiments, Went, Eversole, Marshall, and Hess found that a large quantity of air had to be moved through the greenhouse to remove heat and water vapor sideways, using much larger blowers and exhausts than expected. At the same time, while technical experimentation continued haphazardly on the greenhouse equipment, Went identified the tomato as the organism that displayed the best results of controlled-environment experiments. He produced a series of papers in 1944 and 1945 that outlined the discovery of optimal night temperature for the maximal setting of fruit (15–20°C) correlated with optimal stem elongation. In the last phase, Went worked with Lloyd Cosper to take his tomato experiments into the field, verifying that night temperature was the "major factor controlling stem elongation."[47] It was textbook plant physiology—identifying under controlled conditions an external cause of growth and development before verifying that cause out in the field.

Even under the limited control offered in the smaller greenhouse trials, Went and Eversole claimed two important achievements. First, constant phenotypes came via controlled genetics and controlled environments. As they noted, "whereas tomatoes grown in ordinary greenhouses sometimes grow better, and sometimes poorer after being sprayed with a sugar solution, plants grown under rigidly controlled conditions responded in the same way when they had been subjected to the same conditions." In a nod to claiming standards comparable to other sciences, Went and Eversole pointed out that this fact was not surprising, only that "the amazing thing is that research workers with plants had long been satisfied with lack of reproducibility in their experiments."[48] In other words, the air-conditioned greenhouses were both a comment on the standards of experimentation with plants and an observation about the moral economy of plant scientists, circa 1945. Went stressed that the necessity of experimental reproducibility had not been crucial to the workings of plant scientists, and so implicitly signaled that they must hold to other measures of scientific standards, though these went unnamed.

Second, and more important, controlled experimentation disassociated the "various climatic factors influencing the development." Individual physiological responses could be regulated via the discovery of various optimal conditions. There, Went and Eversole used the example of corn, long thought to require warm nights for best growth. By exactly regulating the temperature, the pair discovered that corn required warm days but cool nights to achieve maximum growth. For Went and Eversole, the discovery of a plant's growth and development dependency on temperature held important consequences for agricultural experimentation via field trials. Given the interrelated variables of climatic data and variety of plant, the pair argued that before trials could even begin "in the field we have to know all qualifications."[49] These early results, replicated time and again in various experiments in various phytotrons, became standard features of undergraduate textbooks in plant physiology illustrating the biological effects of climate. In Arthur Galston's widely used text, for example, students saw the growth curve of peas in relation to temperature, perhaps being asked on an exam to identify value of the maximal point.

As the Second World War drew to a close, many were optimistic that the new technology of the war would be powerfully deployed for science

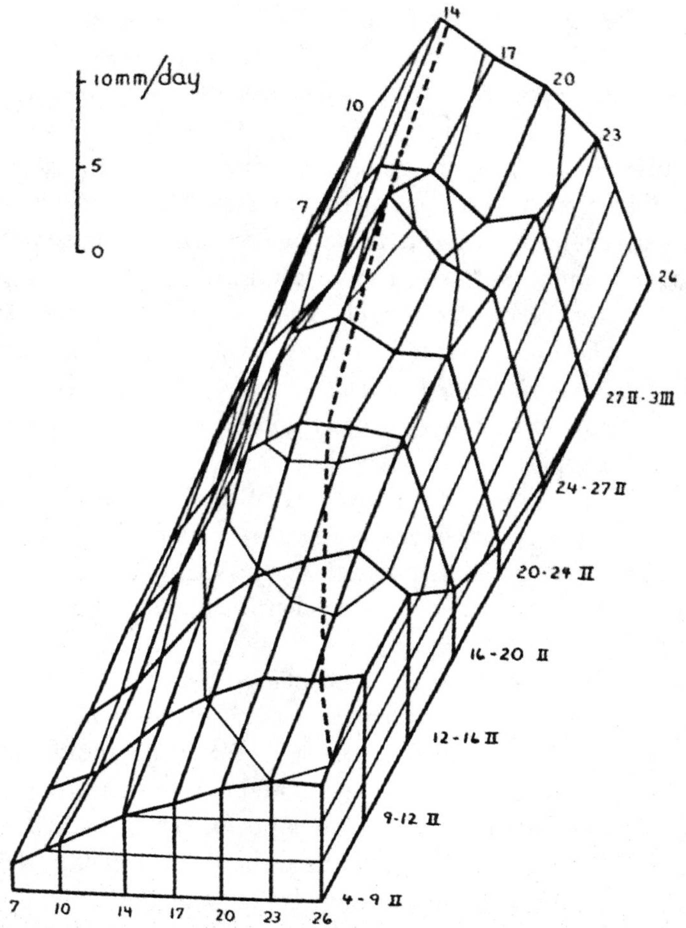

Figure 1.3. "Relationship between growth rate (vertical axis), different constant temperature (abscissa) and age of Vinco pea plants. Shift of optimal growth rate indicated by the dotted line." From Went, The Experimental Control of Plant Growth, figure 21. Reproduced with permission from John Wiley & Sons, Ltd.

in peace, notably at the University of California at Berkeley when the new 184-inch cyclotron first operated in November 1946.[50] At the same time, many were pessimistic about feeding the innumerable hungry masses in Africa and Asia, which gave rise to various plans including the controlled production of algae to be made into food to avert disaster.[51] In Southern California, those on the staff at Caltech proposed their own grand technological dreams, including Henry Borsook's isotope project, Albert Tyler's ultracentrifuge project, and Eversole and Went's new air-conditioned greenhouse project. Building on their wartime studies of the optimal night temperatures of tomatoes and corn, Went and Eversole drew up design blueprints about mid-1945 for "a complete set of air-conditioned rooms and greenhouses [to] supply [the] information necessary for application in the field."[52] As postwar expectations of the revolutionary potential of new technology mounted, Went's proposal starred as the first item that needed a firm decision the day after George Beadle became chairman of Caltech's biology division in early November 1945.[53] Undoubtedly, as the historian Lily Kay noted, Caltech's postwar students heard the almost constant refrain of triumphant technology brought to bear on resolving biological science.[54]

Just before Beadle arrived at Caltech, Went had persuaded Caltech's president, Robert A. Millikan, that his air-conditioned greenhouse laboratory would pay impressive dividends. Millikan subsequently asked Beadle's opinion of the whole business, and whether he believed that the biology division as a whole was "thoroughly sold on this form of expansion."[55] Beadle had misgivings, but not so much about the science of controlled experimentation as about the proposed size of the facility. Over a December dinner with Millikan, Beadle admitted that he had been less than impressed by Went's proposal, particularly the scale of the construction and operation that Went envisioned, but then said that his "enthusiasm [was] growing."[56] Millikan's overtures to Caltech's patrons required assurances from his biology chair, whose recommendation was no doubt tempered by the knowledge that without Went's proposal the prospective funds "would probably not be available for other things."[57] Perhaps with the choice between a substantial grant for Went's project, and no grant at all, Beadle simply chose the economically rational course. On the other hand, Beadle seems to have been taken genuinely aback by the seemingly outlandish size of the project and remained skeptical. Then again, as Beadle himself was only newly recruited from Stanford to Caltech, he may have felt it politically unwise to belittle Went's grand plan, especially in front of an enthusiastic Millikan, and conveniently found unexpected ardor.

Whatever the reason, with Beadle's assurances, Millikan approached one of Caltech's most established patrons, Harry B. Earhart, to fund Went's vision. Millikan and Earhart had maintained a long correspondence since before the Second World War, and the Earhart Foundation

Figure 1.4. From right to left: former Caltech president Robert Millikan, Frits Went, Kenneth Thimann, Caltech patron Harry Earhart, biology division Chairman George Beadle, and Caltech president Lee DuBridge. Photo 10.24–164. Courtesy of the Archives California Institute of Technology.

had contributed regularly to the institute for just as long. The two men were political birds of a feather. Both stressed a return to free-market economies after the war, became vehemently anticommunist, insisted on less government management and control, and a greater place for private citizens and corporations, a position Millikan had held since just after the First World War.[58] Millikan sided with Joseph McCarthy, while Earhart condemned the "raw dealers" and declared Herbert Hoover and Dwight D. Eisenhower "perhaps the two outstanding Christian statesmen in the world today."[59] Just before his death, Earhart looked on vice-presidential candidate Richard Nixon as "vitally interested in rescuing 'Rome from the barbarians.'"[60]

Plausibly, the elder Earhart saw his privately funded plant research laboratory in contrast to the government-managed science, and trusted that inquiry free of governments would best benefit mankind in the years ahead. In considering his support for Went's laboratory, Earhart sought assurances on both its scientific prestige and its promise that a private philanthropist would contribute toward humanity and he pointedly asked Millikan for his frank assessment of Went's program. Millikan said that he believed Went's air-conditioned greenhouses held the "No. 1 position" for prospective research projects at Caltech, with aviation the only possible rival.[61] Indeed, Millikan declared his "conviction" that "this Went project was the most vital research project that can be attacked with fairly immediate prospects of doing something for the benefit of mankind."

> I want to say quite categorically *now* that there is no project on my horizon that can be handled with funds of the magnitude which we have discussed that is comparable in importance with this project. Agriculture is the biggest industry in the world and this new attack furnishes the possibility of doing something the world over which will get away from the terribly expensive cut and try methods which agriculture has always had to use. It should enable us to determine before-hand by inexpensive laboratory experiments what kind of new crops and especially new plant hybrids are best suited to a particular area and a particular season.[62]

Millikan may have been stroking a major institute patron, as no doubt university presidents sometimes must, but his enthusiasm seemed utterly unreserved. "If it succeeds as I expect it to do it should bring results of greater significance to the future of mankind than I can see even in any projects in the field of the utilization of atomic energy. For this Went project has its feet tied to the ground as the other products of overstimulated imaginations have not."[63]

For Millikan to suggest in the post-Hiroshima world that *any* laboratory might surpass the expectations of nuclear physics seems an incredible claim, even more so because Millikan was himself a Nobel Prize laureate in physics. Yet Millikan had decided that a controlled environment laboratory for plant science might outshine atomic physics. The true promise of Went's project rested, for Millikan, on grand social and natural engineering ideals. Went followed Millikan's altogether conservative vision of private enterprise and technological innovation to address the problems of hunger and political instability in war-torn nations.[64] Combining the outlandishness of Went's plan, the conviction of Millikan's boosterism, and Earhart's fantastical finances, an emerging postwar conservative optimism saw private philanthropy build a facility at a private scientific institute for the revolutionary tailoring of crop species to climate to maximize yields. Just a few years later, an entirely unrelated project saw biologists working with the Atomic Energy Commission (AEC) irradiating a large open area from an intense radioactive source pursing the production of viable mutant plants via the new technologies of the atomic age.[65] To men like Millikan and Earhart, a private phytotron appeared conservative compared to a more liberal, government-funded gamma field.

The Earhart Foundation officially granted $200,000 to construct the "Earhart Foundation Botanical Laboratories of the California Institute of Technology" in the middle of January 1946. By the time it opened in 1949 it was the largest laboratory on the Caltech campus. Occupying some 7,000 square feet, the laboratory had six major greenhouses kept at stable day and night temperatures lit with sunlight and nine constant temperature rooms lit by artificial light. All rooms were air-conditioned from below, with air blown through the floor and circulated around the room. In the basement too were the wind rooms, the radioactivity room, and the studio: growth and development was charted using photographs over time. An emergency 30 kilowatt generator maintained electrical supply for the air compressors and lights in addition to mains power. All soil, plants, and people entering were sterilized and the air itself was passed through carbon filters to remove the "smog" that damaged plants. As we shall see in chapter 2, the discoveries of the chemical nature of smog and successful filtering methods were among the phytotron's first triumphs.

Over the next decade, Earhart contributed nearly a half million dollars to Went's vision. As the building took shape and costs ballooned, Earhart generously doubled the amount to over $400,000, and then, five years later, with operating costs spiraling rapidly upward, the foundation once more donated $50,000 for maintenance. Over the same period a number of patrons also underwrote the new laboratory. Some supported the technological vision: in 1951, Beadle and Went launched a campaign to enroll light companies in a "light research" project. Audaciously, they approached General Electric, Westinghouse, and Sylvania Electric with a $30,000 proposal to develop commercially efficient systems of plant growth under artificial light. The proposal suggested that newer bulbs had made some progress but also that plant physiology now revealed how "the plant itself has been very inefficient in converting visible light into the chemical energy of its photosynthetically-produced organic compounds."[66] Westinghouse politely declined, but General Electric, after inquiring what plants might be commercially viable (tomatoes, beets, strawberries, African violets, and orchids, Went proposed), donated $10,000 to support the "work at the Earhart Plant Research Laboratory on the effects of light on green plants."[67] Other grants flowed from research projects: in 1955 the phytotron got a $75,000 grant from the National Science Foundation to support visiting research fellows and graduate students,[68] and later the next year, the Rockefeller foundation donated $111,900 for research on chemical climatology to extend Arie Hagen-Smit's smog research.[69] The phytotron nearly doubled in size in 1960 when the Campbell Soup Company added another building to support its large research program on growing tomatoes in the American Southwest.[70] Such efforts confound the later portrait of Beadle deriding Went's efforts, or any suggestion that Went was an "albatross" to the biology division.[71] Rather Went's laboratory was among the earliest examples of the famed postwar expansion of science, such as when Linus Pauling secured $1.5 million from the Rockefeller Foundation for physical chemistry and his own Nobel Prize, and the Jet Propulsion Laboratory was founded to build missiles and, later, take men to the moon.[72]

FROM RESEARCH LABORATORY TO PHYTOTRON

Since the unofficial name of Went's laboratory, the phytotron, became its cause célèbre, it should be emphasized, as the institute took pains to continually emphasize, that the official name was the "Earhart Plant Research Laboratory." The foundation's generosity, like all endowments showered on Caltech throughout the Cold War, meant that the name of the edifice became paramount. Names always matter, but at Caltech in 1945 the debate over the name revealed everything. Early on, the foundation itself wanted "The Earhart Foundation Botanical Laboratory." Caltech's new president, Lee DuBridge disliked "botanical laboratory," George Beadle said to Went, because "it doesn't imply a broad enough scope." He asked Went what he thought about either "'The Earhart Foundation Laboratory of Plant Biology,' or 'Earhart Foundation Laboratory of Plant Science'?"[73] Went's mid-1948 trip to Ann Arbor to visit his patrons, the elder Mr. Earhart and his son, brought the topic of the name to the fore again; the trio agreed that "Earhart Plant Research Laboratory" seemed both apt and brief.[74]

Thimann gave the dedication at the opening of Went's new laboratory, in a mark of the new laboratory's disciplinary place with plant physiology.[75] But almost immediately, it seemed to some to be far more than just a laboratory. Surely it needed a much grander epithet, thought a pair of biological wits over morning coffee. Bonner and Wildman rechristened Went's shining statement of modern botany and plant physiology a "phytotron." Bonner recalls the story:

> the Earhart Plant Research Laboratory . . . was called an environmentally controlled greenhouse. But my first postdoctoral fellow [Sam Wildman] and I, sitting around about 1950, having coffee, decided it deserved a better or more euphonious name . . . we decided to call it a phytotron—phytos from the Greek word for plant, and tron as in cyclotron, a big complicated machine. Went was originally enormously annoyed by this word. But Dr. Millikan took it right up saying, "This edifice financed by Mr. Earhart, is going to do for plant biology what the cyclotron has done for physics," and he christened it a phytotron.[76]

If Went was annoyed with Bonner and Wildman, it quickly passed; more likely is that Went's supposed annoyance was just a product of an often-retold story by Bonner. The name "phytotron" actually emblazoned everything Went thought about his science and his new scientific facility. He embraced the layers of meaning the new name gave him, and became a prophet for the idea that the phytotron was a model of the physical sciences for the botanical sciences. In what became a near statement of faith, he wrote as early as 1950 that "the cyclotron . . . fulfills about the same function in physics as the phytotron does in the plant science."[77] As phytotrons spread globally over twenty years the mantra that phytotrons would "dissect the mechanisms of the plant as the cyclotron had the atom" was incanted religiously.[78]

As the historian of science Robert Proctor romped, "names are signals in some sense, like medals on a chest or eyespots on a peacock's tail."[79] It matters that the new plant laboratory was called a "phytotron" and it matters too that "phytotrons" spread out across the world. A quarter century later, the plant physiologists Robert Downs and Henry Hellmers in North Carolina still looked back to Went's phytotron, where "the basic principles of physics involved in controlling the environment" were laid down.[80] In other words, the term *phytotron* was not just a local quip, but rather a global coat of arms. Over coffee, playfully and jokingly—they first suggested "thermophotophytotron"—Bonner and Wildman had nonetheless hit on exactly the image of science many plant physiologists so desperately desired. Even while clearly sensitive to the large donation from the Earhart Foundation that had made the first controlled-environment laboratory possible, Went diplomatically suggested that the name merely derived from the combination of "the Greek *phyton,* meaning plant, and *tron,* which has come to mean device."[81] Yet, within a matter of months after being coined, Went had specifically told Caltech's own magazine that the "similarity between the term phytotron and such terms as betatron, synchrotron, cyclotron, and bevatron is intentional."[82]

That intention was evident even in Went and Eversole's initial proposal of late 1945. There the pair admitted that their conception of a controlled-environment laboratory rested on the alignment of the standards of the physical and biological sciences. Much of the impetus for their first phytotron had been, their proposal said clearly, "that especially physicists were reluctant to assist botanists in solving the physical problems connected with the life activities of plants." Physicists had not assisted or could not assist botanists because, quite simply, most botanical work was "poorly defined and practically uncontrolled."[83] Went subsequently embraced the cognomen "phytotron" considering it an award for the control of the experimental environment in botany and plant physiology that finally equated the biological and physical sciences. The Age of Biology was a time of physics.

To appreciate the power of the idea that a phytotron was a cyclotron for plants is to begin to understand how life sciences changed as biologists became technologists. In part, the creation of the first phytotron and its evocative name underscores plant physiologists' continuing nervousness about their status at mid-century. Reviewing the work on the green pigment in leaves as early as 1916, I. Jorgensen and W. Stiles argued that it was "This [research on photosynthesis that] is the prospect that plant physiology is developing into an exact science, utilizing the experiences of the fundamental sciences, physics and chemistry, but nevertheless a science, exact and independent, with its own working principles and methods, directing and stimulating the development of the applied sciences, agriculture and horticulture."[84] When the first comprehensive history of the discipline insisted self-consciously that "Plant physiology is an experimental science and is, therefore, based on experiments," it in effect highlighted the criteria by which science was legitimated, namely, by experiment.[85] The search for experimental command dominated the journal of the Society of Plant Physiologists, *Plant Physiology*, which counted "techniques, methods, analyses" and "environmental responses, stress" as two of the top three areas published during the 1930s.[86] A few years later, Bonner, similarly thought it necessary to open a paper concerning the role of plant physiology in agricultural progress with the assertion, "plant physiology is commonly regarded as a science."[87] Such a claim suggests that one could not simply assume physiology was a science. Bonner, at least, had to say it.

The physicists suffered no such queasiness. Physicists have long regarded all other science and technology as derivatives of physics. The theoretical physicist David Dennison, for example, stressed that "investigations which started in pure fields have become the foundation for new special fields," including satellite communications, the atomic bomb, and the X-ray techniques that led to the discovery of the structure of DNA.[88] Consequently, physicists have long ribbed their biologist colleagues. No less a figure than Vannevar Bush wittily recalled that during the heyday of radar development at the Massachusetts Institute of Technology (MIT) "scientific personnel became so scarce they even took in biologists and made radar experts out of them."[89] Or the story told in good humor by Bonner late in life about how, when he had been accepted into a biology department, his own father thought he had "gone crazy." Bonner said that to most people, "Biology was just a bunch of facts and no science; nothing rigorous about it." Bonner claimed that the physicist Willie Fowler had questioningly teased when they were both graduate students, "Biology? . . . how are you ever going to make a science out of that?"[90] One has to be especially careful not to accept at face value such ripping yarns told late in life, nonetheless contemporary moments offer evidence that the biologists were often the butt of jokes: Caltech's undergraduates annually roasted their famed institution, but saved their sharpest wit for the biologists. Hilariously, one year's Caltech Christmas pageant saw a student searching for his career through each of the institute's divisions in turn. In one scene, the chair of chemistry assures him that it is a "brilliant decision, son; a brilliant decision" to choose chemistry over other fields:

> CHEM: I shudder to think what might have happened to you. You might have decided to become a biologist. Then you would have been lost to science.
> STU: But sir, isn't biology a science?
> CHEM: Oh, come now, son. . . . Running rats and breeding fruit flies may be rather sporting, but let's not call it science. The only science the biologists know is a little organic chemistry, and they don't know much of that. . . . The biologists are nothing but a well-publicized gang of beggars. [All they do] is solicit funds, build buildings, and rob chemists of our subsidies.[91]

Incidentally, it is a line that has never gotten old: as the rest of Andy Weir's crew of astronauts learn that he is still alive on Mars, they write in good humor about the extra room they all have on the spaceship because they left him behind, however, they do have to do his tasks, "but it's only botany (not real science)" his astronautical engineering crewmate chuckles.[92]

Out in the wider world of science, the barbs were just as sharp but not all good-natured. As Bentley Glass prepared the new biology curriculum for American high school students in the late 1950s, he remained conscious of the "lowly esteem in which the biologist is held in comparison with [even] the chemist or physician," and his Biological Sciences Curriculum Committee vigorously fought to establish their credentials in the new technocratic society as equivalent to the physical sciences, even as the biologists suffered charges of "incompetence" from the physicists.[93] Likewise, the influential patron of science at the Rockefeller Foundation's Division of Natural Sciences, Warren Weaver, "shared with the physicist-biologists a view of biology as an underdeveloped subject, rich in potential but shackled by unscientific habits and traditions," according to the historian Robert Kohler. Significant to the development of postwar life science, Weaver consequently identified promising areas of biology to lavishly support, notably biophysics and the early molecular biologists.[94]

Perhaps it was physicists' assuredness that Went had in mind when he privately noted "the awe in which biologists hold physicists" in his diary during the opening of the Australian phytotron in 1962.[95] The appropriation of the suffix "tron" was part of that general awe. Moreover, the awe was widespread, including, for example, worries about their science's "ability to measure up to other kinds of hypothesis-testing science" from ecologists earlier in the century. The twentieth century saw a long campaign to justify ecology as a science by specifically parsing off applied ecology into conservation (leaving basic science as ecology), creating and defending larger theoretical entities such as ecosystems, and appropriating metaphors from the physical sciences, notably Harold Odum's ecosystem circuit diagrams.[96] These connections are especially salient because, as Sharon Kingsland has pointed out, the experience of ecology helped shaped Went's first phytotron at Caltech.[97] Identically, the historian of biology Betty Smocovitis explicitly argued that the emergence of the "evolutionary synthesis" in the middle third of the twentieth century was based on a coherent theoretical core, the parsing off of more natural-history-oriented evolutionary studies, and "successful adoption of experimentation in evolutionary practice through mathematical modeling" all of which permitted "biology to par with the physical sciences."[98] At Beadle's memorial years later, he was remembered as a visionary leader who sought "to bring to biology the insights of chemistry and physics."[99] And of course, the molecular biologists insisted that their successes, as the historian Doris Zallen argued, came via the marshaling of the "conceptual and quantitative tools from the physical sciences" not to mention the physicists themselves.[100] These works of history have replicated the labor of the sciences to bolster and legitimate a new science by establishing the explicit connections to physics. As Soraya de Chadarevian observed, "historians have been obsessed with the contribution of physicists in the origin of molecular biology."[101] Moreover, as biologists themselves have appropriated the methods of physics, so they have also appropriated its assumption that physics reigned as the queen of the sciences. "Ask Aristotle: what fundamentally *exists*?" the present-day evolutionary biologist Armand Marie Leroi has recently written, "he would not say—as a modern biologist might—'go ask a physicist': he'd point to a cuttlefish and say–*that*."[102] In other words, physics did not become the idealized science because it somehow *is*, it has been given that image since the time of Isaac Newton because other sciences appropriated its methods, adopted its language, and most of all when anyone asked "what fundamentally exists?" everyone answered "go ask a physicist."

Plant scientists told their disciples to go ask physicists as they labored to understand the mechanics of the developing plant. "The desirability of a basic understanding of physics and chemistry [for ecologists] need hardly be emphasized," said the Duke University ecologist Henry Oosting.[103] Late in his career, Bonner still said, "biologists should learn enough biochemistry, chemistry, physics and mathematics so that they can profit by interaction with biochemists, chemists, physicists and mathematicians, and that through such interaction occasional great insights may be brought forth."[104] By the mid-twentieth century, plant physiology texts chided earlier researchers' "ignorance of physics and chemistry," and noted how "more accurate conclusions might be arrived at by exercising greater care in the carrying out of experiments."[105] The plant physiologist, Erwin Bünning, later lamented the delayed work on phototropism, noting that "with better training of botanists in basic physical laws," long years of error might have been avoided. During the war, Bünning had starred alongside Pascal Jordan to debate whether organisms worked on Heisenberg's uncertainty principle.[106] Subsequently, Went also directly employed the quantum mechanical arguments of Niels Bohr, Pascal Jordan, and Erwin Schrödinger to argue in the early 1950s that the commonly assumed range of variability in genetically uniform organisms "is not due to statistical fluctuations of the numbers of molecules on which development depends, but is largely caused by inconstancy of and irregularities in the external environment."[107] Advanced physics including quantum mechanics and special relativity had become part of the scientific world of plant physiologists, botanists, and ecologists. When Robert Emerson moved from Caltech to the University of Illinois, it was on the condition that the university hire a physicist or physical chemist to work on his Photosynthesis Project.[108] Oosting concluded as had Emerson that the physical sciences set the standards of experimental control and that increasingly accurate measurements needed to be employed throughout experimental biology.[109] Quite explicitly, Sterling Hendricks, who would coauthor an article with Went advocating a national phytotron, said in his memoirs that "the benign efforts of so many in research, even including something so distant as my work on plants, has origins in . . . two simple expressions . . . the photoelectric effect . . . $E = \phi N h\nu$" and "the expression for the equivalence of mass and energy . . . $E = mc^2$."[110]

At the same time, the explicit integration of the sciences of botany and physics by the cognomen "phytotron" built on an explicitly interdisciplinary culture of science at Caltech. Millikan had always stressed an interdisciplinary organization of science during the initial growth years of the institute in the 1920s and 1930s. As the historian of science Robert Kargon argued, the story of the rise of astrophysics at Caltech is one in which the branches of physics, chemistry, mathematics, and astronomy were all gathered together into a single division at the institute, their progress linked.[111] Subsequently, botany and plant physiology as much as molecular biology were all built on a foundation of interdisciplinary cooperation. The members of the biology division were aware that Rockefeller Foundation support in the early years had secured the "most intimate cooperation with outstanding departments in physics and chemistry" and their "atmosphere of research" that would "bear [the] logical fruit in the biological science."[112] Beadle's close alliance with the chemist Linus Pauling helped build molecular biology at Caltech and forged a powerful interdisciplinary effort between the chemistry and biology divisions.[113] To Went and others, Caltech was at the vanguard of a broad movement to unify disciplines. Caltech of the 1950s glorified the idea that its leading research reputation was, with the exception of MIT, built on bringing the physical to the biological sciences. In a candid letter to DuBridge, Warren Weaver at the Rockefeller Foundation thought "it rather natural to think of the fact that CIT and MIT are two of the leading places where the techniques of the physical sciences are being effectively applied to biological problems."[114] "Phytos . . . for plant, and tron . . . , a big complicated machine" was then as much a description of Went's plant laboratory as Weaver's description of science itself.

For Went and Eversole, technology served to unify disciplines. In their original application, Went and Eversole described how only environmental control produced reproducible conditions that would enable other scientists to work with botanists to expand knowledge "into the borderfield between botany and physics, a practically virgin field."[115] Their plant laboratory was in part a desire to establish common standards of practice to enable research across disciplines, but was also a move toward a larger idealistic goal of unifying science. Botany, far from a coherent discipline, Went unabashedly told the International Botanical Conference in Stockholm, might be unified via experimental control of the environment: he "hoped that the new development of air conditioning greenhouses will form a new tie between botanists of all denominations and produce a new more unified science." Into that field, wearing the heraldic device of the "phytotron," all manner of life scientists, Went said, including "Plant Physiologists, . . . Morphologists, Experimental Taxonomists, Anatomists, Geneticists, Agriculturalists, Horticulturalists, Biochemists and others" could adventure forth "to find now unexpected possibilities."[116]

The goal of unifying the biological sciences was also very conservative. Variously, molecular biology, biochemistry, and evolutionary biology all heralded themselves as uniting the various disciplines of biology in the era after the Second World War.[117] The larger point is that one has to appreciate how a claim to unify science serves a key legitimating act for any new theory, research program, or technological innovation. Smocovitis, for example, argued that by "disciplining evolutionary biology—the *fulcrum* of the biological sciences—[the evolutionary biologists] were to act as unifiers, negotiators of the location of biology, preserving the whole of the positivistic ordering of knowledge."[118] However, while there is no doubt that the evolutionary biologists *claimed* to unify the biological sciences, so too did the molecular biologists, and the biochemists, and so too did Went from his new phytotron. Explicitly, one of the expectations of his new phytotron, Went said, would be that the future would see "the converging, instead of the divergence, of all botanical disciplines."[119] In other words, the ambition to unify the "botanical disciplines" reveals the scientists' underlying assumptions about science: claims to authority were legitimated as much by a scientist standing before his new technology as by claiming to work between disciplines; perhaps mindful of Fowler's graduate school quip, Bonner used to say that he "roams around in the field of biochemistry and plant physiology."[120]

In short, the phytotron was an image of where that array of sciences known collectively as "biology" seemed to be headed. Consider the position of a new student coming to Caltech in 1950. They would read the *Catalogue*, the public face of the institute, to learn what was expected from their postwar majors in biology:

> At the present time biology is one of the most rapidly expanding fields of modern science. In recent years theoretical and practical advances of the most spectacular kind have been made in our knowledge of living matter. This is especially true of those branches of biology in which it has been found possible to utilize physical, chemical, and mathematical methods in the investigation of biological phenomena. A strong demand for physio-chemical biologists now exists and qualified men will find excellent opportunities for careers. . . . Because of the pre-eminent position of the California Institute in both the physical and biological sciences . . . [and] the foundation in the physical sciences received by all students at the Institute, emphasis is placed on the physicochemical viewpoint in the study of living

> systems [whether in] animal biochemistry, plant biochemistry, bio-organic chemistry, animal and plant genetics, chemical genetics, immunology, biophysics, mammalian physiology, comparative physiology, plant physiology, virology, and experimental embryology.[121]

Here is Caltech's biology division's understanding of modern biology at mid-century. Any new student, the catalogue's purported audience, may well have noted the stress laid on the "physio-chemical viewpoint" as unifying an array of specializations including biochemistry, genetics, and physiology. It was a viewpoint that existed in other institutions of biology, including Johns Hopkins, where for example a reenergized biology department under Benjamin Willier equally embraced "the application of physio-chemical methods" to underpin new multiple appointments in physiology, genetics, and biochemistry.[122] At the same time, the catalogue advertised and acknowledged that "biology" contained a large number of interests, techniques, and specializations, not to mention experimental organisms. Biology was disciplinarily diverse: among the staff of the biology division, Went and Eversole were "plant physiologists," whereas Bonner and Beadle were "biologists," but Delbrück a "biophysicist." At the same time, biology was methodologically diverse with an abundance of practices and organisms: when Caltech's biologists assembled their promotional materials they listed no less than twenty-six "fields" of biological interest, collected photographs of sixteen instruments and laboratories, and put a half dozen organisms including *Neurospora, Oenothera, Drosophila,* maize, rats, and guinea pigs on the cover.[123] Quite simply, "biology" around mid-century celebrated its heterogeneity, and Went proclaimed that his new *tron* could well control the experimental environment for any organism.

A FUTURE OF COMPUTERS, HIGHWAYS, AND PHYTOTRONS

Only days after the Victory in Europe Day, the American secretary of war Henry Stimson was warning his new president Harry S. Truman about the dangers inherent in food shortages. Fearing that Western Europe would be "driven to revolution or Communism by famine," from its very beginning the Cold War connected food and political stability.[124] Harry Earhart shared those same concerns, and took action. In the immediate postwar world, he saw Caltech as the dominant institution in science and technology, dedicated to interdisciplinary cooperation, determined to build large technological facilities, and armed with a powerful group of biologists for whom plant physiology was a prominent and noted discipline. He supported one plant physiologist to move swiftly toward establishing a new laboratory for plant physiology and a new style of biological science centered on the control of the physical environment of growing plants because it was an established experimental fact that "temperature and light have an enormous impact on the growth of the different plant parts."[125]

Went's modernist solution to a clear scientific problem, namely, his phytotron, meshed exactly with Earhart's conservative response to what he saw the postwar world facing. In short, technological control offered solutions to both scientific and social problems: Went repeatedly anticipated that the methods his phytotron developed "for analyzing the complex interrelations between organisms and their environment, will be helpful in an analysis and better understanding of our social and economic system."[126] Went believed that any array of complex interactions required, as the wartime experience of systems engineering dramatically showed, the governance of the entire system, not merely individualized components within the system.[127] Social valence was added to political valence in the immediate environs of Pasadena. Went's phytotron represented a glorified technological modernism that promised a new prosperity to the masses. Went embraced a modernist vision in his phytotron (and then even more overtly in his later Climatron at the Missouri Botanical Garden). When Went toured European botanical institutions in 1950, he specifically noted that Europe's greatest hindrance lay in "the existence of so many old buildings. Not only is the maintenance cost of an old building out of proportion, but it stymies the imagination."[128]

The gleam of a modern science of controlled experimentation heralded the reality of progress in plant physiology as much as Le Corbusier's model for towering skyscrapers remade Paris or the freeways of Robert Moses remade New York. With a similar vision, Went believed his phytotron swept clean traditional botanical practice and replaced it with measurable climatic variables, photographs of growth and development, and above all an electronic computer. The computer, purpose-built to run, monitor, and record all environmental conditions in every room, both oversaw and displayed the control over the biological environment, as visitors, reporters, and fellow scientists frequently noted. His electronic

Figure 1.5. "Frits Went at the control panel for the Earhart Plant Research Laboratory." Photo 10.24–136. Courtesy of the Archives California Institute of Technology.

computer permitted the new experimental plant science, Went claimed: "Electronics has brought us a new era of measuring, and of instruments, which in the hands of technicians produce wonders of accuracy," Went gushed.[129] Occupying a prized central space and taking on totemic status, the computer stood as an icon of the modernist turn Went pursued for plant science (fig. 1.5). It was the first computer in a greenhouse. It would not be the last.

CHAPTER 2

AT WORK IN THE CALTECH PHYTOTRON

> The uncertainty in conclusions reached in biological experimentation has led physicists and chemists to distinguish themselves as working in the "exact" sciences.
>
> — **Frits Went, 1956**

AMONG THE charms of a small, intense university like the California Institute of Technology (Caltech), its annual Christmas pageant, written, directed, and performed by students, provides a lighthearted window into how the local community understood science. In the 1949 show, Frits Went's tomato research received no less than an entire song, normally the amount of space that went to an entire discipline. A classic comedy roast, the effort of the students to compose and perform the song speaks to their admiration for their famed physiology professor, his clear obsession over growing tomato plants, and Caltech's new phytotron. Titled "Why Don't You Grow Right?"—sung to "an ailing tomato plant,"' a no doubt smooth crooner serenaded the plant and audience:

> You say something's ailing you down in your root
> But I thinks you'se lazy, you don't wanna fruit!
> Why don't you grow right like some other plants doo-oo?
> Metabolize; take in some CO_2

You've had plenty vitamins and minerals too
And lots of attention from the phytotron crew
Why don't you grow right like some other plants doo-oo?
Metabolize; take in some CO_2

In order to boost up your primordial
I'll give you a shot or two of auxin-a
To make you grow right like some other plants doo-oo.
Why don't you fruit—like some other plants do?[1]

Such a performance suggests that Went was the toast of Caltech in the late 1940s because he had finally cracked how to make plants "grow right" by building a phytotron. For the next decade, the Earhart Plant Research Laboratory, the phytotron, was one of the jewels of botanical and plant physiological research: in 1955, the chairman of Caltech's biology division, George Beadle, wrote to the original Earhart patrons, "scientists from all over the country and the world continue to flock here to work in this unique laboratory."[2] Twenty years later, however, plant research had completely disappeared at Caltech, and the phytotron itself had been demolished.

The life and death of the first phytotron, 1949–71, neatly bookends many of the most dramatic changes to the shape of biological knowledge in the past sixty years. This chapter describes how biological knowledge involved the technological labor to determine the exact set of conditions that produced a defined and measurable plant, in other words, a plant "grown right." Went's overall aim was the elimination of variability from fluctuating environments to produce rooms of identical plants (fig. 2.1). Variability, Went announced in *Science* in October 1950, presented "some of the most important problems in biological research."[3] Later, in a 1956 issue of *American Scientist*, Went reiterated that the lack of control over variability was "usually the greatest handicap in biological experimentation." In these major scientific publications, Went said that "one of the most important results obtained in these air-conditioned greenhouses and growing rooms is the extent to which biological variability can be reduced."[4] Previously, the study of heredity had concentrated on establishing the often-broad variations across types since the late nineteenth century.[5] Surveying his own field of evolutionary biology nearer the end of the twentieth century, for example, Ernst Mayr noted how "most nat-

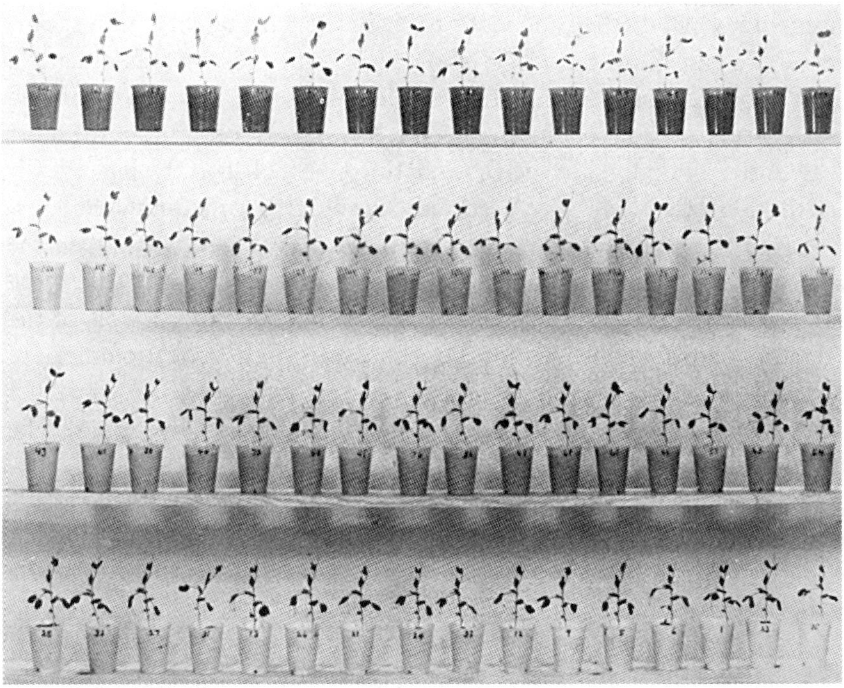

Figure 2.1. "Uniformity of Pea Plants." From Went, *The Experimental Control of Plant Growth*, plate XX. Reproduced with permission from John Wiley & Sons Ltd.

uralists . . . stressed that variability is a normal attribute of populations and that what characterizes populations is indeed the kind and amount of variability."[6] In contrast, Went understood variation quite differently. Variation was not a normal attribute of experimental organisms, but a sign of a lack of standardized control over the experimental subject. To be grown right was to be grown under controlled conditions. To be identical meant that plants possessed identical genes and had been grown in identical environments. Moreover, only with identical plants could the research agenda of plant hormones, metabolism, flowering, or nutrients go forward. Consequently, under the glass of the first phytotron in the 1950s and 1960s, a set of practices emerged to produce a plant grown right leading to a style of science of the phenotype that saw biological knowledge equated technological control.

In addition, because the phytotron established a claim to right-grown plants, it could also evaluate how and why plants grew wrong, most immediately when they appeared sickly in the haze by then almost permanently engulfing Los Angeles known as smog. Smog was already a serious health and environmental problem in Los Angeles and elsewhere by the 1940s, for both plants and people. As we shall see, some of the most influential work done in the first phytotron established the biologically harmful effects of smog and led directly to curbs on automobile emissions. In retrospect, it was unfortunate that Went never leveraged the smog work to gain more authority for his phytotron. If we accept the famed sociologist of science Bruno Latour's insights on the role of the laboratory in the wider world, Went thus missed a critical moment to solidify the fact of the environment's role in biology and to construct the phytotron as the essential laboratory where other such facts could be discovered.[7]

Instead, Went sought the production of the right-grown plant because only with identical plants could botany and plant physiology become experimental sciences as the "exact" sciences were. Claiming to be a "biologist," and speaking around the height of his influence in biological circles, Went talked about the geneticists, physiologists, agronomists, and entomologists then researching plants, before announcing a new "group of scientists," the "Plant Climatologists." Plant climatology had been a difficult subject, he explained, because of technical limitations to every scientist's ability to grow plants "under strictly controlled environments or synthetic climates." However, it was a subject whose time had arrived because his Earhart Plant Research Laboratory permitted "for the first time" the desired experimental conditions. Until that point, Went concluded, "the uncertainty in conclusions reached in biological experimentation has led physicists and chemists to distinguish themselves as working in the 'exact' sciences."[8] No longer. Accepting that the definition of an organism was a product of its genes and environments and that plant breeders and geneticists had established control over the genes, environmental control was the next frontier toward establishing an experimental basis for plant science.

The struggle over the acceptance of environmental control was a contest, Went believed, that would decide whether biology became an exact science or remain just a bunch of facts. The environment was made into an experimental object by being broken into discrete variables such as temperature and light and then building technological systems to make each variable reproducible. Those systems specified and made measurable each individual component of the environment. Remarkably, however, the labor to determine the exact set of technological systems that produced controlled environments also meant that the phytotronists developed an unexpected but growing awareness of feedback in the daily practice of operating the phytotron.

Feedback, in turn, had two significant consequences. First, feedback offered deeper insights into the shape of the environment, or what we might consider the nature of Nature. Was there, the phytotronists began to ask, actually any possibility of keeping each element of climate controlled without disrupting another? A debate subsequently erupted over whether the phytotronic environment represented a "normal" environment, and indeed a debate over what a "normal" environment might even be. While it continued into the 1980s, the high point saw a Caltech phytotronist defend the idea that normal nature was at best complex and perhaps even entirely chaotic, leading to speculation that what biology really needed was a "chaosotron."

Second, feedback undercut the idea that a biological science of the environment might be reductionist, even as reductionism rapidly became the dogma of other life sciences fixated on genes, especially molecular biology. Dismissing the claims of his molecular biologist colleagues, Went told his classes in the 1950s that while the cell might be reduced to chemistry, "the tree has graduated from molecular biology."[9] He went so far as to declare molecular biology trivial because it involved only molecular interactions for which "one does not need to know much more than chemistry, quantum physics, thermodynamics."[10] Such vitriol may now sound extreme, but it suggests that Went believed the soul of biology was at stake with the rise of molecular biology. In contrast to the emerging reductionism of life to molecules, Went and the phytotronists believed biology to be a science of genes and environments.

THE PHENOTYPE AND THE PHYTOTRON

The immense problem of the complexity of "the living organism" was multiplied by the "complex physical system [of] climate," Went lectured his undergraduate students throughout the 1950s. The only way forward, he advised those budding biologists, was to work with "genetically homogeneous material analyzing one factor at a time: light, temp."[11] After 1949, he pointed out to his students that this is exactly what researchers did in the phytotron. In one of his own papers coauthored with Marcella Juhren and William Hiesey, Went demonstrated how the responses of different species of bluegrasses of the chromosomally complex genera *Poa* could be studied "under controlled conditions with but a single known external variable."[12]

From the undergraduate lecture to the research journal, the technology of the phytotron solved biology's two-body problem: genotype + environment = phenotype. It was the Danish plant physiologist Wilhelm Johannsen who defined the distinction between phenotype and genotype. Johannsen's "genotype," as the historians Staffan Müller-Wille and Hans-Jörg Rheinberger observed, "referred to whatever remained identical in living beings through generations and was therefore amenable to experimentation, just like the molecules in chemistry and the atoms in physics."[13] Plant and animal breeding for agriculture and sport had encouraged and promoted study of the genotype. Classical Mendelian geneticists, focusing on the production of new characters in evolution, simplified their work through the assumption that the phenotype approximately expressed the genotype (i.e., genotype ≅ phenotype). For geneticists, the range of variation across a single population was irrelevant because any environmental differences operated on genetically identical organisms.[14] Twin studies became popular because twins shared identical genetic material and approximately identical environments, or different environments in the case of separated twins.[15]

But in his original work, Johannsen had identified not one but five distinct types of "variability" within any identifiable species. His work was important for the research programs of botanists and plant physiologists: Johannsen had distinguished the various results from hybrids governed by Mendelian genetics and mutations from what he termed "individual and fluctuating variability" and the "often conspicuous differences . . . which develop under strongly diverging external conditions," in other words, the environment.[16] In practical terms, the search for genes was only a subset of a larger biological program. At least in France, the historian Christophe Bonneuil argued that applied plant physiology, botany, and agriculture commonly relegated the gene to secondary importance until perhaps as late as the 1970s.[17] Likewise, the existence of the gene mattered little to the studies and standards of plant improvement originating in Sweden, and to the subsequent botanists, plant physiologists, horticulturalists, and breeders in a host of agricultural settings, research stations, and plant breeding industries. None of these scientists were overly concerned with either the pursuit or discovery of the "gene" but rather employed empirical breeding techniques to generate huge numbers of "lottery" varieties from the later nineteenth century onward.[18] Still, the growing obsession with genes gave rise to experimental programs every bit as much a part of the atomic age as the phytotron, namely, the shotgun approach of Ralph Singleton in the "gamma field" of the American National Laboratory at Brookhaven in pursuit of viable mutant plants via the irradiation of a large open area from an intense radioactive source.[19]

Quite simply, such successes meant that the twentieth century was awash with attempts to rationalize the time-consuming and expensive selection process because potentially valuable breeds might be identified. The controlled environment of the phytotron offered an immediate solution to the selection of new breeds. James Bonner would tell a meeting of the National Academy of Sciences in 1959 that he considered the Campbell Soup tomato program "one of the great uses of a phytotron" because it specifically connected the facility to plant breeding work. When the president of the Agricultural Research Institute asked if the geneticist "would get more for his money?" with phytotronic research, Bonner answered that "the greediness of the geneticist for large numbers [of plants to be tested for climatic adaptation] has to be balanced in the phytotron by the number of dollars he has." While the cost per square foot of growing space was nearly six times the cost of the same space in a field, Bonner noted, the geneticist only needed one-tenth of the plants in a phytotron. Moreover, "he saves time. . . . in the phytotron screening program he will be able to screen through F_4 in 18 months. In the field he uses up at best four years, more if the kind of summers are not right."[20]

Its boosters believed that the phytotron fulfilled the dreams of generations of biologists. By holding an organism's genes constant, breaking the environment into its components, and then varying only one component at a time while holding all the others constant, phytotrons made the phenotype experimental. For plant physiologists, half of the equation, the problem of standardizing the genotype, was largely resolved by the 1950s: Went chose "a good uniform variety" of tomato, while Harry Highkin used peas because "it is easy to obtain inbred homozygous strains."[21] Highkin, who came to Caltech in the 1950s, initially as a research fellow courtesy of a National Science Fund (NSF) grant,[22] spent his first year selecting pea varieties' "most significant quantitative characters and most convenient methods of measuring." His second year was then occupied with developing the "culture of the selected varieties" toward the overall "physiological, genetical and evolutionary studies."[23] The genotype seemed well understood and under control.

The real problem, plant physiologists insisted, was to create an experimental science of the other half of the equation, the environment, to accompany the assured science of genes and breeding. It was accepted as an underlying fact that genetically identical organisms grown under differing conditions gave often radically different experimental outcomes. Significantly, the demonstration was already a standard pedagogical exercise at Caltech by the 1940s. The "Bean Test" experiment for new graduate students served to demonstrate that "many biological assays unfortunately are so sensitive to slight changes of conditions that the results of different tests will be inconsistent although the tests will be consistent within themselves." At Caltech, Went, and fellow physiologists, Arie Haagen-Smit and James Bonner, regularly conducted the Bean Test experiment in the Biological Assays course. Students learned to explode one core assumption of genetics work: the assumed claim that the experimenter held identical experimental, reproducible organisms on which one might begin to assuredly experiment. The experimental procedure served to convince the student that not only would the usual culprit, "slight differences of [experimental] technique," produce inconsistent results, but so to would the beans themselves "because the supply of beans is not controlled and the selection of proper individuals is largely intuitive."[24] The evident conclusion was that every experimenter had to ensure that the variability of the bean stock was itself not the experimenters' undoing. And at least at Caltech, graduate students in botany and physiology soon learned that the tests on which they had come to rely were often anything but reliable because the same genotype grown under different conditions possessed a different phenotype that corrupted the assays' results.

The power of the phytotron was its capability to assure physiologists that they held phenotypically identical organisms via identical genes and growing environments. Pedagogically, doctoral dissertations opened with standard demonstrations, such as that of Went's student, Jean Paul Nitsch. Nitsch invoked the argument from the biological assays course about the proper reproducible standards of both experimental procedures and plant materials. Notably, Nitsch devoted the first chap-

TABLE I

THE EFFECT OF ENVIRONMENT ON THE FLOWERING OF THE SMALL GHERKIN (<u>CUCUMIS ANGURIA L</u>.)

Flower types found on the vines 83 days after sowing.

Temperature		Length of day	Type of open flowers
Day	Night		
30°C.	30°C.	16 hours	no flowers
30°C.	30°C.	10	no flowers
30°C.	30°C.	8	no flowers
30°C.	26°C.	8	no flowers
30°C.	23°C.	~~8~~ 16	male
26°C.	23°C.	16	male
26°C.	23°C.	8	male + female
26°C.	20°C.	10	male + female
26°C.	20°C.	8	male + female
23°C.	23°C.	8	male + female
23°C.	17°C.	10	female
23°C.	17°C.	8	female(1)

(1) Some occasional male flowers opened in this series.

Figure 2.2. "Table 1," from Nitsch, "The Role of Plant Hormones," 7.

ter of his dissertation to demonstrating that growing conditions bear significantly on the development of organisms. In his particularly nice example, Nitsch planted gherkins in the phytotron for eighty-three days. Nitsch specified as part of his training the control of every condition from nutrients (the gherkins were watered with standard Hoagland's solutions twice a day) to insulating against outside contaminants and pollination, to what conditions the gherkin seedlings were grown in. Nitsch experimented with seven different combinations of day and night temperatures, and tabulated his results, as shown in figure 2.2.

The results, which are given first in the overall argument of Nitsch's dissertation, are presented as the clearest demonstration of the effect of temperature on development. The experiment showed that a difference of just 3°C in the daytime temperature combined with 6°C at night caused the distinct shift from all female to all male flowers, that the higher the daytime and nighttime temperatures, the more male flowers gherkins produced, up to an absolute limit of 26°C night temperature when no flowers are produced regardless of day length.[25] As Nitsch summarized, "in contrast to previous work, these experiments show that environment has a profound effect not only on the number but also the type of flowers produced."[26] In his fine dissertation, Nitsch displayed the new "experimental biology" in phytotrons. An experimental result like Nitsch's served as a scientific and pedagogical demonstration of the necessity of establishing controlled environmental conditions. For Nitsch "previous studies" lacked scientific legitimacy because they had been unable to use standard plants that were both genotypically and phenotypically reproducible. In short, the message from Nitsch to the community of plant scientists was that standard genetic organisms tested against standard reproducible climatic variables in phytotrons would reveal the right phenotype.

Pedagogy begat practice. An experiment from early 1955 illustrates fully how plant scientists created the "right grown" plant for use in experiments by eliminating variability from their experimental organisms. Bonner and his student, Mary Lou Whaling,[27] planned fifty-three experimental trials that would take place on six different varieties of *Avena* (oat) seedlings, both hulled and not hulled, measuring between two and four centimeters at planting over five months. The experimental series sought to reveal the effect of the method of planting and the length of the coleoptile (the sheath around the shoot tip in grasses and cereals) on the initial growth rates. Across six varieties, the experimenters tried various concentrations of growth hormones and temperatures and measured the results in terms of growth. After some thirty experiments conducted at controlled temperatures between 5°C and 25°C and with concentrations of growth hormones IAA (indoleacetic acid) between 17.5 mg/100 ml and 70 mg/100 ml (or 1×10^{-3}M) and of 2,4-D between 22.1 mg/100 ml and 88.4 mg / 100ml (or 1×10^{-3}M), the pair began experimenting with inhibited growth hormones like DCA and 2,4,6-T. In addition, there were several seemingly outlying experiments such as the attempt to slow growth by adding between 1 ml and 5ml of $CaCl_2$ (calcium chloride) to the control solution.

Such research stemmed directly from Bonner's two-decade-long search for growth hormones to regulate or inhibit growth, stem elongation, or flowering. Such a research program required, of course, a standard set of plants all grown right so as to conclude that the growth or flowering was indeed the product of an added hormone. Consequently, before any standard solution could be used to determine growth effects, standard plants needed to be grown right. Bonner and Whaling generated standard stocks of experimental organisms by taking genetically uniform seeds and growing them for ninety-six hours in the phytotron's "red room" at 25°C. Of those plants, only those that had grown between 2.75 mm and 3.25 mm in length from the first node to the tip of the shoot were selected. The rest were discarded. The temperature of 25°C was not chosen arbitrarily, but rather represented the temperature at which the least variation in growth rate occurred. Critically, previous experimentation had been necessary to establish this point of minimum variability in growth. As the associated chart at the beginning of the experimental notebook displayed, at 30°C, 3 cm nonhulled plants in a standard solution showed growth rates anywhere between 0.22 mm/hr and 0.32 mm/hr.[28] Likewise at 15°C, the pair saw growth rates of between 0.12 mm/hr and 0.14 mm/hr. At 25°C, however, the shoots only grew between 0.22 mm/hr and 0.23 mm/hr.[29] To a known minimal error, the shoots were thus experimentally reproducible.

This experimental practice had deep consequences, its practitioners argued. On the one hand, the production of strictly controlled environments in the first phytotron had distinct epistemological implications, namely, that the plant physiologists claimed to possess secure knowledge about the growth and development of plants because they now held genotypically and phenotypically identical experimental organisms that came from a set of practices. The experimental practice is another excellent example of the schema developed by Hans-Jörg Rheinberger, although the phytotronists did not possess his descriptive language. While the technical object of plant science was a reproducible standard plant, the more important epistemic object was the means of that reproduction, namely, the set of experimental practices to create standard plants. In this case, the epistemic object of phytotronics was what came from both the creation of a phytotron as a technological complex and the creation of the twofold procedure.[30]

On the other hand (and to plant biologists the more immediately important), such an experimental practice solved the problem of Lysenko, that Soviet crusader against Mendelian genetics. As Went explicitly argued, what had allowed the Lysenko "controversy to develop" was "a lack of experimentally sound evidence as to the effects of environment on organisms . . . due to the technical difficulties involved in experimentation on the effects of environmental factors on the growing plant." By the 1950s, Went noted, "the genetical background of variability is now either well understood or is under investigation," and certainly no one inside a phytotron rejected genetics as Lysenko had done, nor did they support the claim that traits could be passed on to the following generation.[31] As the Cold War intensified during the Korean War, the persecution of geneticists in Russia had made saints of them in America. In contrast, students of the biological environment suffered as seemingly passive fellow travelers of Soviet science: "the environment has either been neglected or has been make a political issue," Went acknowledged, because the only way for many scientists to avoid being seen as even tacitly supporting communism was to ignore the study of the biological environment. In a remarkable moment, Went suggested that while politics had corrupted genetics in the worker's paradise, politics had likewise corrupted the environment in the home of capital. The whole distasteful affair had impeded research into the conception and measure of the "environment" in the growth and development of plants, Went wrote in 1957, and had made it difficult to persuade people that the lack of evidence was not a failing of the botanists or plant physiologists, or a theoretical absence, but simply a technological shortcoming. Once plant scientists took reliable genetic stocks and produced homogeneous phenotypes under controlled conditions, Went argued, the necessary requirement of "*reproducibility*" of results followed.[32]

Reproducibility was a standard by which the phytotronists measured their claim to establish facts. Reproducibility took a number of forms, namely, the reproducibility of the experimental plants, reproducibility of different variations under controlled conditions, and finally reproduction of results by other groups. In the wake of Bonner and Whaling's experiments on coleoptile extension growth, for example, Bonner could directly compare results with Noel Kefford in the Australian group in Canberra. Kefford wanted to know "what Pasadena vermiculite has that distilled water doesn't have." The production of standard plants meant that one major variable of botanical and agricultural work could be eliminated, allowing research to concentrate on environmental causes of growth and development. As Bonner said to Kefford, "you would do a public service to write up the results that you have up to the present time to discuss in details the time course as a function of the different environmental variables and to end up with a specific discussion of the initial growth rate in high auxin concentration. It should be possible, I would think, to find out without too much further difficulty the reason for the difference in initial growth rates with Pasadena vermiculite-grown plants as opposed to Canberra vermiculite-grown plants."[33] Briefly, from a single plant, through a range of environmental and chemical variables, the experiment migrated outward taking it and its experimenters all the way to Australia. Of course, such claims to the reproducibility of results lent legitimation to the scientists and their knowledge because the normative expectation is that if knowledge can travel it must be truly part of a universal idea.

Bonner and Whaling's experimental run in the mid-1950s was just one example among hundreds of the mature phytotron at work. Now masters of standard plants and operating under standard conditions, scientists could engage in comparative work to discover the causes of differential growth and development. Broadly, biologists continued to search for mechanisms:[34] Went, Thimann, and Bonner pursued hormones in the 1930s and subsequently they and many others pursed the mechanism of the environment in the 1950s. In its first decade, the phytotron hosted research on photosynthesis and the utilization of light energy by plants. It enabled physiological study of the climatic response of crop plants, particularly, Went noted, beets, tomatoes, strawberries, and corn grasses.[35] Between 1952 and 1954, Luis Gregory placed the maximum yield for Kennebac potatoes at 20°C day and 14°C night temperatures, the same conditions for eggplant. G. C. Camus found that tobacco varieties not only grew at remarkably different rates, those rates changed over the age of the plant: for example, the Cuba White variety begins life with an optimal night temperature of 30°C, which shifts down to 14°C. At the same time, Camus also noted that light intensity had an effect on new leaf formation, nearly an extra day separated new leaves grown at an intensity of 500 ft-c. (foot-candle) less than optimal. Lloyd Evans found that broad beans reached their maximum growth of shoot length in just fifteen days at 26°C but it took seventy days at 7°C. Meanwhile, Albert Ulrich, sponsored by the Sugar Beet Development Foundation, revealed that while the weight of the beet increased with day and night temperature to a maximum at 23°C, the sugar content was highest at a night temperature of 4°C (nearly 12 percent) decreasing steadily with increasing temperature, a remarkable conclusion.[36]

Went was especially excited to join with a pair of plant pathologists from the University of Wisconsin from 1953 to 1957 and successfully gained American Cancer Society funding for an experiment in tissue differentiation in trees, "galls" via insects. Under Paul Kramer, a stand of standard oak trees was grown in the phytotron, as his collaborators developed tissue cultures from normal and abnormal (gall) tissues. The first year of Went's grant produced only negative results, with a lack of abnormal tissue induced on the trees, and even a lack of insects successfully integrated into the growth rooms of the phytotron. The use of the phytotron's controlled spaces was justified on the basis that nature was uncontrolled: "wild populations are too uncertain as to availability and freedom from parasites."[37] In April 1956, Went could suddenly report success. After finding "the ideal investigator," an entomologist named William Hovanitz who could obtain adult insects, "extracts from abdominal glands from these Pantania flies have produced gall-like swellings for the first time."[38] Went wrote to the American Cancer Society hoping to add an entomologist to the staff of his phytotron for the project, noting: "I think that our success is just based on improved technique."[39]

Another major area of research was the "ecology and physiology of the native vegetation, a counterpart to the investigation of crop plants, in which evolution becomes an experimental problem."[40] Here Went worked alongside Alberto Soriano on desert germination. Early on, the pair proudly announced how the controlled laboratory experiments revealed that a certain total amount as well as density of rain were necessary conditions for full germination of seeds from desert plants.[41] Two years later, Went sought a dedicated staff member working on "experimental ecology" from Caltech president Lee DuBridge.[42] What seems equally interesting is Went's focus, already apparent in the early 1950s, on an experimental science of evolution. Importantly, however, such an experimental science would be the culmination of basic physiological problems (photosynthesis), followed by studies of the genotype and physiological relationships to the environment. Only after these discrete studies could the overall ecology and physiology of evolving organisms become an experimental reality. Air-conditioning provided environmental control and

Figure 2.3. From left to right, Frits Went, James Bonner, and an unnamed researcher comparing growth rates in the phytotron. n.d. Photo 10.24–21. Archives. California Institute of Technology.

permitted, Went argued, the production of reproducible plant material necessary to conduct reproducible plant experiments. Moreover, "each environmental factor can be controlled separately, [and] the effect of these variables on plants can be individually investigated." Lastly, "the sum total of the naturally occurring fluctuations in environment (which we call climate) can be measured, providing an experimental basis for plant climatology and ecology."[43] The 1950s thus saw an eclectic array of projects operating in the phytotron as Went sought to maintain a vast array of technological hardware and expertise.

Unexpectedly, however, it was the discovery of the chemical nature of Los Angeles smog as well as the first studies of its effects on plants that became one of the most famous research projects to emerge from controlled-environment work in the phytotron. As early as the 1940s, Los Angeles was famous for its orange haze known as smog, trapped between the geography of the city's surrounding mountains and the firm inversion layer of Southern California. The growth of automobile culture and the petroleum industry appeared as obvious causes, though it took several years to demonstrate that smog was primarily the result of two chemical processes. On the one hand, sulfur dioxide (SO_2) from petrol engines created one sort of smog, but so too did the oxidation by ozone of unsaturated hydrocarbons. By 1949, the Los Angeles county Air Pollution Control District director, Gordon Larson, determined to understand and battle smog, had met with Went and Beadle. Initially, Los Angeles County hoped for a six-month project costing just over $3,500. Went hoped to engage the county for at least a year. At the same time, though keen to accept the county's money, Went was reluctant to have smog "leaking into the phytotron!"[44] Went asked Sam Wildman—the one who had coined the name "phytotron" with Bonner—to provide input because, he thought it was hardly news that "smog damages plants."[45] Went missed the mark. In fact, over the next decade, mainly via work done by Arie Haagen-Smit, much of the significant work on the photochemistry of smog and its biological effects emerged from the phytotron, including some of the first research that concluded that smog was "probably concerned with eye irritation." Subsequently, Haagen-Smit led efforts to establish emissions standards in California to curb both the health and environmental effects of smog.[46]

Though the deleterious health effects of smog usually made headline news, the larger environmental impact of SO_2 and the creation of ozone from hydrocarbons in bright sunlight were equally important. Beginning in 1950, Haagen-Smit in the Caltech phytotron "cooperat[ed with] the University of California Riverside Citrus Experiment Station" and the Los Angeles Air Pollution Control District to determine the extent of injury to plants.[47] Given the size of California's agricultural industry, plant injury was a serious topic. What particularly drew attention was a previously unknown form of leaf damage known as silver leaf that caused the underside of leaves to develop a metallic sheen. Haagen-Smit led a five-man research team that took spinach, endive, oats, and alfalfa plants, grew them under controlled conditions in the phytotron and then fumigated samples with a variety of gases, gasoline fractions, unsaturated hydrocarbons, and ozone. It was probably Went's reluctance to have smog throughout the phytotron that caused Haagen-Smit to use for his project "two adjacent gas rooms in the basement" kept at a steady air temperature of 22°C and relative humidity of 60 percent, and lit with "fluorescent tubes and tungsten bulbs." Both rooms were fed outside air. One room thus became a fumigation chamber, where various samples of plants were fumigated with various gases and gasoline compounds. In contrast, the chamber next door was the "control" chamber where the air went through a series of filters.

As the team's significant paper of 1951 makes clear, smog research in the phytotron required a series of "initial experiments" to "determine what kind of filters would remove [phytotoxic agents in smog] from the polluted air." They tested water scrubbing columns, tap water with suspended carbon, absorbent cotton, and even an "industrial gas mask." Settling on "activated carbon canisters," they could fully remove smog from the air and hence test the growth and development of plants under controlled conditions. Then, in both rooms, identical plants were grown under identical conditions, except, of course, air quality. Following standard phytotronic practice, the plants themselves were grown initially under controlled conditions in a "plant-growing chamber in a smog-free atmosphere under controlled conditions of temperature, humidity, light intensity and photoperiod until they reached a suitable size for fumigation, 10 days for oats and 30 days for all the rest."[48] Subsequently, Haagen-Smit's team tested fourteen industrial gases, and various gasoline plus ozone combinations. Their plants showed typical silver leaf damage from 1-α-Pentene and, damningly, "gasoline" as well as "gasoline plus ozone." Thus, smog definitively caused damage to plants and risks to health.

Haagen-Smit would later recall that when people went to the top of the Empire State Building in the 1950s, they could barely see the street below through the dirty brown haze in contrast to the clear views of a generation later. He saw that, alongside the famous simultaneous elimination of pesticides like DDT, the successful attack on smog signaled a profound ecological awareness and the beginnings of a large social and political movement. Bonner later said that "the story of the study of smog is in large part the story of Arie Jan Haagen-Smit single-handedly fighting the resistance of the American Automobile Industry." Initially, few wanted to act but by the end of the 1950s he would host "vice-presidents from the Ford Motor Company in [his] office."[49] In the 1960s, appointed by then governor Ronald Reagan, Haagen-Smit chaired the influential Motor Vehicle Pollution Control Board establishing standards of emissions throughout the industry, including getting NO_x exhaust emission control devices added to cars through an amendment to the 1977 Clean Air Act of Congress.[50] In 1968, he was appointed chairman of the Air Resources Board in California, where he suspended Volkswagon sales for ten days until the company produced its certificate of compliance. Better air quality and tighter emissions standards were a direct result of the controlled environment experimental proof of gasoline's harmful effects on living organisms. Incredibly, the modern clear, breathtaking views of the length and breadth of Manhattan from the Empire State Building are owed in no small part to the first phytotron.

CYBERNETIC PLANTS

Central to the operation of any big science facility, yet almost invisible in published work, are the operations managers. Managerial control was a critical facet of the control of the complete environmental system. The phytotron's staff of "plumbers, electricians, refrigeration and air-conditioning specialists [to] check, oil, clean and adjust all machinery at regular intervals" adopted the standards of the plant physiologists, noting that "only with such continued care can the complicated machinery of a 'phytotron' be kept in working condition."[51] Within the phytotronic system, the plant physiologists had exactly the kind of manager the Cold War era prized: George "Pret" Keyes was the true master of the phytotron's "intricate design [and] control systems." He managed the crew of fifteen staff to move, regulate, feed, water, and care for plants on a grueling seven-day schedule.

Above all, Keyes oversaw Went's obsessive decontamination process. Controversially, "all persons, materials and air entering the Earhart Plant Research Laboratory have to be decontaminated, sterilized or filtered."[52]

The regime of control had to cope with an immense amount of potentially corrupting material ("100 tons of gravel" per year), persons ("18,000 annual entries"), and air treated ("about 1,000 tons per day"), suggesting the extent of the phytotron's operation. At the same time, Went aimed to eliminate from his sealed environment foreign pollens, seeds, pests, and diseases inadvertently carried by almost unwelcome visitors. The regime also involved control of the researchers themselves and included locked doors, sterilized jumpsuits, and continuous hand washing and hair combing.[53] As Went and Keyes sought to control people like plants, feedback appeared in another guise: as Went noted, "once a person becomes aware of the fact that he may be a vector of insect-spreading, it is amazing to observe how many aphids or other plant pests are carried on one's clothing."[54] In other words, the very act of entering a space to control it was also the mechanism by which corruption penetrated the facility of control. Consequently, regardless of role, everyone went through the decontamination procedure; technical status no longer conferred social status as everybody changed into standardized and sterilized clothes inside the laboratory; Went called them "laboratory clothes." Laboratory clothes or jumpsuits were reminiscent of the full body overalls of engineering clean rooms (fig. 2.4). Colored white like medicine's white coats and scientists' labcoats, the phytotron's jumpsuit stood as a powerful symbol of a moral economy of a cybernetic science predicated on control.

Regimes of sterilization, washing, and changing clothes established control over the researcher as technological systems established control over the experimental climate. One result was that Went no longer emphasized the need to establish model organisms in the face of superior experimental control. As we saw in chapter 1, Went's laboratory began conservatively as an outgrowth of the scientific culture created by Thomas Hunt Morgan's geneticists at Caltech who, as Robert Kohler's foundational work demonstrated, created knowledge through a standardized organism, *Drosophila*, and associated experimental practices. The successful research program of the 1940s biologists spread far and wide and created a variety of reproducible "model organisms."[55] Copying Morgan, Went's earlier experiments in the smaller greenhouse had also pursued a model organism, a "botanical *Drosophila*" Went said, repeating the earlier phrase of E. B. Babcock working at Berkeley.[56] However, sometime in the first few years of the phytotron's operation, Went no longer stressed making his favorite plant, the tomato, into a model organism.

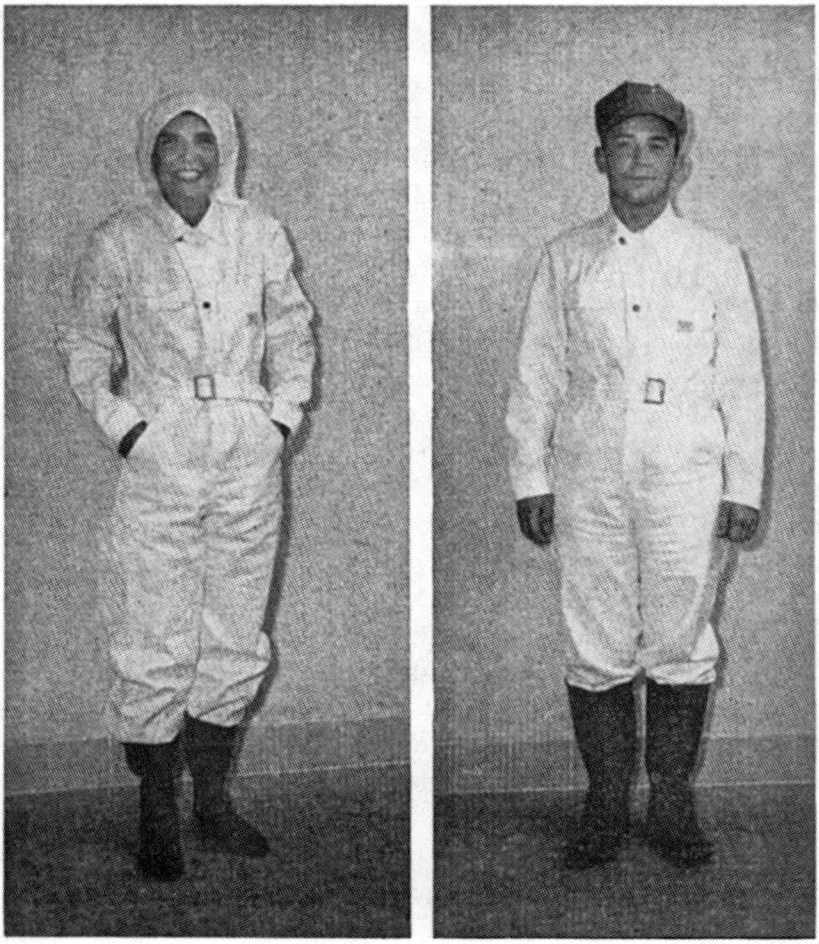

Figure 2.4. Phytotron Jumpsuits. From *Engineering and Science* (June 1949), 6.

Instead, plants were selected "based on their suitability for the solution of specific problems."[57] Studies of photoperiodization, for example, used the tomato or strawberry, and investigations into vernalization more often used peas, in contrast to germination's focus on common desert plants. For Went, no one specific plant solved the key problems of botany and physiology because the specimens were not standardized via known genetics and known growth environments. In other words, Went laid claim to the reproduction of *both genotypically and phenotypically* identical organisms—any organism, not just flies—establishing causation for traits from either genes or geography. Of course, we now know that a generation later, one model organism, *Arabidopsis*, did come to dominate plant science at the same time as ecology's medium-size experiments using model systems in the inevitably named ecotrons.[58]

Back in 1949, however, believing himself free to use any organism, Went paid considerable attention to improving the phytotron's technology. Went damned practically everything in old greenhouses—from their patterns of ventilation, to their rate of air flow, to their methods of shading glass via whitewashing (recommended "only by its cheapness"). Even the traditional arrangement of plants in "University greenhouses and demonstration collections" exemplified old and misguided botanical research. Because they required numerous and diverse genera and species, the institutions arranged their collections "according to taxonomic systems." For Went the technological future pointed to new arrangements of experimental systems for plants along "ecological and climatological lines." Went's "ideal arrangement" was "a number of separate greenhouses kept at different temperatures."[59] Consequently, the operational range of the climatically controlled rooms and cabinets in the Earhart Plant Laboratory was 0–40°C. But Went conceded that realistically 10–26°C was "the most effective range" for plants. As he explained, "in growing plants the right temperature means the difference between success and failure."[60] Of some fifty-seven closed spaces, only two fell outside this range, one at 4°C and another at 30°C. In fact, the Earhart Laboratory's technological systems could do no better than plus or minus 1°C; even as the facility went into operation its technological limits became apparent.

While temperature, humidity, pressure, and day length were all variables able to be controlled within known tolerances by the end of the 1950s, another fundamental component of the biological environment, namely, "light," remained a more troublesome variable for decades. By the 1950s, it had been determined that light is the basic energy source of all living things. Light is the energy component of photosynthesis that transforms carbon dioxide and water into oxygen and organic matter.[61] In Went's earliest experiments, and then in both the Australian and the French phytotrons, the issues of the proper sources of light, its distribution in a room or chamber, and its control and measurement were never satisfactorily solved nor consensus reached. Went believed that the physicists had once again provided the ready-made conception of the biological action of light. In the early 1960s, Went recalled hearing a talk that noted how physicists considered "light intensity the brightness, or what the plant physiologists call light intensity, is really the flux density, which is really the integration of the intensity over the solid angle to be considered." For Went, the talk solved "the question as to how to measure intensity: with a spherical or a flat meter . . . The light which a particular leaf receives can be best measured by a flat surface light meter . . . whereas for a plant as a whole the hemispherical meter is better."[62]

From the 1950s onward, the determination of the proper sources and the proper measurement of light for experimentation preoccupied many plant scientists. The early major unit of illumination, candlepower, was defined as the illumination given out by a "British standard candle which was made of spermaceti, weighed 1/6 lb, 1 inch in diameter, and had a wick made to burn at the rate of 120 grains per hour."[63] Spermaceti, an oil from the heads of sperm whales, was well-known for producing clean and bright candles, perfect for illuminating those ballrooms so favored by Jane Austen's heroines, but disastrous for whale populations, which were hunted to near extinction by the middle of the nineteenth century.[64] No botanist used candles for experimentation, of course, but the unit foot-candle remained in widespread use: Went continued to use foot-candle, as did many early phytotronists. However, in the growing twentieth-century industry of artificial bulbs and lamps that became fundamental components of the environmental systems in phytotrons, the standard unit of light flux was the "lumen." In contrast to candlepower, one lumen was the light flux falling on a spherical surface of 1 ft^2, one foot from a point source of one candlepower. In other words, the lumen was the measure of a one-square-foot segment of a total sphere of illumination coming from a light source. With the advent of arc-filament lights, fluorescent tubes, and sodium vapor lamps, among others, phytotronists often found it easier to measure the output of the light source. A fluorescent tube, for example, has an illuminance of about 4,000 lumens. However, the light source at the center of the one-foot sphere could also be expressed as a light intensity, or the candela, which was subsequently defined as the luminous intensity of a monochromatic source at 540×10^{12}Hz (or $\lambda = 555$ nm) that has a radiant intensity of 1/683W/sr.

All that is to say that the emerging controlled experimental environment for the life sciences was, as for physics and chemistry, critically aware of the need to establish uniform standards of units and their measurement. Indeed, debates over units are one visible element of biologists' struggles to specify and then measure the phenomenon in question. Moreover, these debates were never really settled. By the 1970s, both the specification of what should be measured under the variable "light" and how it should be measured were still being questioned. The Committee on Growth Chamber Environments of the American Society for Horticultural Science took up the topic of the measurement of light to new extremes in the 1970s. The first chapter of the horticulturalists' growth chamber manual offered a brief physics lesson in light, including the quantized nature of light. The author, the plant physiologist J. Craig McFarlane of the new Environmental Protection Agency's Environmental Monitoring and Support Laboratory in Las Vegas, offered two systems of measurement for light: photometric and radiometric measurement. Photometry measures the visual quantity of light received by "a standard human eye." Radiometry, on the other hand, measures the radiance of a light source falling on a given surface area.[65] In short, photometry measures when light is received; radiometry measures when light is generated.

In addition, as technologist biologists pointed out, light actually had two components to be specified and measured. First, there was whether to measure the light given out by the light source, or to measure the light falling on a leaf or plant. Second, one then had to fix on a unit of measurement. Part of the specification and measurement of the biological environment under controlled conditions became the process of determining the units of light as an environmental variable. Light, like growth and development, is the sum of two independent variables. On the one hand, it was a function in nature of daylight length. In Caltech, but also in the Australian phytotron and the New Zealand phytotron, artificial lights supplemented natural light. As daylight waxed and waned, artificial light maintained a constant illumination. This was important in Pasadena, but crucial for New Zealand's phytotron because the town of Palmerston North "is not a sunny place," William Laing remembered, it had "lots of cloud and the day length varied too much for consistency."[66]

Even more crucially, and the subject of intense study for decades, "light" is also a sum of all the intensities of all the wavelengths produced by a source, with energies changing in relation to those wavelengths. Technologist biologists like Went and Sterling Hendricks took pride in knowing and using the equation for light's dependency on wavelength: $E = hc/\lambda$.[67] Indeed, measuring plant growth and development at specific wavelengths of light was among the phytotronists' earliest programs of

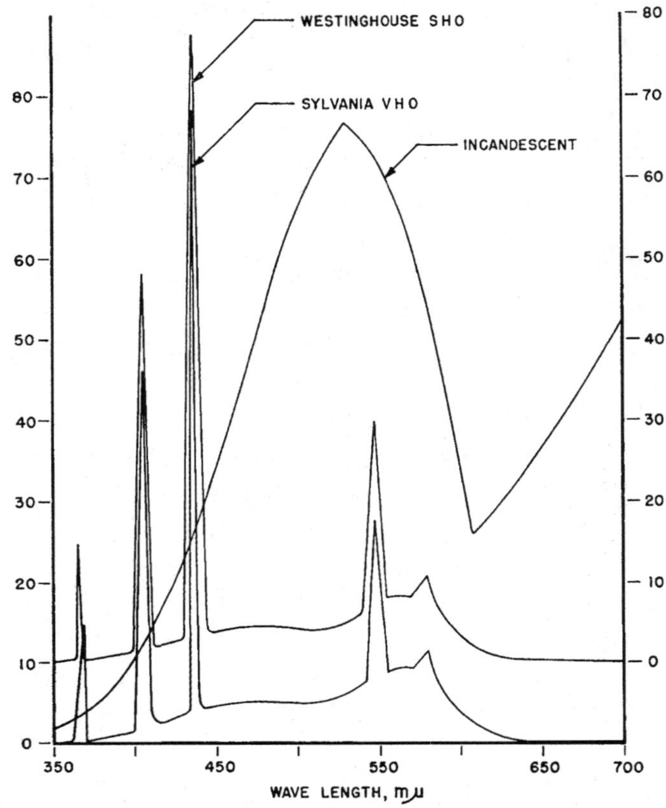

SPECTRAL EMISSION CHARACTERISTICS

Fig. 7. Spectral emission characteristics of VHG fluorescent lamps.

Figure 2.5. "Spectral Emission Characteristics" and "Phytotron Light Intensity Distribution," in the Washington State University air-pollution phytotron ca. 1961. From Adams, "An Air Pollution Phytotron," 473, 474. Reproduced courtesy of Air & Waste Management Association.

research. Hendricks became famous as he diligently grew plants under monochromatic (i.e., a single wavelength of light) sources. Subsequently, most phytotrons generated the spectral characteristics of their light sources themselves. In New Zealand's phytotron plant scientists went to great efforts to use a mixture of lamps that were "carefully calibrated and arranged to be uniform between rooms."[68] In smaller phytotrons, like the three-chamber unit at Washington State University, however, researchers used newly standardized "spectral emission characteristics" of incandes-

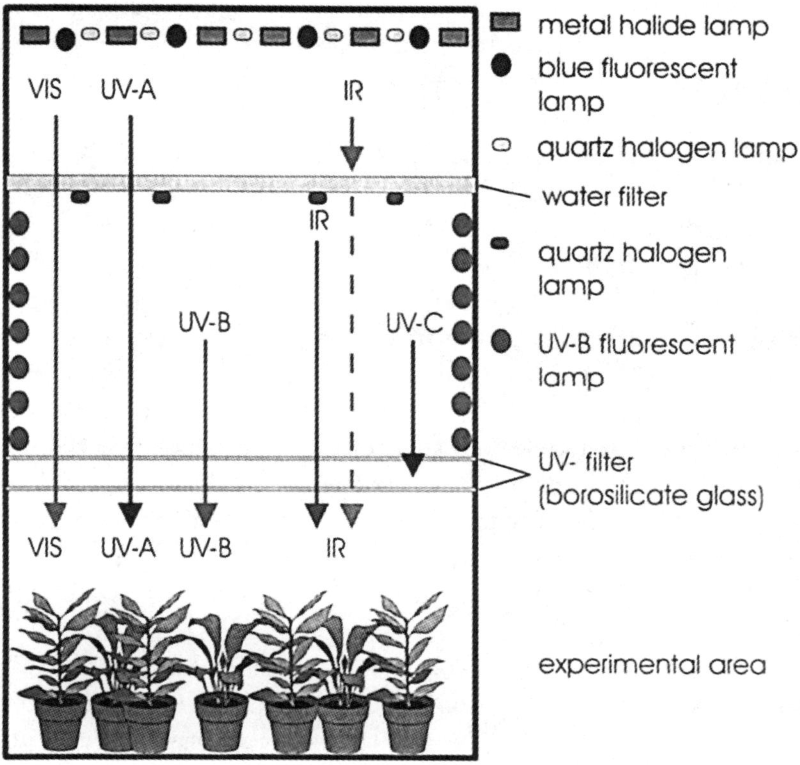

Figure 2.6. "Schematical outline of the lamp and filter configuration of GSF sun simulators" and "Spectral contribution of the individual lamp types . . . 1. Metal halide, 2. Quartz halogen lamps above the water filter, 3. Quartz halogen lamps below the water filter, 4. Blue fluorescent lamps 5. UV-B fluorescent lamps." From Thiel et.al., "A Phytotron for Plant Stress Research," 458.

cent and "very high output" fluorescent lamps. Westinghouse's new "very high output" fluorescent lamps specified that they had major spikes in intensity at 400 and 440 nm, the latter over eight times as intense as the broad profile from 350 to 700 nm. The incandescent lights had a radically different profile, however, steadily increasing in intensity from 350 to 550 nm, before dropping off around 600 nm (fig. 2.5).[69] By way of comparison, the phytotron of the Helmholtz Zentrum in Munich, Germany, built in 1996 incorporated a "sun simulator" by combining a variety of lamps with a variety of filters both across the ceiling of the chamber and along the walls to match the seasonal and diurnal patterns of natural light (fig. 2.6).[70]

These were all efforts in the continuing work of technologist biologists to specify and measure the multiple properties of light for use in biological experimentation. Seeking to standardize units, the American Society for Horticultural Science Committee on Growth Chamber Environments noted the experimental certainty that different wavelengths of visible light generated different growth and development in plants from several monochromatic studies, but questioned the proper units of measuring light. McFarlane advocated that light should be the "measurement of the incident quanta between 400 and 700 nm and is typically expressed in terms of nano-Einsteins per square centimeter per second ($nEcm^{-2}s^{-1}$)," effectively a measurement of the number of photons of light per unit area per unit time.[71] Subsequently, the Crop Science Society of America supported McFarlane's use of the nano-Einstein as did newer phytotrons such as Agriculture Canada's Research Station in Lethbridge, Alberta, which offered detailed instructions to researchers on how to convert "quantum sensor output" measured in micro-Einsteins (rather than nano-) into the more familiar radiometric units of watts, or the photometric units of lux.

Units matter, not least because to be universal they must be readily convertible: the Lethbridge phytotron warned its researchers, for instance, that regardless of whether the researcher was measuring light generated by a source or light received onto a surface, the conversion from photon units is "complicated," particularly because the "spectral distribution curve of the radiant output source must be known." Moreover, even then "the accurate measurement of W_λ [the total radiant output of the source] must be known," a "difficult task, which should not be attempted without adequate equipment and calibration facilities."[72] In other words, the specification of biological light continued to plague plant scientists at many institutions for many years. Robert Downs, for example, rewrote the North Carolina State University (NCSU) phytotron's standard handbook as recently as 2004 to include a long section on the proper measurement and units of illuminance. According to Downs, "intensity refers to the light source and provides little information about the amount of light received by the plants." Likewise, "illuminance measurements have very little real meaning in plant science because of the great difference between the spectral sensitivities of the human eye and the plant photochemical systems." Downs suggested "photosynthetic photon flux density" as the proper unit of light measurement because "photosynthesis is a quantum process [and thus] the most realistic measurement of the radiation used by the photosynthetic system is a number of photons within a specific waveband." Downs, like McFarlane, agreed that the band was 400 and 700 nm, but he specified the units as μmolm^{-2}s^{-1}, where the "mol" is the amount of some substance. But here Downs encountered a problem. SI (International System of Units) rules, he noted, disallowed the inclusion of a qualifier, such as the qualifier necessary to determine what the substance was, which the mols signified; or, as he rewrote the unit, mol quanta m^{1}s^{1}. He thus instructed any potential visiting researcher to the NCSU phytotron to "describe the measurement as a photosynthetic photon flux density, or PPFD, of some number of μmolm^{-2}s^{-1}."[73]

Complicating the specification and measurement of an environmental variable even further, the ongoing debate over the proper units was determined from the "light" from just a single source, whether it was one lamp, tube, or candle. Each room in a phytotron, though, comprised dozens or even hundreds of light sources. As the facilities evolved, lights were mounted above, to the side, or even in the growing floor. This added another layer to the complexity of phytotrons. One the one hand, researchers had to ensure that no part of a growing room received a different level of light—there could be no dark corners or bright floors. On the other hand, banks of lights might still leave shadows if plants overlapped one another. In the phytotron at Washington State University, for example, researchers were supplied with three-dimensional cross-sections of the light intensity distribution in each chamber.[74] In later cabinets and rooms, the walls were often stainless steel or even mirrored to reflect light evenly to all parts of the three-dimensional space. "Uniform illumination," one text recommended, "may be achieved by spacing the lamps further apart near the center and closer together at the sides of the room." Even then, different rooms would reflect light differently, while the lamps themselves would change characteristics as they heated, aged, or even simply accumulated "dust."[75] As Went was fully aware from the physicists, the act of creating a measurement changed the very conditions of the measurement itself.

FEEDBACK AND THE PHYTOTRON

Through the struggle over the specification and then measurement of just one variable of the environment—light—the new technologist biologists grew to appreciate the difficulty of defining the plant itself. The technological control of the biological environment was a complex and messy business, made worse because even once variables were specified the phytotronists discovered that every component of climate also interacted with every other variable. In other words, feedback appeared constitutive of the environment, the plant, and life itself. Feedback, in turn, implied a more cybernetic meaning for the "environment," as the technologist biologists sought to gain control over plants via control over mechanical and electronic systems. It was here that computers offered a solution to the complex control and measurement of numerous interdependent variables. The arrangement of the phytotron was controlled by an electronic computer, which Went thought vital for consistency and keeping records of the many environmental variables. The role of electronic computers, Went anticipated, would "in the future contribute significantly to the development of biology" because biologists realized that "the highly complex interrelationships between constituent parts, whole organisms and environment require equally complex mathematical treatment."[76]

The plant physiologists in their first phytotron were not alone in beginning to grapple with feedback. Throughout the Cold War era, feedback as a principle came to dominate fields like artificial intelligence, electronics, cybernetics, and computing.[77] Feedback emerged from the demand for control itself and the technological imposition of control through interconnected systems, and considerable engineering experimentation always accompanied the phytotronists as it did with the cyberneticians. Feedback was not anticipated in the original design of the first phytotron but was a product of action in action; it was a product of the technological construction of biology. As one textbook, evocatively titled *Physics in Botany,* insisted, "to be really effective, and to meet the demands of modern biological methods, an automatic system must exert very close control of the variables in the system. . . . At the highest level so far reached, machines are available which are able to scan their own product, and then, using any variation in the product, regulate their own working. Thus effects are made to act back on their causes so as to preserve stability and this is called the *feed-back principle.*"[78]

Lighting became the key feedback battleground. Illumination possessed the twin variables of length and intensity, but nature confounded the ambition for control. Although the control of temperature to known limits stabilized early on, values for light, as Went noted, proved elusive. Where early experiments in temperature at least suggested optimal ranges, experiments in different open greenhouses, in different seasons, and different locations had made it "impossible to determine the optimal light intensity required by a particular plant, using natural daylight." Consequently, the earliest results achieved by Went in his new controlled environment showed the differences between growth rates under shaded greenhouses and artificial light. As Went pointed out, open field crop plants often required light intensities of 5,000–10,000 ft-c for optimal growth, but photosynthesis saturation in any individual leaf occurred at 1,000 ft-c. He could not offer much explanation for the difference, but speculated that for any leaf to gain, on average, the necessary 1,000 ft-c saturation, the whole plant required much higher intensities over longer periods.[79]

Modernist to the core, Went declared that sunlight, though cheap, was highly variable and "only available during the day."[80] To overcome that limitation, the first phytotronists celebrated that the fluorescent tubes seemed "almost ideal" as sources of controllable artificial illumination. Went covered nearly the entire ceiling in fluorescent tube panels. Though available for illumination for any period, fluorescent tubes function with the highest efficiency, Went informed his readers, near 50°C—much lower and light intensity begins to vary, and near freezing, "the lamps may not start at all."[81] Thus his phytotron now had two demands on its temperature control system that directly fought against each other. On the one hand, the plants needed to be kept at a constant temperature of between 10°C and 26°C but, on the other, the fluorescent tubes needed to be carefully maintained at 50°C. At 10°C the tubes would barely start; at 50°C the plants would burn.

The solution was to separate the fluorescent tubes into panels sealed against the environment of the greenhouse. From above, ducted air passed separately through the light panels heating the fluorescent tubes to control for light intensity. Directly below, but isolated physically from the temperate system of the lights, were plants kept about 30–40°C cooler. The demands of each system necessarily affected the operation of other systems. As the room below was increasingly cooled, the lights above would have to be heated; as the light intensity fluctuated from inconstant temperature, the absorbed heat of the plants declined, raising the ambient temperature of the room and necessitating more cooling.

It is a microcosm of the Cold War: the cycle of control and stability generated feedback, which came from the interaction of the parts of the system itself. The insoluble paradox of the technology was that tighter control of light intensity gave greater temperature variability at the border of the heating and cooling systems. Even by 1980, Downs still complained that "although light sources have been separated from the growing area of plant growth rooms for many years, the thermal characteristics of present and potential barrier materials are poorly understood."[82] Unresolved but important, the phytotronist's experience of feedback illuminates much about the era, dominated as it was by the desire to control interconnected variables. Like technological systems locked into feedback relationships with each other, so too were the experiments and the experimenters. Experiments could no longer be built, operated, and maintained by solitary researchers, but required teams of scientists, legions of technicians, and droves of staff. In later phytotrons, such as Pierre Chouard and Jean Paul Nitsch's Gif-sur-Yvette phytotron, these systems dominated almost completely. "Sunflowers, wheat and many other plants have been grown from seed to seed without ever experiencing natural light," they announced proudly, making the sun an expendable part of the phytotronist's totalitarian ambitions.[83] It was environmental control that would give a standard plant. In effect, the phytotronists declared nature to be abnormal, chaotic, and capricious. The reliance of botanists and agriculturalists on the sun for lighting displayed their prescientific attachment to a science that was similarly abnormal, chaotic, and capricious.

For Went and Nitsch, the removal of the uneven solar cycles and its replacement with measurable artificial light sources created the practices to produce a "normal" plant. Like the point masses of physics, the "normal" plant did not actually occur in nature. In nature, a plant suffering randomly because of the vagaries of the weather might grow unusually, wither unexpectedly, or die without warning. As one speaker commented at a late phytotronics conference, "the plant in its natural environment proceeds from crisis to crisis, and the grower attempts to reduce the frequency and extent of these crises."[84] Because no condition or growth pattern could be reproduced, a substantial uncertainty continually remained. The controlled environmental space of the phytotron addressed the growth and development of "normal" plants, normalized because they were grown under controlled conditions. Went's first controlled environmental space was thus analogous to Ivan Pavlov's famous physiological factory. Pavlov surgically altered his experimental dogs to permanently access "normal" physiological processes. As the historian Daniel Todes argued, Pavlov's used animals that were surgically transformed into normal scientific subjects became the center of his celebrated laboratory work, the basis of his replicable experiments, and the source of his authority to both physicians and scientists.[85] For Pavlov as for Went, the proper definition of the "normal" organism was central to the correct experimental method for studying it. In both cases, the organism in the state of nature was not natural.

The phytotronists faced such challenging epistemological issues squarely. In probably the most enlightening example, in 1960, around the height of the phytotron's status in biology, the Caltech physiologist Harry Highkin gave a seminar at Cold Spring Harbor, New York. Highkin's presentation explained the "effects of different, specific environments on the heritable system of an organism," in his case pea plants. During the following question-and-answer session, recorded in the symposium's journal, one member of the audience balked at the suggestion that constant conditions were truly the opposite of periodically changing conditions, remarking that "Truly constant light, temperature, etc., are entirely foreign to the normal experiences of living organisms." It cut to the heart of the phytotronists' ambitions: it was the study of the "normal" plant that justified the elaborate research instruments. Moreover, the phytotron did not reduce the complexity of nature, but rather the complex technological systems permitted every environmental variable to be controlled and adjusted. But Highkin's questioner had hit on an awkward point for the phytotronists: if knowledge of normal plants could not stem from constant conditions because that did not represent "normal" nature, then the phytotronists would have to change their conception of the experimental environment itself. Highkin diplomatically agreed with his questioner, but it was clear that the criticism was not entirely foreign to him since he had a ready retort: "I think we do recognize the significant differences between environmental factors which are kept constant, varied periodically, or varied aperiodically. The first two we can easily control in the 'Phytotron.' The latter—aperiodic control of the environment—is most difficult, and I don't know of any laboratory where this has been done. It has been suggested that a 'Chaosotron' be built where just such experiments would be conducted, i.e., controlled chaos."[86] Here was a truly revolutionary moment for biology: true nature was chaos. To replicate and experiment scientifically on nature required, in fact, controlled chaos machines! We do not know who suggested the construction of a "chaosotron," but while modernism reigned, any complexity, even controlled chaos, was not an insurmountable problem.

As modernism faded, however, chaos, or at least complex functions, became the new reality. Perhaps, if "normal" nature was actually chaotic then maybe this meant that, even if genes were linear, a "normal" plant itself was chaotic because any plant was some product of its genes and environment. Phytotronists, among others, began to grasp an epistemology of complex probabilities, perhaps even chaos. Mathematicians and meteorologists started to work on the problem of weather as a chaotic system, with its strange attractors, nonlinear dynamics, and unpredictability.[87] At the same time, the noted theoretical physicist turned population ecologist Robert May began his study of model ecosystems by "assum[ing] an unvarying, deterministic environment," but added complexity halfway through because "real environments are uncertain, stochastic. . . . Equilibrium is not the constancy of the physicist, but rather an average around which the system fluctuates." For May, who became Lord May and later published work from Britain's Ecotron, "we obviously can no longer speak of *the* population N(*t*) at time *t*, but only of its probability function."[88] The anticipated research program of the first phytotronists based on a simple linear equation that an organism was the sum of its genes plus its environments had become the study of complex functions. The subsequent struggle to specify and measure a complex science of biology would help end the optimistic "Age of Biology" that once promised direct causes of growth and development from the deterministic identification of specific genes and environments.

THE DEPARTURE OF WENT

Went moved to the Missouri Botanical Garden in 1957. There he built the Climatron, as we shall see in chapter 3. Contrary to any idea that plant physiology in phytotrons was a moribund research agenda in contrast to the rising star of molecular biology, Went's departure from Caltech rested almost exclusively on a growing clash between personalities and finances. Far from perceiving any failings of the phytotron as a facility, the Earhart Laboratory received another major grant from Campbell

Soup to nearly double the controlled environmental space available in 1958. What is clear is that Went suffered an increasingly antagonistic relationship with Beadle, the chairman of Caltech's Biology Division. An early issue was over the title of "director" of the Earhart Laboratory. It was a sore point with Went, especially after Beadle reminded him at least twice formally that the division did not recognize directors even in physically separate buildings or laboratories.[89] The director and chairman's relationship turned uglier still when Beadle pointed out that plant physiologists "have been guilty of doing some pretty poor work." Went denied that Beadle's assessment had any relevance for his operation, noting instead that "only with the work of Bonner and Wildman" could there be "any criticism of plant physiological work at Caltech." Staggeringly, Went charged that in their work on auxins, the pair had "been more consistently wrong than anyone else I know." Once more, Beadle attempted to talk Went back from such personal attacks, reminding him that everyone makes mistakes. Beadle calmly wanted to note that the increasing demands on the space in the Earhart Laboratory implied that Went could select only "the very best."[90]

By the end of the 1950s, Went's relationship with his Caltech colleagues had completely disintegrated. Went confided that he felt Beadle and Bonner had treated him like a "small boy."[91] He left Caltech, he said, because he "did not feel that the Biology Department had much to do with biology any more."[92] His diary confessed his belief that Beadle's approach, "like Bonner's, is a desk approach, with no connection with the living organism any more. They have built up imaginary schemes according to which organisms might be explained, but for that one has to close his eyes to most of the facts of Biology."[93] Went may have accused Beadle of having an "armchair approach" to biology, but it seems clear from the minutes of their meetings that Went played a small role within the Biology Division itself during his time at Caltech. He did not often attend meetings and rarely had any business to propose or discuss. His annual fellowships were awarded without debate. But when it came to other shares of the divisional pie, like allocations of graduate students, Went and plant physiology consistently lost badly to biophysics. Of the five teaching assistantships for 1955–56, for example, three went to biophysics students, and one each to immunology and plant genetics. Two plant physiology graduate students, Stanley Burg and Vernon Burrows, received graduate standing, but sat on the reserve benches to get teaching assistantships.

The only other two applicants for plant physiology were simply not recommended.[94] The following year, the Biology Division's screening committee recommended no student of plant physiology.[95] All this took place in an era where graduate assistantships often went unfilled. According to an early NSF survey on financial support available for graduate students released in August 1953, some fields found it difficult to fill graduate student places. According to the survey respondents, many departments, holding to rigorous standards, simply left assistantships vacant rather than accept anyone other than outstanding candidates. At the same time, as much in biology as in astronomy, money lured many potential disciples away from science. Many departments fretted about a world where the "higher salaries of industry and business are too much for the prospective graduate student to resist."[96]

There was also scandal. The details are scant, but questions emerged in the 1950s over the determinations of the chemical structure of "auxin a" and "auxin b" by the plant physiologists back in Utrecht, Fritz Kögl and Hanni Erxleben.[97] Went and Haagen-Smit, themselves both graduates of Utrecht, had worked with samples sent by Kögl in the 1930s. But something was amiss. Just before the war, it fell to Haagen-Smit to sort out the issue that no one else could positively identify Kögl's samples. After sending some seven samples to Caltech, the last one appeared active, but one of the Caltech staff members, already suspicious of the whole business, tested Kögl's "auxin a" sample and found it to either be indoleacetic acid (IAA) or at least heavily contaminated by indoleacetic acid. That Caltech staff member was Kenneth Thimann. Recent historical reconstruction of the issue by Peter Karlson has portrayed Kögl as an easily led administrator, and he argued that "the whole [auxin a, b] story was faked by Dr. Hanni Erxleben."[98] Nearly thirty years later, using mass spectrometry on the preserved original samples of auxin a and auxin b, J. A. and J. F. G. Vliegenthart finally definitively declared auxin a and auxin b as "non-existent."[99] In one last twist, however, Bonner wrote to Karlson in 1982, saying he had heard not only that Erxleben misled her boss and plant physiologists the world over, but also that she was rumored to be a Nazi agent planted in Holland who had disappeared after the Second World War![100]

Scientific scandal, fake hormones, and Nazi spies aside, Went himself seemed to be little involved in the drama. He remained credited with isolating the first auxin, and was perhaps saved by his insistent requests that Kögl and Erxleben send reproducible samples to Caltech. Yet Bonner would also claim to Karlson that, even after he and that "agent-provocateur" Sam Wildman had demonstrated that indole-3-acetic acid could be extracted from shoot tips as early as 1946, "Went insisted until the last time he ever spoke to me . . . that auxin a and b are the true stuff."[101] In reality, the whole thing disappeared by the 1950s. Plant hormone workers abandoned the unproductive work of isolating and identifying real or imagined auxins, and simply moved on. Haagen-Smit, according to Bonner, went silent on the whole sordid episode (including three expensive years with William Bergen at Caltech collecting huge quantities of urine from lavatories trying to produce auxin a), but recovered his reputation by leading smog reduction efforts. Near the end of his life, after Wildman had also energetically chased the details of the auxin a story, Bonner declared, "Kögl dead and I hope we can declare auxin 'a' dead and we can continue with our previously announced policy of never thinking of it again."[102]

On a campus of scientific giants, friction between academics is nothing new, and the existence of auxin itself was never questioned. Instead, what seems to have most disturbed Went was his inability to solve a near insurmountable dilemma. In the creation of the phytotron, Went hoped he would concentrate on fundamental research, but the ever-present demands of soaring maintenance costs demanded never turning away "potentially or actually" plants of primarily "economic importance." Went insisted that "the ultimate aim in studying plants in the Earhart Plant Research Laboratory is to get an overall picture of basic plant behavior," but he simultaneously remarked that from coffee to orchids, and tomatoes and strawberries and sugar beets, a number of "private and public organizations" financially supported the phytotron's operations.[103] Phytotronic research was expensive: after two full years in operation, the annual cost of the laboratory had rocketed to somewhere "about $120,000 a year," Went estimated.[104]

In short, the reality of the phytotron as a scientific instrument remained at a distance from its rhetorical identity as a basic science instrument, and it increasingly frustrated Went. On the one hand, Went emphasized that the full development of "basic physiological studies . . . which of necessity embraces an appraisal and integration of genetical, physiological, and ecological factors," was "so complex that to date satisfactory experimental approaches have not yet been developed," hence

Figure 2.7. Caricature of Frits Went by Hans Gloor, Biology. n.d. Photo 10.24–6. Archives. California Institute of Technology.

the need for "a new instrument" like the Earhart Laboratory. Even after Went made such claims, however, a trio of plant physiologists working on grasses with Went readily admitted that agriculture would be the real beneficiary of "basic physiological studies."[105] On the other hand, even by mid-1955, "weed" research occupied nearly one-fifth of the "trucks," while beans and peas, seemingly more standard scientific objects, occupied about two-fifths. The remainder of the space contained a variety of economic crops like trees, roses, tobacco, soybeans, and orchids. Indeed, the

Figure 2.8. Caricature of Max Delbrück by Hans Gloor, Biology. n.d. Photo 1.3.1–25. Archives. California Institute of Technology.

importance of the case of orchids lies in the fact that Went actually used the orchidists quite scurrilously as a means of supporting his expensive facility until patronage could be secured to support basic science.[106] Went insisted that he wanted a science to rival the "exact" sciences and to create a "Theoretical Botany" akin to "Theoretical Physics."

Once again, the wit of students pinpointed the underlying struggles of Caltech's biologists. Fundamental research was the currency of their moral economy; as the chorus of the Biology Division's 1949 Christmas Follies had sung out, "Truth with a capital T is more for us."[107] Yet it was only the molecular biologists who appeared to leave the business of applications behind and become the image of the physical scientists. Transparently, in the mind of the biology graduate student and amateur cartoonist Hans Gloor, Went appeared as mad Doctor Frankenstein and Max Delbrück was already the armchair theoretical physicist (figs. 2.7 and 2.8). In the poses struck in the cartoons, Went was still trying to make his tomato plant grow right, which dramatically contrasted with the purely thinking molecular biologist Delbrück using just equations scribbled on a blackboard. In such caricatures, there was no notion of the historical reality that so much early molecular biology was built on the promise of wondrous medical cures or that research in the phytotron had already identified smog as a threat to health. The reality did not matter: what mattered to both the molecular biologists and the plant physiologists was to become what they envisioned as ideal "scientists," theoretical, mathematical, and pure.

SUNSET ON THE FIRST PHYTOTRON

The Earhart Plant Research Laboratory came under the new leadership of Anton Lang. Supported by Bonner, who became the chair of the Biology Division when Beadle moved to Chicago, Lang oversaw the construction of the new wing of the phytotron, built and paid for by the Campbell Soup Company to develop tomato varieties that would produce commercial crops in warm climates. The subject of plant physiology waned at Caltech in the 1960s. In practical terms, Bonner took over all teaching of plant physiology at the institute after 1960 but he became, officially, a biologist. Plant physiology disappeared from the list of Caltech professors' specialties in the Caltech catalogue the following year with the death of Eversole.

After the mid-1960s, the phytotron declined as Caltech expanded. The acquisition of the rest of the block occupied by the phytotron around 1962 opened up a major phase of the development of the Caltech campus, which would nearly double in size over the next decade. The plans for the new chemistry building, as the Biology Division soon learned, ignored the shading effects on the adjacent greenhouse. Bonner and Lang, the last remaining researchers, mounted a defense of the phytotron, but the chemistry building went forward as planned. Ironically, the expansion of chemistry signaled the slow death of the first phytotron. Lang, supported by Bonner, valiantly preserved some of the research program, and the pair worked to recruit new patrons in the form of the Agricultural Research Service and then the United States Department of Agriculture (USDA). Clearly desperate, Bonner suggested that the Agricultural Research Service entirely take over the running and operation of the phytotron.[108] Lang seemed despondent about the facility's future, offering to meet the USDA director of Science and Education, Nyle Brady, "in my office or in the phytotron (ruin)."[109] The rest of the Biology Division looked hungrily at the Earhart Laboratory's space, facilities, and research funds. In late 1964, Bonner prepared a short synopsis of the phytotron's then precarious position within the Biology Division for Brady's tour, noting the "growing resentment on the part of the staff concerning the magnitude of the Earhart operation."[110] Several members of the Biology Division, including Delbrück, specifically requested a tour of the laboratory before they made up their minds about what do to with the facility, and so Lang and Bonner put together a little informational tour for their own colleagues, especially the more "phytobiophysically illiterate members of our staff."[111] The tide had turned against the phytotron at Caltech, however. The last act was when Lang secured himself a new position later that year at Michigan State University complete with an offer from the Atomic Energy Commission to fund a new Plant Research Laboratory, comprising greenhouses and numerous controlled-growth cabinets. Like the Caltech phytotron, his new Plant Research Laboratory combined multiple disciplines "biochemistry, biophysics, genetics [and] microbiology" to determine the "influence of radiation upon function and development of plants."[112]

As bulldozers razed the world's first phytotron in 1972, a Caltech administrator rationalized it to his president, "Progress simply made the facility obsolete."[113] "Progress" saw that the valuable space was quickly rebuilt into the developmental biology laboratory completed by 1974 under Robert Sinsheimer. Sinsheimer was a member of the new breed of biophysicists who had transferred from physics to biology at Massachusetts Institute of Technology (MIT) during 1948–49, after having worked in the MIT Radiation Laboratory during the war. His first appointment as a biophysicist was at Iowa State University in 1949, before he was brought to Caltech under Delbrück in 1953.[114] He would chair the new Caltech biology as it fostered, like Stanford's, the new gene jockeys of 1970s and 1980s biotechnology.[115]

Plant research disappeared at Caltech for a generation.[116] Circa 1970, Bonner explained that biology at Caltech had been compressed into just three definitions, molecular biology ("biophysical and biochemical studies upon nucleic acids, proteins"), cellular and developmental biology ("analysis of cellular function"), and neuro- and psychobiology ("the nervous system as a principal integrative component of higher organisms").

By the 1970s, biology students at Caltech could not take even a single class devoted to plants, and the Biology Division offered just one physiology course, enrolling but a lone student that year.[117] Bonner, by more or less outlasting such vagaries, could rightly claim to be the sole remaining plant biologist at Caltech: as he told a visiting speaker even in 1981, "the level of competence in plant biology here [at Caltech] is approximately '0' (except for me) but very high in everything concerning molecular biology, genetics, etc."[118] As Bonner noted, by the early 1980s the plant sciences once more began to attract funding and students on the back of genetically engineered crops and the whole genetically modified organism revolution. But "plants," at least as the earlier plant physiologists had defined them did not return to Caltech; the subject and facilities for environmental control were notably absent, completely swept away by "plants" holding the promise of gene splicing and recombinant DNA.

In short, the new developmental biology laboratory that replaced the phytotron signaled the changing meaning of biology at Caltech, namely, that biology would become a science of genes instead of a science of genes and environments. Molecular biologists like Max Delbrück, Gunther Stent, and James Watson constructed a story of the mythical historical inevitability of molecular biology's victory in the late 1960s.[119] Yet, even in 1970, Beadle was still happy to be the keynote speaker for the dedication of the University of Wisconsin Biotron, the most complete climate-controlled biological laboratory space in the world, as we shall see in chapter 6.[120] Beadle spoke at length on the topic of "life sustaining environments: the need for knowledge," implying that he remained interested in climate controlled research as well as convinced of its importance, perhaps echoing the various claims of Buckminster Fuller about "Spaceship Earth," which had gained considerable currency around that time.

Over nearly twenty years, the first phytotron had hosted an effort to standardize the experimental plant, smog and cancer research, and tried to grow the perfect tomato. Moreover, it had fostered much of the early community of phytotronists who spread the message of controlled environments for biology far and wide beyond California. Most phytotronists worked in the Earhart Laboratory at one time or another: when Nitsch graduated in 1951 he helped found France's phytotron, eventually the largest in the world; Lloyd Evans guided Australia's Otto Frankel through Caltech's phytotron for the first time, and eventually himself became the director of the Australian phytotron in 1962, as we shall see in chapter 4; Henry Hellmers and Paul Kramer had already moved on from Caltech to found the cooperative phytotrons at Duke and North Carolina State universities, the subject of chapter 5. They all took from Caltech a clear sense of the power of doing biology under controlled conditions. Newly establishing standardized experimental procedures to generate standard plants spread far and wide so that by the mid-1970s a group from the American Society for Horticultural Science proposed to work on the "repeatability or standardization of chambers conditions and procedures" in the Biotron at the University of Wisconsin-Madison. Remarkably, they worked toward "a biological method" that involved determining the "standard cultural procedures and environmental settings" for controlled-environment facilities to then establish "a baseline growth curve for a number of common horticultural plants (lettuce, marigolds, tomatoes, and birch)."[121] Though subsequently the building of phytotrons went into decline, Evans believed that Went's "broader vision for research in phytotrons remained, that it could explain how plants, both wild and cultivated, whether in pure crops or complex plant communities, respond and adapt to 'climate' as a whole." Both Pasadena and Canberra attained this vision, Evans believed, but "by the time computer simulations of crop growth were developed most phytotrons had reverted to the humbler role of providing standard plants."[122]

CHAPTER 3

THE CLIMATRON

> Just as the cyclotron, bevatron and synchrotron belong to the most ambitious research tools the physicist has created, the Climatron undoubtedly is the most modern and advanced research and demonstration tool of the horticulturalist and botanist.
>
> — **Frits Went, 1960**

SHE LOOKED beautiful. She glowed for the cameras as she sat on her high-backed chair. Her off-the-shoulder dress revealed an elegant neckline, her gloves stretched regally well above her elbows, and she wore a tiara and held a scepter in her right hand. She was Miss Carolyn Lee Neidringhaus, Saint Louis's Veiled Prophet Queen for 1958 (fig. 3.1). Civic functions and receptions would follow for the young lady over the next year, with the first stop in early October being an audience with children in the Shaw House at the Missouri Botanical Garden. The house of Henry Shaw, another grand nineteenth-century philanthropist, sat at the heart of the Missouri Botanical Garden, formerly Shaw Gardens, as Shaw himself represented the heart of a particular American culture of civic pride, engagement, and charity in Saint Louis, the first westward stop across the United States.[1] Shaw's Garden donated a spectacular bouquet of orchids for Miss Neidringhaus that year, more resplendent than those of years past. George Pring at the garden, whose pride and joy was the orchid collection from whence the bouquet had come, received due praise at the annual Veiled Prophet Queen's ball, a highlight on the Saint Louis social calendar.[2]

Figure 3.1. 1958 Veiled Prophet Queen. From *Missouri Botanical Garden Bulletin* 47, no. 2 (1959), 39.

Attending the ball was the newly arrived director of the Missouri Botanical Garden, Frits Went, who noted privately in his diary that on this evening he had witnessed "the most elaborate social function" he had ever seen.[3] Scientists like Went recognized that debutante balls had a social function but rarely viewed their own professions in a similar way. To their patrons (and their historians), however, scientists, their facilities and discoveries all have very distinct social functions. This next chapter in the history of phytotrons is all about the social function of the

Figure 3.2. The Climatron at the Missouri Botanical Garden. Author's photograph.

science of controlled environments. The chapter explains that between 1959 and 1961, the people of Saint Louis themselves raised $700,000, the largest private grant since Henry Shaw's initial endowment establishing the Missouri Botanical Garden a century earlier, to fund Went's climate-controlled display greenhouse, inevitably called the Climatron. Using the insights of the historians Karen Rader and Victoria Cain that the display of nature describes the social function of science,[4] the case of the Climatron exposes not only the worldview of science at the height of postwar modernism but also the worldview of a declining American cities' elite. As we shall see, the postwar cultural elites of Saint Louis fought to maintain their declining city, both by renewed commitment to traditional ceremonies like the Veiled Prophet Queen ball and by investing heavily in remarkable modernist buildings, the Climatron among them.

When the Climatron opened in 1962 it showed the technological control of nature via a public display of four whole plant ecologies unified under a single geodesic dome. The public literally walked into the Climatron past a Honeywell computer. As far as Went was concerned the necessary social function of the Climatron was to gather support for a future and grander controlled-environment laboratory. In itself it did little overt scientific work, but as the Climatron attracted paying visitors, it was supposed to underpin the creation of the infrastructure of the next broader research program into the environment as a category of biology.[5] Went regarded the Climatron as the next, though explicitly intermediate, step between a phytotron for a single plant holding one set of climate variables constant and a full ecotron that could replicate a set of organisms and environments. Officially, Went's job took up the "challenge" of restoring the garden to "the forefront of botanical institutions, to make it again fully effective as a horticultural center, and to increase its usefulness for the public and for the future of Botanical Science."[6] From his first tour in late 1957, Went saw an institution in decline, and commented on shabbily planted beds, poorly run work, and crumbling buildings that could not be repaired and were better bulldozed. The dilapidated greenhouses, he said, required complete replacement not least because "in 10 years the old type of greenhouse will be completely obsolete."

The answer was obvious to Went: Missouri needed modernism, badly, and the garden's external evaluators concurred. "To this modern world," the report read, Went offered "an interest in ecology, climatology and technology."[7] As we saw in chapter 1, Went had come from modernist California where two things indelibly defined progress: air-conditioning and aluminum.[8] Technology had met social function in late 1945 when a philanthropist, Harry Earhart, sought political stability by funding Went's new controlled-climate botanical laboratory, subsequently dubbed the phytotron. Consequently, Went's vision of modern botany at the Missouri Botanical Garden would begin with "new houses, new in design, in engineering and in function" to replace the rusting past, which, as Went noted in his diary, would necessarily include "a certain amount of air-conditioning . . . especially in this climate."[9] The centerpiece of these new buildings would be the new climate-controlled space of the Climatron.

In contrast to Went's vision, the conservative cultural elite saw the Climatron as a modern edifice of control to rally around. By the 1960s, Saint Louis's urban core had decayed, as in so many American cities of the era, while its suburbs flourished. Hoping to attract new industry and business, Saint Louis, just as Detroit, Trenton, and Philadelphia, cleared land to permit new grand buildings to appear but these structures also destroyed declining neighborhoods.[10] To the elites, it was a modernist re-

Figure 3.3. The Missouri Botanical Garden's Climatron interior as Technological Eden. Author's photograph.

invention of the city, and architectural historians have noted that by the 1970s guides to Saint Louis for visiting conferences and professional organizations featured the Climatron alongside modernist spectacles like Frank Lloyd Wright's Kraus House (1955), the Abbey Chapel (1962), Eero

Saarinen's Arch of Westward Expansion (1958), and the Lambert Field Main Terminal (1955).[11] The baby-boomer generation embraced grand technologies of human progress such as early space flight, Buckminster Fuller's geodesic domes, and commercial jet aircraft that displayed ever-greater technological control over human environments. Indeed, Saint Louis's cultural elites actually built what the era's modernist thinkers theorized, specifically modern technologies as "an authentic expression of the machine age, and a necessary agent for progressive social change," as the historian Thomas Misa argued.[12] A necessary stop on any tour, the Climatron attracted conferences of mathematics teachers, for whom Went gave talks and tours, and a recommendation from the Art Education Association equating it with Saarinen's Arch as "environmental art," though only the "ladies section" of the American Water Works Association made the trip.[13]

In the early 1960s, the Climatron was thus the product of Went's public remaking of plant science and a public elite's remaking of their city via an architecture of control. It was, however, ultimately an unstable alliance. After only five years as director, Went ignominiously resigned from the garden, even though he had amassed much of the funding to build his new laboratory. He had perhaps overplayed his image as a "modern" plant scientist; certainly his scientific staff did not support him when he faced off against a resurgent trustee of the ancien régime, Henry Hitchcock. Went lost his head just as assuredly. The crisis delayed any new construction until the end of the 1960s, when under the new director, Peter Raven, the new, though much changed, laboratory space finally got under way. Unlike his phytotron, however, the Climatron remains a centerpiece of any tour of the garden to this day.

DR. WENT GOES TO MISSOURI

> From Eden to Gethsemane, the Bible is a book of Gardens. It is with a garden that Genesis begins, and with a vision of trees bearing fruit that the Book of Revelation ends.[14]

The story begins in a public garden in Missouri. Henry Shaw planted his gardens along a stretch of pastoral land adjacent to a public avenue in 1859, a public and private institution he himself would live and work in until his death. Having made his fortune from the astute importation of commercial goods for the budding frontier town via the Mississippi River and then exporting lead back to England the same way, Henry Shaw represents one of the self-made icons of the United States in the nineteenth century. His garden, endowed to public trust in Saint Louis, functioned as a scientific establishment beyond mere personal satisfaction or commercial gain. Shaw received advice from Sir William Hooker at Kew Gardens, father of Joseph Hooker, as well as Asa Gray, the already renowned American botanist at Harvard. Gray directed Shaw to the local botanist George Englemann. The three men, Gray, Hooker, and Englemann, brought into contact via Shaw's botanical ambitions, guided the early growth of the Missouri Botanical Garden.[15]

A century later, Shaw's Botanical Garden struggled to preserve the legacy of Shaw both financially and scientifically. When Went moved from the California Institute of Technology in Pasadena to the Missouri Botanical Garden in Saint Louis he believed he was switching from one scientific institution to another, albeit one at the height of its prowess to another struggling against further decline. Privately, Went's initial survey concluded that the garden had no horticultural research, possessed "no other botanists of standing" except Edgar Anderson, and carried no research students.[16] Anderson was still an important figure, renowned for his work on hybridization and botanical research via populations, but by the late 1950s his productivity was clearly waning.[17] In contrast, Went's star shone brightly. On meeting the new chairman of the garden's trustees, Went reveled in the impression that the Saint Louisans "were prepared to do practically anything I wanted to make me accept" the directorship, eventually offering him a professorship at Washington University, alongside community and trustee support for his plans.[18]

Went accepted the appointment as director of the Missouri Botanical Garden on May 1, 1958. Over the next five years, Went would labor to reestablish the Missouri Botanical Garden as a leading botanical institution by preserving its important library and arboretum as well as innovating a new appearance for the housing and display of a tropical plant collection. Initially, he considered simply shifting over to Washington University's school of botany the scientific elements of the garden, especially the herbarium and the library. He thought "it obvious" that Washington University should be linked to the botanical future of the garden, not least because Shaw's trustees had, by and large, little interest in any scientific mission, or the herbarium or library. As a modernist experimental scientist, Went believed he could restore Shaw's legacy by removing the scientific facilities from the decorative functions of the garden's trustees: he certainly believed that the future "trustees of the garden will not have the necessary competence in guiding its scientific development."[19]

In contrast, a number of Shaw's trustees, composing the elites of the Saint Louis community led by the prominent Saint Louis citizen Robert Brookings Smith, thought the Shaw endowment should fund shows, displays, and attractions to promote the cultural and civic life of their city. Brookings Smith graduated from Princeton, became a partner in a brokerage firm, and later began his own business. In addition to being a trustee of the Missouri Botanical Garden, he was also a trustee of Washington University.[20] Facing the trustees, the garden's new director made a point of stressing that while various fundraising campaigns might support the maintenance of the gardens and displays, Shaw's endowment itself should be "largely used for research and development, and not for regular garden maintenance." Both parties accepted the notion that the income from the Shaw endowment defined the type of institution the garden would become. For a scientific institution, as Went pointedly explained to Brookings Smith, the "endowment money can buy completely new approaches and imaginative development which can be provided only with the greatest difficulty through public agencies."[21]

In other words, as Went's diaries make plain, he already did not understand the public nature of the institution wooing him, or appreciate the function of that first social gathering he witnessed. Went too easily dismissed the new social environment of Saint Louis and was perhaps too readily swayed by trustees' assurances of support. In large part, Went had initially underappreciated the social function of the garden because he believed the Missouri Botanical Garden was only a temporary stepping-stone on his way to greater scientific renown. He insisted that he would only move to Saint Louis for one or two years, because unbeknownst to either the garden's trustees or Caltech, Went had also entered negotiations with Yale University, which he visited *immediately* after his initial reception in Saint Louis. His longer-term goal, in fact, focused on replacing Paul Sears as a professor of botany at Yale, but because Sears would not retire for two more years, Went took no "official" action, and left Yale assuming that he would remain "considered the main candidate."[22] The garden's directorship, then, enabled Went to leave

Caltech early, and put behind him what he saw as the disappointments of George Beadle's negative opinion of him personally and professionally. At the same time, because of his authority within the botanical world Shaw's trustees gave him carte blanche to build another great controlled space, but he worked believing that a sinecure Yale professorship lay in his near future. For the moment, however, Went was firmly in Missouri.

One example gives the flavor of the state of the garden as Went saw it, as well as the type of solutions he envisioned. The garden possessed another, far larger property at Gray's Summit outside Saint Louis, and there stood another arboretum, all but abandoned. At first sight, Went thought the space readily convertible into classrooms or laboratory work in biology or ecology for educational purposes.[23] Beyond those general terms, it rapidly became clear that the problem was not finding a use for the Gray's Summit property, but finding a unique use that would bring people and money. Though the arboretum could be a field laboratory, Went acknowledged, "it probably would not be self-supporting." It could also be a campsite "but the principal youth groups are well supplied." As a display area for formal gardens, it would merely replicate what "has been done in several places elsewhere in the United States." The only idea Went saw as feasible and productive to appeal to the touring public involved building "a historic village reproduction" "with museum-display of buildings, décor, and furnishings" telling "the history of old Missouri."[24] An odd suggestion from a botany professor, no doubt, but Went at least appreciated that any scientific mission required funds from elsewhere, and because science is expensive the quaint historic village seemed a viable solution. And Went's particular concept of an air-conditioned, controlled-environment greenhouse was especially expensive.

The Missouri Botanical Garden trustees had a vision of a general reworking of Shaw's Garden for several years, at least since George Moore retired as director in 1953 after over forty years in the position.[25] Five bleak years passed for the garden between Moore's retirement and Went's arrival. The end of Moore's directorship severed the last link holding the garden to Shaw's original nineteenth-century dream, at least so said the consultancy firm hired to propose new directions. Moore's long tenure had been a "period of adjustment from the Age of the Universal Man, who knew something about everything, to the Age of the Specialist, who works long and hard attempting to learn a lot in a limited field."[26] Since Went was appointed, we must assume that such sentiments appealed to a group of Saint Louisans for whom Shaw's Garden represented a long philanthropic tradition of civic duty. It is unclear from the sparse trustees' minutes how the cultural elite of Saint Louis made its decision to emphasize the scientific identity of the garden as a way to halt its decline. But the appointment of someone of Went's stature within the botanical community signaled the trustees' seriousness.

The trustees perhaps saw in Went a glamorous, professional scientist. He was, after all, a former president of the Botanical Society of America, a current professor of plant physiology at Caltech, the director of the Earhart Plant Research Laboratory, and the discoverer of auxins. No doubt, Went's spectacular approach contrasted markedly with that of Edgar Anderson, "slugging away at a big detailed job; which for the most part is pretty unglamorous."[27] Went brought major scientific results and had a considerable reputation even as far away as New Zealand: "authorities of the caliber of Dr. Went" carried considerable weight in physiological, botanical and forest circles.[28] The cultural elites of Saint Louis certainly opened their checkbooks for their new scientific star: at some point immediately after Went's appointment, Brookings Smith could report receipt of $10,000. The anonymous gift from an unsolicited donor signified "very tangible approval of our program for rebuilding the Garden to its former prominence in the botanical world [and] an indication of how much a public spirited citizen of St. Louis appreciates our contribution to the scientific, educational and cultural needs of the city."[29]

In any event, there was little possibility that Went could return to Pasadena. Louis Levin, the deputy assistant director of the National Science Foundation's Biological and Medical Science division, received Went's first grant from Missouri cautiously, even skeptically. Rumor had reached him, evidently, that Went's tenure in Missouri might be brief, and that Went was already contemplating returning to Caltech. Went assured Levin that any such notion was dead: I have "burnt my bridges behind me," Went said.[30] And burned his bridges he had. Writing to his old colleague Henry Eversole and Eversole's niece, Lucy Mason Clark, Went mentioned that he no longer felt able to live in Pasadena, especially after "Dr. Beadle's evaluation of me as a scientist and as an administrator." With the publication of *The Experimental Control of Plant Growth*, summarizing the initial efforts at controlled greenhouse experimentation, he told them it was time he moved on to other challenges. Missouri would only be a year or two, he hinted, before he might build "probably on a larger scale [a] new set of air conditioned greenhouses and controlled temperature rooms [to] house not only plants but also small animals, and plant diseases."[31] The next year, Went and Sterling Hendricks supported the need for an American "Biotron" in no less than the journal *Science*.

For a scientist who did not relish administration or claim any ability at fundraising, one scarcely understands why he would move to Saint Louis at all unless he felt even more forced out of Caltech than his diary reveals. What is true is that when Went arrived at the Missouri Botanical Garden he encountered a larger, more diverse organization than what he had left behind in California. In 1958, the garden employed a staff of nearly sixty, including gardeners and laborers, as well as the staffs of the public relations and director's offices. The Research and Instruction side of the institution had ten members, led by Edgar Anderson, who had been with the garden for some twenty-five years as the curator for useful plants in the economic house. The arboretum, the herbarium, and the orchid department had another ten people, most of whom had been at the garden for less than five years.[32] George van Schaack ran the library and herbarium "in addition to," Went was startled to learn, "mathematics teaching." Taxonomist Robert Dressler was in charge of the *Annals*, the major scientific publication of the garden. Though Went later said he regretted keeping people in their positions, it must have been far less trouble to keep Hugh Cutler as assistant director, and allow George Pring to continue to obsess over his orchids at Gray's Summit. Likewise out in the gardens, Louis Brenner and Lad Custak continued caring for most of the horticulture.[33]

If the garden's people remained unchanged, its buildings underwent a rapid and dramatic transformation. Circa 1958 when Went arrived in Saint Louis, nearly every building of the garden's infrastructure was nearing collapse. So "extremely bad" was the citrus-alcove house that it "would have to be dismantled." The outside of the six orchid houses all needed work, and the growing houses seemed merely "average-to-bad." Went hardly knew how the Fern Palm Economic houses were still standing. Only the Linnean House and the experimental house had been recently refurbished, testament to a new impetus toward reestablishing the garden's scientific credibility.[34] Went saw only the tired rust of a bygone age in his first months: "The value of the existing garden elements such as walls, arbors, etc., and the importance of many of the existing old trees was discussed," noted the minutes of a meeting about the old dilapidated Palm Houses. "It was generally agreed," the minutes continued understatedly, "that there is very little value in much of this and we should not restrict our concepts to retaining existing garden elements and plant materials."[35] Emboldened by Brookings Smith, Went advocated the removal of the palm houses entirely and their replacement with a brand-new structure. Whether the palms or the palm house itself was the more sacred cow to the older trustees is unclear, but their removal clearly displayed the new order.

Went and Brookings Smith took down a venerable section of the old Shaw Gardens, but they would meet stiff resistance in other areas. Certainly, the garden would hire no other physiologists to augment Went; any new researcher "must not be a physiologist but might be an algologist," Went noted to Levin at the National Science Foundation.[36] Went hoped maybe to secure a geneticist. Instead, though Went long had been dismissive of taxonomy, the subject continued to dominate the garden's efforts. Moreover, the established research staff did not concur with Went's lackluster assessment of the scientific efforts at the garden. They claimed that the Missouri Botanical Garden's Henry Shaw School of Botany's association with Washington University made it a true center for "horticultural research" rivaling Cornell University. Furthermore, the steady publication of scientific work through the garden's *Annals* gave the institution a worldwide audience for its local scientific work.[37] Likewise, the garden's herbarium remained central to their work in taxonomy and morphology. Overall, the research value of the garden stemmed from the close association between "a first-class herbarium and library" and "horticultural and botanical garden work."[38] In contrast, Went regarded taxonomy and morphology as nearly as antiquated as the rusting Palm Houses, and noted that the research spaces possessed "little space for students and scattered facilities."[39]

In the heady days after Sputnik's launch, Went, Cutler, and Brookings Smith determined to drag Shaw's Garden from the nineteenth century into the twentieth century in short order: "we must get a grid system installed through the entire Garden" to direct plants, Went insisted.[40] High technology would lead the way. The future required, the consultant report read, a "phytotron and greenhouse near Climate Structures, Research Center and Entrance Facilities."[41] Via a vision of the modern botanical gardens, the phytotron at the Missouri Botanical Garden was born. Well aware that a new, well-publicized "research program would be of great benefit in arousing potential donors to the importance of the research aspect of the garden," Went, Cutler, Brookings Smith, and Rohrback seized the moment to build not just a phytotron but an even grander icon of modern science.[42]

FOUR SEASONS IN ONE DAY

For Went, the American National Science Foundation (NSF) offered the best hope for the kinds of funds necessary to rebuild any modern botanical garden in Missouri. At the end of 1958, soon after Went arrived, he and Cutler formally applied for $300,000 to rebuild the garden's greenhouses, its arrangement of plants, its public education program, and its museum program. The grant noted the addition of Went and Norton Nickerson as new scientific staff. Nickerson had been a fellow at the California Institute of Technology in the summers of 1954 and 1955, and then a botany instructor at Cornell until 1958. He received one of the early research grants for work on "lazy" corn and gibberellic acid. The NSF awarded $250,000 on June 30, 1960. The NSF read about a scientific institution with crumbling facilities but notable potential. Went and Cutler dwelled especially heavily on the area of education and training, noting that in the new facilities "public education will be stressed." The familiar call of a lackluster public school system justified the appeal of federal funds to a private institution for public ends. "If grade school children are to get any education in biology, this will have to be done through museums, zoological and botanical gardens, and we would like to be among the pioneering institutions." The pair even suggested supplying food made from plants in the garden such as "tamarind cookies, dates, olives, ginger" to the garden's restaurant.[43]

At least publicly, Went sailed into 1959 convinced that the endowment could be shifted to "scientific" concerns. New money could be raised easily from new grants, while the local Saint Louis community could support the traditional maintenance of their garden. As the budgets forwarded to Went indicate, the idea of shifting garden maintenance onto the consuming public seemed reasonable and necessary. Orchids and the public garden maintenance consumed twice as much money as the entire scientific side, which included research, instruction, the herbarium and library, as well as publications.[44] If the garden could be effectively divided into two complementary halves, Went believed, both a public space and a scientific institution might flourish. With confidence, Went promised Levin that he possessed the "assurance of the trustees that within a month they will have underwritten the funds necessary for rebuilding our Palm House. In its place will be erected an air-conditioned plastic greenhouse, which will probably bear the same relation to ordinary display greenhouses, as the Earhart Greenhouses bore to the regular research greenhouse."[45] With that same assurance, Went made a two-month tour throughout South America at the end of 1958. When he returned to Missouri he found definite progress on the central new greenhouse plan. As he noted in his diary, the trustees welcomed "a geodesic dome of 175' diameter and 65–75' height" "(to be named something like plantosphere, or sylvarium, or floradome)."[46]

Inevitably, the *name* of the facility "was coined specifically."[47] This time, it was the architect, Eugene Mackey Sr., who devised the name *Climatron,* evidently responding to Went's idea that he wanted a climatological laboratory. Informed readers of the garden's *Bulletin,* no doubt anxious to visit the new structure, read erroneously how Went had "coined" the name to emphasize the climate control functions of the new house.[48] And so the pattern of his earlier phytotron repeated itself: though he did not father it, Went once again wore proudly another tron badge. Moreover, the embodied symbol of tron moved readily between California and Missouri; even Went's analogy was identical to his earlier phytotron: "Just as the cyclotron, bevatron and synchrotron belong to the most ambitious research tools the physicist has created," Went wrote, so "the Climatron undoubtedly is the most modern and advanced research and demonstration tool of the horticulturalist and botanist."[49]

THE DOME OF HEAVEN

By May of 1959 the garden's board had raised a substantial sum for the Climatron. New and old met in the Climatron. Went the scientist advocated the new, and the garden's trustees accepted that they lived in an age of science to whose icons the general, paying public would flock. Certainly, the newness of the design would be celebrated far and wide: the garden's *Bulletin* assured its readers, "After all, we were building not just a new greenhouse for Shaw's Garden—we were inventing a radical, new kind of facility, never before tried anywhere in the world, for a radically new approach to the growth and display of tropical plants."[50] Yet the

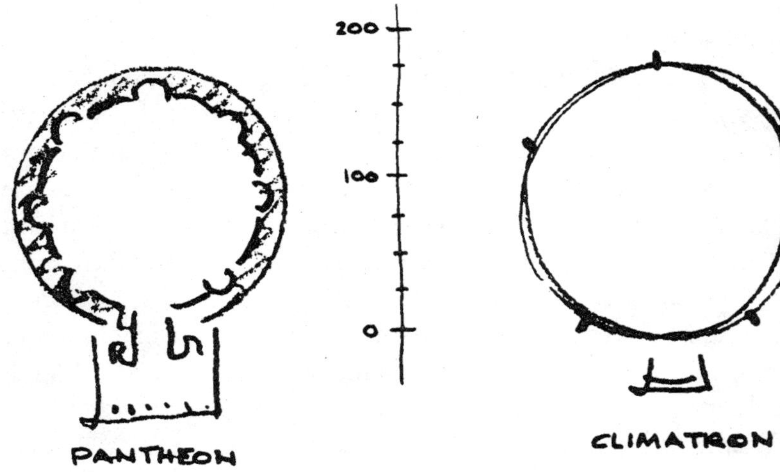

Figure 3.4. Pantheon: Climatron. File "Reynolds Award." Record group 3/2/6, box 21, folder 4. Archives. Missouri Botanical Garden.

architectural firm of Murphy and Mackey sought inspiration in classical forms. Among their early sketches for a single columnless space, they made a direct comparison between the Pantheon in Rome and a dome for the Missouri Botanical Garden. The "Dome of Heaven," all hoped, would stand in "striking contrast" to the "old-fashioned" palm house.[51]

Under a single dome, Went's radical new space took shape rapidly. By August, the name and the major architectural elements had been settled. A single unified ecological space would house four different climate areas all under a "geodesic climatron." To Went, the design promised that visitors would "experience the various seasons of the year and . . . see the effect of temperature, precipitation, altitude and latitude on plants as they relate to various geographical regions of the world."[52] The plans called for five areas (rather than the eventual four): Autumn, Cold, Spring, Summer, and Desert. The "Cold Climate" area would display, for example, "the affect of perennial winter. Northern latitudes or above timber line. Tundra. Mosses-lichens-grasses-sedges-rhododendrons-Labrador tea."[53] Went was overjoyed as he witnessed the fulfillment of his grand display of public science: "Let me say," he said to the engineering firm, "that the . . . climate controlled facilities is exactly what I had in mind."[54]

So what did Went have in mind? Fortunately, Went's evolving notions

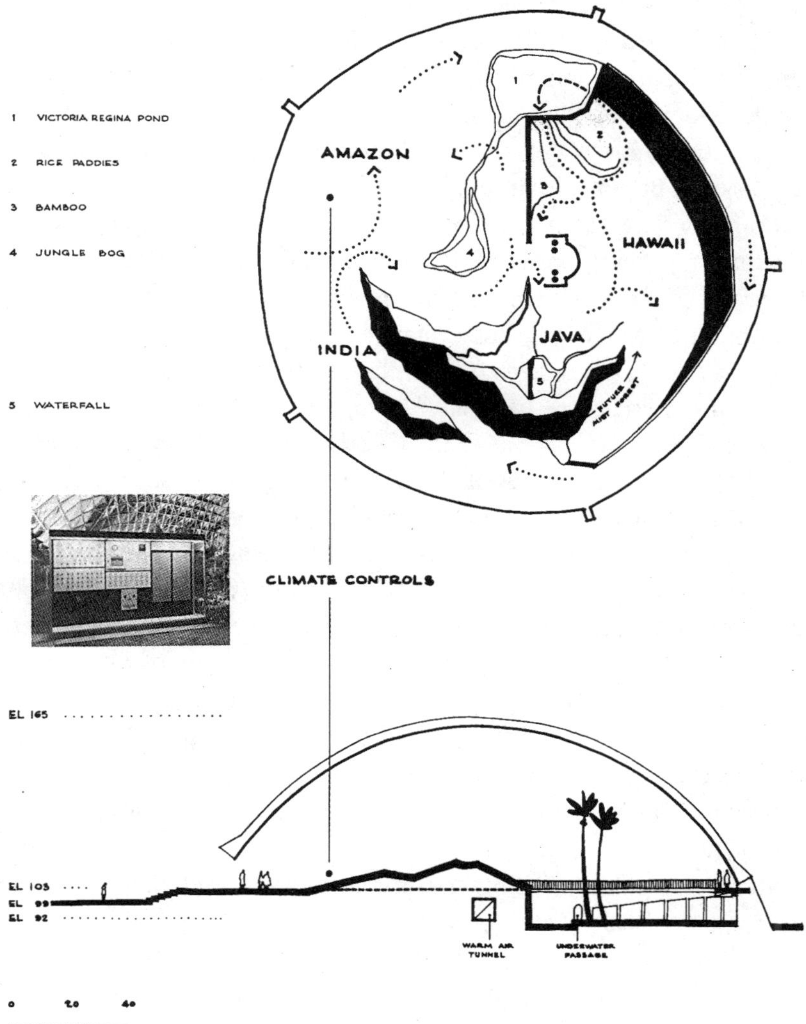

Figure 3.5. Interior plan of Climatron. Archives. Missouri Botanical Garden.

of his new phytotron, the Climatron, took shape in long drafts of handwritten notes buried in his personal papers. His notes convey the impression of a centerpiece able to attract attention for the garden, and, one suspects, himself. The new Climatron would be unique as a greenhouse, utilizing air-conditioning once more to "maintain a number of different climates in a single enclosure." It would be unique in allowing visitors to tour all the climates of the world in a single hour. It would be unique as the "first in a series of new greenhouses to be built to replace the obsolete and deteriorated old ones."[55] As the world entered an era of unique firsts in space, uniqueness, Went hoped, would also stem the tide of decline at the Missouri Botanical Garden.

Construction began rapidly. The Buckminster Fuller geodesic dome of aluminum tubing was erected in only six weeks. On reflection, however, the architects considered the plastic skin inadequate, or at any rate no manufacturer would guarantee that their product would last any more than two years. The trustees faced a choice of sacrificing the Shaw palms or erecting a structure likely as not to need replacing in a few years. Modernists from Robert Moses to Le Corbusier would have noted proudly that "the immediate present—and the palms—were sacrificed to the future," as the garden's *Bulletin* declared.[56] With the palms left out to die, Went and the architects saw the passing of the outdated palm motif. For them, the new structure stood "in striking contrast to the old-fashioned Palm House it replaced," and became "the principal feature of a series of improvements . . . overlooking the principal landscape of the 100-year-old Botanical Garden."[57] In characteristic 1950s style, the past and the present made way for the future, a shining, aluminum, geodesic future, now covered with the material of the future: "plexiglass." The excitement of creating an entirely new world gripped the garden: "We were building not just a new greenhouse for Shaw's garden—we were inventing a radical, new kind of facility, never before even tried anywhere in the world, for a radically new approach to the growth of plants and display of tropical plants."[58]

Hexagonal geodesic segments enclosed a tropical space. Once the plastic skin was attached the following year, a vast air-circulation system moved hundreds of thousands of cubic feet of air per minute through dense tropical foliage. Inside the Climatron, crossing from one side of the dome to the other, the air became noticeably cooler, at least ten or fifteen degrees; continuing on farther, the air became more humid. At its extreme, in the southeastern section, the environment would "closely resemble the Amazonian lowland." Likewise, should a visitor return at nighttime, they would feel noticeably drier, as if in regions "such as Western Pakistan and India." Only under those strange conditions could sugar cane and mangos grow in abundance. Went specifically looked to illustrate that high humidity was not, in fact, a requirement of a successful greenhouse: "this fallacy has been exploded by experiments in the Pasadena phytotron, where it was shown that temperature rather than humidity is significant for plant growth."[59] In the Climatron, Went employed the lessons his California phytotronic science had shown him about maximizing plant growth. Humidity was connected to the difference between day and night temperatures; less difference gave high humidity and greater difference lower humidity. By specifically adjusting the daytime and nighttime temperature of the sections of the Climatron, humidity and growth could both be controlled at will.

"Only a *visit* to the Climatron can make one realize what a modern greenhouse can be, and that the era of the musty, overheated, overshaded and overplanted greenhouse is doomed," Went insisted.[60] Some sixty thousand Missourians agreed in just the first month, and paid to wander through the modern greenhouse. We can gain much of the flavor of those early turns through the Climatron from the long pages of notes Went contributed toward information brochures, tourist guides, and educational handouts:

> The night blooming jasmine (*Cestrum nocturnum*) is one of the most interesting plants of the Climatron. Formerly it was in the old Palmhouse, but it did not do well, partly because the air was too stagnant, partly because it was too dark, and partly because it did not get the right light treatment. In its native habitat (West Indies) this shrub flowers abundantly in October, and much less in May. It behaves in the same manner where it is growing in Southern California. The reason for this flowering behavior is that *Cestrum nocturnum* is a so-called long day—short day plant. This means that no flowers are produced unless it first is subjected to a number of days when it receives more than 12 hours of light per day (long day), followed by 2 short days. In the Climatron, which is illuminated every Friday, Saturday and Sunday until 9 or 10pm, the plants *always* receive at least 3 long days every week, and therefore they get always a succession of 3 long days plus 4 shorter ones, which is apparently sufficient to make them flower.[61]

As much as the garden itself, the plants also profited via the Climatron. Under controlled conditions, and benefiting from close study, a heretofore unusual flower bloomed consistently year-round giving a wonderful scent in the tropical environment. By sacrificing the palms to the future,

Figure 3.6. "The nerve center of the Climatron," The Control Panel sited near the front entrance. From *Missouri Botanical Garden Bulletin* 53, no. 6 (1965), 2.

modern technology and science achieved another success. Went underscored how, by tearing down the stagnant past that could not conclude the causes of growth and development in plants and replacing them with controlled spaces able to experiment with the environment, wondrous results necessarily occurred. The lesson seems plain enough. If a pleasant flower could be made productive all year-round in the Climatron, the imagination soared considering the possibilities for economic crops. Perhaps not coincidentally, those economic crops lay just ahead of the visitor: "to the right, across the path, are rubber tress (*Hevea brasiliensis*) growing" so large that by 1965 they "should produce locally-harvested rubber"; sugarcane (*Saccharium officinale*) lay behind the papayas, and "several species of *Citrus* are growing all coming from China"; then mango tress, avocado trees. "Always on display are *Cattleya*," the corsage-type orchid. Curving around the visitor encounters the *Eucalyptus globulus* from Australia, "one of the most important trees in the world!"[62]

In the Climatron, the architecture of modern technoscience reached new heights. "On Sunday evening, Oct 1, 1960, with a distinguished audience of leading scientists, government officials, industrialists and many other friends of the Garden in attendance, the Climatron was dedicated to American Science, to the people of American, and especially to St. Louisans."[63] To hear the president of the National Academy of Sciences, Detlev Bronk, give the dedication speech, the audience would have walked past the Climatron's "Control Panel" (fig. 3.6), situated on the left immediate as one entered the facility—verily, the computer in the greenhouse. Barely had the facility opened to the public when it received the R. S. Reynolds Award for 1961, having won the international competition hosted through the American Institute of Architects. The citation noted, "the tropical lyricism of the botanical displays seems so successfully carried out by the architects of this structure, that it must be a marvelous experience for the visitor."[64] Even more so than the scientists themselves, the architects privileged imagination over reality for evidently the awarding committee had not been through the Climatron.

In many ways, of course, the Climatron did for the Missouri Botanical Garden precisely what it was supposed to do: it increased visitors and generated income. The people of Saint Louis and elsewhere flocked to the garden to see the new spectacular greenhouse, to walk under the waterfall, to go into the tunnel and take photographs of the geodesic structure. Inside, they listened to music, and became for a short while a part of nature once more. "A spectacular feature planned by Dr. Went is the lighting system which gives the effect of an artificial sun and moon. Banks of high-intensity lights revolve slowly to give the progressive visual effects of sunrise, daylight, sunset and moonlight. The lights make a complete revolution once every five minutes, about the time required to stroll around the Climatron's 550-foot periphery."[65]

According to the figures collected by Went, "the steep rise in attendance in 1960 and 1961 has to be attributed almost solely to the building and operation of the Climatron."[66] Attendance records were broken in the first months after the Climatron's opening, a point emphasized continually over the next two years by the director's office. Yet, even under the guise of success, the director's office reports dwell on the expected yearly income of the garden, and in particular the part played by the Climatron.

> The success of the Climatron is showing to what extent Saint Louis has been waiting for the type of renovation which is now going on in the Mis-

> souri Botanical Garden. Not only in the amount of recognition [and here Went had inserted local, national, and international recognition] we are receiving for the Climatron unprecedented for a venture which has cost under one million dollars (compare this with the costs of our space program!), but it has caught the imagination of St Louisans, who are flooding to Shaw's Garden in ever-increasing numbers (in 1960 more than twice as many people visited the garden than in the 1949–1958 era).[67]

During the planning phase, Went estimated that the Climatron would generate a figure of $100,000 per year. In the nine months following its opening, it looked on track, and the summer months still lay ahead. Went expected that at that peak time, the Climatron would more than pull its weight in attracting people to the garden. "Whereas greenhouses during summer tend to get too warm for comfort, especially when they are not shaded, the Climatron air circulation system keeps the temperature comfortable even on warm days."[68] His expectations proved correct, and the Climatron saw even more visitors willing to pay the extra twenty-five-cent entry fee (a hotly contested issue) for the next several years. By mid-1961, the Climatron had found another niche in the ecology of the garden, evening visitors. By remaining open until 9:00 or 10:00 p.m., the Climatron generated further income on Friday, Saturday, and Sunday nights.[69] Somehow, the Climatron really was the Bevatron of botany to the Saint Louis public.

FOREVER AUTUMN

The future dominated Went's considerations of his scientific work, and his tenure at the Missouri Botanical Garden was marked by his constant appeal to the morrow. Went regarded the Climatron merely as a stepping-stone. The future would see an even larger laboratory built from the tokens of the paying public's visits to his Climatron, its aura and its turnstiles cementing the foundation of his new research center where experimental work on evolution might finally take place. As early as 1959 Went noted in his diary, "I have decided that the main theme I am going to work on, and which I will be developing at the garden is Experimental Evolution." What he meant by the phrase "experimental evolution" entailed an incredibly broad program of the study of the growth and development of life in whole ecosystems. "I would probably want to work on the interrelationships between plants, which probably are largely chemical, on the germination controls, on the effect of jungle trees on each other and on young plants in general, on experimental desert associations, to be recreated in the desert greenhouses, on transmission of characters between species by non-sexual means." Whole jungles and whole deserts were just two parts of the whole world that Went had in mind. He also said grandly that "it should be possible with our air-conditioned greenhouses to study the distribution of temperature and photoperiod response in the plant kingdom."[70] As Went explained to James van Sant, "It is not the number of hairs or the flower color which controls the survival of a species, but its response to prevailing temperatures, light, humidity."[71] Went saw the entire plant kingdom as coming within his grasp within his new climate-controlled laboratory. His diary's entry concludes, "The Garden would become famous for this work."[72] None of this research came about because Went would soon resign as director.

The trouble started when, to pay for the future, Went suggested an orchestra. He excitedly told his major patron, Louis Levin at the NSF, of his new plans to build a two-thousand-seat auditorium for the Saint Louis Symphony. "Nothing could be more delightful than a stroll through the garden or flower displays during a concert intermission."[73] The city of Saint Louis sought a new location for its cultural center, and Went became convinced that it should be embedded in the garden. His thinking seems to have been that between a cultural center and the Climatron, he would have amassed enough security to launch his campaign for the large, experimental evolution laboratory space he fervently desired.[74] As he noted to Levin, "I am getting impatient to get my new laboratory space; I have to hold off on research projects the way it stands now. Then I also will be able to take on a number of research students and again receive foreign fellows like in Pasadena."[75] For Went, the new research center seemed more like his own academic institution. Somehow Went's Yale plans had evaporated, but a bolder ambition had taken their place: simply, the ticket sales from the Climatron would fund the scientific efforts at the garden, while the architectural distinction and uniqueness of the complex would firmly establish Went's position in the botanical and scientific world. Over the next two years, Went's plans appeared to be on course. Another NSF grant was awarded for the research center in 1961, and a year later, Went secured an additional $188,000 from the National Institutes of Health, while the trustees independently raised an impressive $200,000. Yet the planned center just kept growing. By mid-1962, Went applied for and received a one-year extension (because the original NSF grant was only for two years). But a year later, with still no final design for the research center agreed upon, Went's doom was nigh.

Went's departure from Pasadena seems dignified in comparison to his acrimonious last few months in Saint Louis. As the story appears from Went's papers, longtime trustee Henry Hitchcock had successfully maneuvered the garden's Board of Trustees so that Robert Brookings Smith, their president since 1958, would no longer stand. In a classic coup d'état, Hitchcock, it seemed to Went, had firmly declined the presidency before reluctantly accepting the position. Went attempted to counter these internal political maneuvers as early as late 1962. In an ultimatum delivered to the trustees, Went proposed "a choice between two—and only two—alternatives": either Brookings Smith stayed on and supported Went, or the president's departure inevitably signaled Went's own end.[76] Suddenly "frank talks" between the director and his new president discussed the "serious lack of communication" between them. Hitchcock accused the former president, Brookings Smith, of being "too authoritarian," running the garden with a deficit, unitarily acquiring funds to build the Climatron, and holding on to pipe dreams of a million-dollar endowment. Went, for his part, approved of Brookings Smith's style of management, heartily backed his grandiose plans, and expected more money to flow into various new projects, especially the proposed cultural center. The Climatron, Went "aired," illustrated how large glamorous projects successfully gained funding while traditional, conservative albeit inexpensive ideas continuously floundered. Added to these distinct differences of vision for the Missouri Botanical Garden were personal animosities that had clearly lingered for years. Hitchcock accused Brookings Smith (and Went by association) of a "secret deal." Went cited letters and minutes to Hitchcock, who "had to see the minutes before he believed." After an hour, Went left convinced "that there is no future for me at the Garden under [Hitchcock] (it would not be *with* him)."[77]

At the next Board of Trustees meeting, Went avoided conflicts with personalities, and instead sought to convince the members of their duties under Shaw's will and the agreement they had made with him back in mid-1958. It was clear even to Went that by then his own vision of the garden's distinct public and scientific aspects differed from the board's conception. Went saw the public face of the garden as requiring a new entrance, a new parking lot, and a new auditorium (perhaps united with the Saint Louis Symphony into a grand Cultural Center) all built on the model of the Climatron. As the minutes relate, "Dr. Went also inquired as to what the Board's feeling was towards the music program, [and] the maintenance of the restaurant."[78] The minutes fall silent on what, if any, their feelings were. Recalling the incident years later, Went condemned Hitchcock for his narrow-minded parochialism when faced with something like exotic food. Yet the restaurant, we know from Went's notes, was but a microcosm of his overall vision. "The restaurant will not be a conventional one," Went had outlined back in 1961, rather "it will explore and demonstrate the relationships . . . between [the] taste of plant products and climate." From the varieties present in the Climatron alone, skilled chefs would make the world's most remarkable dishes. "Thus eating," Went envisioned, "will take on a new meaning, apart from culinary and caloric aspects, and nobody will leave the restaurant without some new experiences and understanding."[79] All this proved too much for Hitchcock, who specifically identified the restaurant as a key issue to oust Went from the garden.

In April 1963, Went pressed the board once more for "the erection of a building to provide research space and house the library and herbarium," and the board "unanimously decided" to make the research building the "number one project" to support scientific work.[80] The trustee's support of the research space makes it difficult to get past the impression that Went and the board saw the garden in two distinct and conflicting ways. To Went, the public and scientific gardens were far more united than the trustees viewed them. The trustees remained devoted to the scientific aims of Shaw, and when outside funding became available, they were more than ready to support purportedly scientific endeavors. For the trustees the public gardens commanded the support of the Shaw endowment; science was external to orchids. In contrast, Went believed that scientific botany enjoyed the support of the Shaw endowment while the public gardens and orchid displays should be necessarily supported by public funds gathered from the people. For Went, the Climatron was not a scientific space but a display of science for the public; however, to the trustees, their Climatron *was* a modernist scientific space of controlled growth and development.

With restraint, Went resigned as director of the Missouri Botanical Garden in October 1963. He wrote to Paul Kramer at Duke, then

Figure 3.7. The geodesic dome of *Silent Running*.

building his own phytotron, that "unless very soon the Trustees of the M.B.G. change completely, the Garden will be finished as a scientific institution."[81]

In fact, the end of Went's time at the Missouri Botanical Garden did not mean, as it had in California, the end of his tron and it certainly did not mean the end of the garden. In Saint Louis, the Climatron continued to be a central attraction for the city, a topic of debate, and in time, received substantial refurbishment funds to be restored and expanded. For the garden, the Climatron succeeded both in drawing new crowds into the old institution and in successfully making "tropical plants grow as well in the temperate zone as they do in their native homes."[82] Went became remembered as the man who built the Climatron, but he did not return to the garden for nearly twenty-five years. His modernist plexiglass vision may not have revolutionized biology, but it undoubtedly saved an august institution in America's scientific heartland. Hitchcock may have preserved the garden for bland food and debutante balls in the 1960s, but by the 1970s, the social function of the Climatron had radically changed. One can only imagine Hitchcock's reaction on learning that the Climatron inspired the movie *Silent Running* and its dystopian vision of humanity's future in space, where the biodomes of the last of Earth's organic matter are being jettisoned into space as waste, and then detonated (fig. 3.7).[83]

To a scientist like Went, the Climatron was always about the future. As Evelyn Fox Keller powerfully noted, representations of science-fictionness in science, "hopeful statements," are "expressions of a kind of intentionality . . . they actively contribute to the construction of future scientific reality."[84] Beginning as a modernist vision of a controlled future as conservative cultural elites struggled to preserve their world, it also publicly displayed a style of experimental plant science where the control of the environment offered to govern plant growth. Moreover, it has been for decades now a place where people go to witness a number of ecologies diverging under different climates and to learn about the future of their planet. Though one walks through the Climatron as a tourist, hopefully one appreciates that people are as embedded in their own ecological niches as every fish, tree, or orchid.

CODA I

THE FINALE OF FRITS WENT

The tree has graduated from molecular biology.
—Frits Went, circa 1951

THE DISCOVERY of auxin and the creation of the Climatron book-ends the life and work of Frits Went. Starting as a laboratory scientist, he became a builder of technological systems and finished as the director of the first climate-controlled ecosystem. Throughout, Went championed the fulfillment of plant physiology and botany as experimental sciences via control of the biological environment. To that end, Went created a new big biology. He became the founding figure for the generations of technologist biologists that went forth from Caltech and created a new biology of environmental control. However, when Went resigned from the Missouri Botanical Garden in late 1963, he was largely cast adrift from plant science, which itself was changing rapidly. Officially, Went took up a professorship of botany at Washington University, but after only a year he left to become the director of the Desert Research Institute in Nevada. He retired from the institute in 1972. In Nevada, to complete the circle of the scientific community, he was replaced by Dwight Billings from the Duke University phytotron, who was the author, back in 1952, of the holocoenotic environmental complex diagram that conceptually specified the physical biological environment.[1] For the next twenty years, Went lived quietly in relative obscurity. His remaining papers from the 1970s and 1980s show a scientist lost, his articles consistently rejected by journals, his ideas dismissed, and his grant proposals turned down. In his twilight years he worked on a strange theory linking the presence of black particulate carbon in the atmosphere to blue skies, and even privately published a small book on the topic.

Before moving on to the stories of two second-generation phytotrons and the Biotron in the final three chapters, I would like to pause in this Coda for a moment and briefly offer an epilogue to the founder of phytotrons. It is the warts and foibles of a scientist like Went that invaluably add to our understanding of the whole character of recent science; just because Napoleon lost to Wellington at Waterloo it does not make him any less historically significant—indeed, precisely the opposite! Went's conception of the study of life assumed control over genes and then worked to establish control over environments, and thus, to him, any real science of organisms had to be built from the pairing of the sciences of genes and environments. What he witnessed over his lifetime, however, was the rise of a predominantly molecular understanding of life and its processes, as did others, notably both Paul Kramer and E. O. Wilson, who believed that the molecular wars ravaged biology above the cell by the 1970s.[2] While scientists after the 1970s would caution molecular biology against its extreme reductionism, as far back as 1942, Went felt it necessary to warn the botanical community that while "we may be able to reproduce any number of partial reactions going on in the plant in a test tube or on a piece of paper," to "consider Life in general" requires "the final fitting of these reactions into the unity of the living plant."[3] According to his biographers, some fifty years later, after witnessing the remarkable changes throughout biology, he died convinced that "the domination of physiology by molecular biology was an impoverishment of biology," and had up to the very end "decried the growing reductionism in biology and was especially disturbed by the increasing emphasis on DNA, to the virtual exclusion of other subjects."[4]

In contrast to the continued fantastical celebrations about DNA, Went saw the contribution of the plant sciences fade, his own reputation reduced to merely the discoverer of auxins, and the greatest indignity of all, his phytotron forgotten. The discovery of auxin remains his famous triumph, and the wider scientific community commemorates it frequently, yet what Went considered his real triumph, his controlled environment successes, are ignored. Went was evidently surprised when, in the mid-1980s, the orchidologist Joseph Arditti encouraged him to recall his contribution in spite of Went's emphatic protestations. "You may perhaps argue that you are not primarily an orchidologist," Arditti wrote back to Went, but Went's work on auxin, the development of the Vacin, and Went's medium for orchid seed germination had all made "significant contributions to orchids," Arditti insisted. "Without auxin there can be no tissue culture, no control of rooting, no studies of flower physiology in orchids."[5]

Went saw himself quite differently than Arditti did. When Went later reflected on the achievements of phytotrons for an undergraduate plant physiology textbook, he dwelled on his auxin discovery but he also reiterated what he believed was the genuine contribution of the development of phytotrons, namely, to make the biological sciences truly experimental. Over the "last 30 years phytotrons have helped to make ecology an experimental rather than a descriptive science," Went wrote.[6] It seemed obvious to Went that his phytotron had made the study of environments equivalent to the study of genes in biology, but had also made the study of biology equivalent to the study of physics and chemistry. One perceives Went's frustration in his later writings and biographies about the fact that such achievements merited no commemoration. He had devoted himself, he said, to following "the experimental method." What he meant by that was on display in his phytotron—the production of standard, controlled and repeatable environments and the production of a standard methodology to produce standard plants for experimental work, as well as the production of the standardized researcher. The practice of science being played out in the phytotron was straightforwardly the product of "the experimental method now so thoroughly entrenched in most branches of science."[7] In other words, control over the biological environment made botany and plant physiology into sciences just like physics and chemistry Went advocated, recalling the familiar refrain that the phytotrons of the plant scientists were akin to the cyclotrons of the physicists. Nor was Went alone in such ambitions: his French counterpart, Pierre Chouard maintained that the "phytotronic method" had revolutionized the science of biology.[8]

While Went labored mightily to recast the epistemological basis of the plant, or even the biological sciences, and was successful insofar as he built several phytotrons and helped create a new scientific community around them, the world of science was also the fight for institutional and disciplinary recognition, grants and patronage, disciples and devotees, and sheer publicity. Went was disastrous as a booster and publicist. Indeed, according to his biographers, Went moved from Caltech to Missouri to avoid the increasing administrative burdens of a prominent scientific and institutional director. While Went "left behind his cherished [Caltech] phytotron and the desert ecosystem he had come to love," Arthur Galston later wrote, "what propelled him outward was a growing disenchantment with the burdens of administering and raising grant support for the phytotron and of coexistence with colleagues who were increasingly committed to reductionist molecular genetic approaches to the problems of biology."[9] In other words, the tragedy of Went unfolded because in an era of big science, bigger budgets, and the biggest instruments, he left one phytotron complex for another even larger one, yet was never able to preserve any of the social levers to move his patrons, secure his supporters, and undermine his opponents.

Of course Went tried to carve out memorials for what he believed was his lasting legacy, namely, phytotrons. In Went's last major presentation before an audience in Paris in 1976, he rambled but reiterated all the fundamentals on which the array of climate-controlled facilities and a scientific community had been built: the "grand botanists" of the nineteenth century, Julius Sachs, Hugo de Vries, and Charles Darwin had each attempted to explain the behavior of plants "in relation to their environment" but they failed, Went said, because "it was necessary to control the environment." In the twentieth century, however, technological developments like fluorescent tubes and air-conditioning had made the study of experimental ecology "doable." Those technologies had made botany and plant physiology into sciences with experimental standards equivalent to those of physics and chemistry. As Went had said before, "not one chemist or physicist would consider obtaining results of experiments under badly regulated conditions or with lesser known equipment." Yet, having left Caltech and having been forced out of the Missouri Botanical Garden, Went seemed as lost as his ambitions to establish environmental control as a standard methodology of biology. His insight that botany and ecology required controlled conditions remained unappreciated, the triumphs of his research program including photoperiodicity "one of the most important factors of the environment," rhythms, thermoperiod, and the causes of flowering unacknowledged. Genetics was now everywhere, he observed, yet it could not discern where the "environment changes the hereditary behavior of a plant." He had continued to work on the biological environment and was fully a technologist biologist: Went's last experiments in a small phytotron in Reno, he said, aimed to reproduce the conditions of the night sky. He told his audience that he had built himself a new phytotronic component, a "celestial radiator," supposedly to absorb longwave radiation coming from leaves to get plants below the temperature of dew condensation. Casting a final long look over modern biology and its by then, in his view, imbalanced priorities, Went asked his audience "where modern genetics would find itself if Mendel had had a Phytotron at his disposal."[10] Where indeed.

CHAPTER 4

THE POSTCOLONIAL SCIENCE OF THE AUSTRALIAN PHYTOTRON

> Australia already has a cyclotron; why do we have this overlapping and duplication?
>
> **— Prime minister's speech at the opening of the phytotron, 1962**

LIKE MICKEY Mouse and the Beach Boys, phytotrons migrated from Southern California to the rest of the world in the 1950s and 1960s. Frits Went's first phytotron at Caltech served as both a model and a training ground for a whole generation of phytotronists who established their own facilities around the world. A global biological science of controlled environments took shape and by the 1970s some thirty phytotrons existed. Prominent among that array stood four great facilities: France's grand phytotron, the massive Soviet phytotron, the Australian phytotron, which is the subject of this chapter, and the American Biotron, the subject of chapter 6. The story of the Australian phytotron sits in-between the stories of the Caltech phytotron and the American Biotron because the goal of total control evolved from its first incarnation as a local facility at Caltech, a private institution, to later national phytotrons like the Biotron that claimed greater control over larger ranges of variables to universalize the new science of the biological environment.

In Australia as much as in California, the transformation of biologists into technologists took place through the explicit appropriation of the physical sciences. The first phytotronists had so successfully managed to broadly spread their message that a phytotron was a cyclotron for biology, that when the Australian phytotron opened in 1962 the then prime minister, Sir Robert Menzies, an imposing man and political figure, and an elegant, persuasive, and memorable public speaker, could tell a humorous story about how his civil servants had at first confused the two: "When CSIRO [the Commonwealth Scientific and Industrial Research Organisation] first of all decided that a request ought to be made to the government to find the money for a phytotron, the proposal fell into the hands of a relatively junior officer in my department who read it and in all good faith scribbled out a note saying 'this is nonsense, [Mark] Oliphant [the preeminent Australian nuclear physicist] already has a cyclotron; why do we have this overlapping and duplication?'"[1] Sitting among the invited guests at the opening of the newest icon of Australian and plant science was the scion of phytotrons himself, Frits Went. Listening assuredly as his Climatron took shape under a Buckminster Fuller geodesic dome, perhaps Went recalled the day over a decade earlier when two wits in his own laboratory had jokingly first coined the term *phytotron* as the audience laughed at the mix-up over exactly what sort of tron they were building. Went should have been pleased because the anecdote unintentionally revealed that the core rationale of the phytotron had been successfully replicated. That night Went wrote in his diary that biologists were in awe of physicists, while years later Lloyd Evans, specifically recruited from Went's Caltech laboratory to help build the Australian phytotron, recalled that Australia's scientific leaders were "enthusiastic because [they] saw the phytotron as linking physics with biology."[2] Indeed, just as the Californians had, the builders of the Australian phytotron aimed to establish total control over experimental conditions of light, humidity, temperature, and nutrients to reproduce in the biological sciences the epistemological certainty of the physical sciences.

To that end, the Australian phytotron substantially expanded the range of possible environments over the original Earhart laboratory with fifteen glasshouses capable of sixteen-hour photoperiods with day/night temperatures ranging from 15°/10°C to 36°/31°C in 3°C steps sat alongside 107 independently controlled day-length cabinets, in addition to frost rooms and dark rooms. The facility's crown jewels, however, were "38 artificially lit cabinets in which irradiance, temperature and daylength and in some cases relative humidity, can be accurately controlled."[3] When it opened in 1962, the Australian phytotron stood as the largest facility yet dedicated to research on the environmental causes of plant growth and development. Over the next twenty years, the facility produced some remarkable insights to the nature of biological processes, including 590 published papers. Notably, it showed "the significant role a phytotron can play in interdisciplinary research," particularly in the elucidation of the varied environmental responses to the so-called C_4 pathway of photosynthesis.[4]

Unlike the privately operated Earhart laboratory, the Australian phytotron was built and operated under the beneficent aegis of the Australian government's scientific arm, the Commonwealth Scientific and Industrial Research Organisation (CSIRO). As this chapter explores, the emergence of the facility parallels the evolving political and social environment of postwar Australia that accompanied and helped construct the changing meaning of the physical environment. As much as the creation of the first phytotron took place within the struggle to specify the focus of the science of biology, the creation of a phytotron in Australia was part of an intense and wide-ranging debate after the Second World War concerning no less than the conception of the country itself. The phytotron was widely regarded as demonstrating Australia's place as a global scientific leader, as was the Parkes giant radio telescope that had opened only a year earlier. Indeed, national pride saw the Australian phytotron renamed the CERES—the Controlled Environment Research Laboratory—shortly before it opened. For the Australians, compared to previous phytotrons, their facility possessed a greater "number of small cabinets . . . each capable of providing a wide range of conditions . . . when greater accuracy of environment control is required": Evans pointed out in a column in the prominent journal *Nature* that a phytotron might have been an American invention but the CERES was the improved Australian design.[5]

The enthusiastic embrace of modernist technology to attack the problems of the world was part of the redefinition of Australia. Modernism, a global movement as prominent in Australia as in the United States, the United Kingdom, or France, was on full display in projects concerned with computing facilities, new telescopes, controlled-environment laboratories, and nuclear power. All emerged as major national technoscientific endeavors throughout the West in an era that demanded control over nature and over populations, and put its hope for that control in modernist technology.[6] Menzies told the crowd at the phytotron's opening that it obviously showcased the "growing application of science, through technology, to the problems, the practical work a day problems, of the world."[7] But it is important to note that the phytotron was only one part of the wide modernist assault on industrial, social, and scientific development of postwar Australia: the year the phytotron proposal came before the Australian government, 1958, the eagerly awaited Lucas Heights, Australia's first and only nuclear reactor, opened in April. It had been made possible by increasing American and British cooperation in sharing nuclear secrets and the utilization of northern Australia for the earliest British nuclear tests. Although only a research reactor used for the production of medical isotopes, Lucas Heights signaled Australia's joining the club of nuclear powers, a marker of status in the postcolonial world.[8] In that same year QANTAS, the national airline, supplied the country with a round-the-world service via pressurized, American-made, tri-tailfin, aluminum, "Super Constellations," a clear signal of a new modernist internationalism.[9] Aircraft, as Nick Cullather pointed out, were similarly potent symbols of the modern in the postcolonial world: famously, in 1966 when Benigno Aquino described to a reporter what the Green Revolution meant to the Philippines, he connected the transformation of crops to that of aircraft: "'Here is the bullock cart. Here is the nineteenth century,' [Then, pointing across the road to a paddy of stubby, dark shoots planted in orderly rows], 'There is the jumbo jet! The twentieth century.'"[10] Clawing to provide economic and food security as a bulwark against regionalism communism, Western governments invested in the performance of phytotrons as much as in nuclear reactors and aircraft.

As a modern scientific technology for traditional agriculture, Australia's phytotron was exactingly threaded between the old and the new ideas of the nation itself. While Russel Ward's *The Australian Legend* argued that Australia remained attached to Britain culturally, Stuart Ward maintained that Australia had rapidly become multicultural via large-scale immigration in precisely this period, and noted that Australian political culture had finally broken away from its firm commitment to British race unity right around the time of Britain's application to the European Economic Community.[11] Along another axis, Daniel Oakman

Figure 4.1. The Australian phytotron. CSIRO publicity photograph a21400. Image courtesy of the National Archives of Australia.

argued that an acceptance of "modernization theory in the post-war years" was part of Australian moves toward maturity not least because "Australia saw itself as a developing nation, requiring external capital and economic stimulation."[12] Yet at the same time, Australia reenacted an imperial vision to take civilization to her own colonies. Paul Hasluk, the minister for territories between 1951 and 1963, viewed the Australian protectorate of Papua New Guinea as a "society still awaiting the full effects both of the techniques and mechanical strength of Western civilization." Australia's duty, Hasluk believed, was to create both a "community and an economy" whereby "primitive beliefs and codes" gradually gave way to a "new order."[13] In fact, the new order would be decolonization. Also in 1958, talks between the Australians and the Dutch, the last remaining colonial powers in Southeast Asia, saw New Guinea gain its independence. At the same moment that a modern future of aluminum aircraft, nuclear reactors, radio telescopes, and a phytotron was realized, Australia's last colonial inheritance from the First World War departed.

In place of imperial re-creations came multilateral agreements. In 1951, Australia signed the ANZUS treaty with New Zealand and the United States, signifying the pressing dual concerns of Australia and the United

States in having to "live in the Pacific."[14] But Australia also entered into the Colombo Plan with its Commonwealth partners seeking comfort via the continuation of white, Western leadership: the Colombo Plan's basic aid philosophy reckoned upon "the application of modern technology and skills to the under-developed and traditional economies."[15] In other words, the major political and social transformation of Australia from an imperial son to a postcolonial regional partner demanded that the phytotron also explicitly strengthen cooperative links between Australia and the Colombo Plan area, especially India, Sri Lanka, and Southeast Asia. Australia's reorientation toward the Indian Ocean is in line with more recent scholarship on the Cold War that highlights the deep connections between third world nations (itself a constructed and loaded term), and brings simple superpower bipolarity narratives under scrutiny.[16]

At the same time, modernist national efforts of science and technology like the creation of a phytotron were solutions to the problems of social order in Cold War–era Australia, as much as they were in postwar Saint Louis. In short, the phytotron served as a universal biological instrument in a multiscience world in the same way that Australia attempted to serve as an international presence in a multinational world, particularly in Southeast Asia. Pointedly, the Australians lauded their phytotron as a "multi-science" institution between botanists, plant physiologists, biologists, chemists, and physicists and later celebrated how they had "brought together an interacting community of scientists whose research encompasses a wide range of plants and experimental approaches."[17] Through visiting scientists gaining valuable scientific and technical expertise from their time in the phytotron, it was hoped that Australia would gain control and even good will in a time of postimperial turmoil.

THE DIVISION OF PLANT INDUSTRY, CANBERRA, 1945–1960

The story of the Australian phytotron begins in Australia's capital, Canberra, where the Division of Plant Industry of the CSIRO looked to the future. Most members of the British Commonwealth created Councils of Scientific and Industrial Research (CSIR) in the 1920s following the scientific successes of the First World War. In centralized research institutions, the United Kingdom, Canada, South Africa, India, New Zealand, and Australia in effect sponsored national science. Whether they were in the new Dominions or colonies, Councils of Scientific and Industrial Research participated in imperial science. As Roy MacLeod characterized the period, science at the imperial center remained influential because it had access to learned societies, journals, and the great scientific institutions of the Empire such as the Royal Society and Kew Gardens, which isolated or small numbers of practitioners on the periphery could not join with fully.[18]

That relationship began to change even before the Second World War. Between 1939 and 1945, along with Great Britain's primacy in global affairs, it was sundered. The economic uncertainty of the Depression raised the appeal of technological and scientific solutions for Australia's struggling primary industries. Australia's leaders remained distinctly pragmatic about sponsoring research, and consequently the biological, forestry, and agricultural research "Divisions" entirely dominated the early years of Australia's CSIR. Global circumstances saw the Dominions' local scientific and technological efforts expand as the waning British Empire called on every morsel of its strength. For example, Australia gained access to early British radar systems two years before the famous Tizard Mission that delivered every important British technological device into the hands of American industry, including early atomic bomb plans.[19] At the same time, there was a powerful move toward scientific independence: Thornburn Brailsford Robertson, a nutritional physiologist, railed against what he saw as the paternalism of England: "I feel that we must make a strong stand against any kind of paternal dictation which would reduce us to the status of lab-boys of British scientific administrators."[20]

Scientific independence assumed a new importance during the Second World War as Australia found itself forced to redirect its national focus from Britain to the United States. As Australian industries built planes and ships to meet the Japanese threat, local physicists developed their own radar systems while local agricultural and biological scientists bolstered food supplies so successfully that agricultural Australia powerfully contributed to the global war effort by supplying local food to United States forces in the South Pacific.[21]

Consequently, having fulfilled its international commitments and expanded its local scientific base, Australia's self-image as an agriculturally prosperous, independent nation swelled. Postwar reconstruction would continue this trajectory: the budget of CSIRO's Division of Plant Industry increased by over 300 percent, from £82,000 to £333,000, and staff levels more than doubled between 1945 and 1952.[22] The expansion of Plant Industry signaled the alliance between agricultural science and Australian prosperity, yet for Australian scientists, sheer size did not equate to scientific strength, nor did it ensure future success. Instead, many Australian scientists sought greater support for the pursuit of basic science from their government patron.

As the 1950s dawned, the new chairman of CSIRO, Ian Clunies-Ross, acknowledged, one weakness of the Division of Plant Industry remained a lack of basic research as the "scientific foundation" of agricultural science. Of the three "basic" areas—Physiology, Genetics, and Microbiology—Genetics had "die[d] during the war years, while Physiology, as a separate section, was only brought into being a year or two ago."[23] Clunies-Ross believed that the maturity of Australian science required, exactly like the maturity of American science, a greater proportion and investment in basic science. And just as in the United States, support for basic science went hand in hand with the expansion of governmental support for science. Later characterizations of Clunies-Ross describe him as "keenly interested in politics, and tolerant of the partnership between science and the state. [He] welcomed growth, equating more with better."[24] Clunies-Ross became known for placing "economic objectives [of scientific research] in the background rather than the foreground."[25] While balancing this generalization with examples of federally supported applied science producing immediately beneficial products (such as the myxoma virus against rabbits in 1950–51 and Frank Fenner's campaign against polio), the historian Boris Schedvin argued that basic science become favored above applied science within the CSIRO.[26] Pragmatic and easily sold to the electorate, applied science continued unabated, but, for many, including the CSIRO chairman and the organization's historian, the maturity of CSIRO as a scientific organization and the maturity of Australia as a nation rested on their commitment to basic science.

The appointment of Otto Frankel as chief of Plant Industry in 1952 numbered among the many significant changes to the staff and character of the CSIRO during these early postwar years. Born in Vienna in 1900, the product of a traditional gymnasium classical education and the First World War, Frankel struggled to locate a place for himself in the postwar German universities. After a brief stint in chemistry, he landed at the School of Agriculture in Berlin in 1922 to begin work on genetic combination of plants, in a way that Frankel himself later likened to chemists' manipulation of molecules. Unemployed again in the late 1920s, Frankel worked as a private plant breeder, then as an agricultural adviser to the Zionist movement, before stumbling across a geneticist's position in the newly established Wheat Research Institute in New Zealand. Moving to the Southern Hemisphere, he established a breeding program—using crosses of New Zealand wheat to bolster mechanized harvesting and baking quality—and emphasized quality testing. Practical matters like baking did not hold Frankel's interest for long, and from the late 1920s until he left for Australia in 1951, he conducted numerous fundamental cytological investigations.[27]

In the years immediately before he proposed a phytotron, Frankel significantly reorganized Plant Industry. The division had been previously centered around either "a 'project' or 'crop,' eg. Agrostology or tobacco, versus 'subject' basis, eg. Chemistry or genetics," Frankel outlined, with only a few "fundamental sciences" and fewer practitioners of fundamental work. With support from Clunies-Ross and riding a postwar wave of support for basic science, interdisciplinary and international science, Frankel implemented changes to the research practices of the major institution of Australian plant biology in 1952. He outlined securing "about fifteen chemists and physiologists" for Plant Industry, indicating that "a development in genetics is foreshadowed," and, remarkably, "a physicist is provided for."[28] At the same time, a new social organization for science would parallel a new interdisciplinary, project-based approach to basic plant science. All researchers began to work in research teams: the "whole idea" provided "a loose framework for small teams of workers to inter-act to the utmost." Envisioning teams of researchers focused around large instruments, Frankel advocated "a grouping on a broad 'faculty' basis" as the "essential" way to "provide effective scientific leadership" for Australian agricultural science.[29] According to his biographer, Frankel asserted his authority and style quickly and successfully. With quantitative genetics and cytology receiving the earliest support, a nationally and internationally renowned evolutionary genetics group emerged over the 1950s.[30]

Frankel's reorganization transformed plant science in Australia. It became renowned as a leading center of expertise. By 1960, after his visit down under, James Bonner, the famed Caltech biologist claimed to the Rockefeller Foundation that "as a matter of fact, we often refer to the Division of Plant Industry laboratories in Canberra as 'the National

Institutes of Health of the Plant World.' It is undoubtedly the largest collection of people doing basic research in the plant sciences to be found in the free world today."[31] Like many scientists in many fields, Bonner actively created international community through correspondence and visits. Indeed, his trip to Australia relied on an earlier tour by Frankel who, back in 1955, had visited several American biological laboratories on a Carnegie grant. Passing through Caltech before returning to Australia, Frankel met Went and Evans and became convinced that Plant Industry required a major investment in a phytotron.

CREATING THE AUSTRALIAN PHYTOTRON, 1958–1962

Australia's phytotron owed much to its American progenitor, not least because it grew from a graft of Caltech's instrument, via Went's postdoctoral student, Lloyd Evans. Evans moved to Caltech in early 1953 immediately after completing his doctoral work on soil structure at Oxford. It would seem that soil science did not quite suit Evans. He had taken some time to settle into the research, as his adviser E. W. Russell noted when writing to Went to secure his appointment. Evans was not the "kind of person to impress on first acquaintance, at least it took me quite a time to be certain how good he really is, but . . . he wears very well."[32] Looking to move from soil science to plant physiology, Evans told Went that he had been very impressed by his phytotron, and that Caltech presented an opportunity for him to return to the field of botanical research. Evans's appeal to Went sounded very much the colonial seeking imperial wisdom. "There is naturally a great field for such work [environmental factors on plant growth and development] in New Zealand, and some is being done in small growth cabinets. We may never have a 'phytotron,' but I would very much like to be able to work in your laboratory for some time . . . in order to become acquainted with the problems and techniques of such work."[33] As it turned out, New Zealand built a phytotron in the 1960s.

As Evans later recalled, his postdoctoral years at Caltech made him "enthusiastic about the potential of phytotrons for analyzing climatic limitations on plant growth and development."[34] Evans adopted the grandeur of Went's vision, became adept at operating the technological complex of the Caltech phytotron, and joined the burgeoning new community of phytotronists. According to Went's report to Evans's Commonwealth Fund patrons, Evans dived into the world of the phytotron swimmingly. Went praised Evans's contribution to the "community life" of Caltech's plant scientists and celebrated Evans's research. "The results which Lloyd Evans has obtained during the last months are very promising," crowed Went. "He is studying the climatic response of beans, about which very little was known until he started his work." Reversing his usual priorities, Went noted to Evans's patrons that a basic understanding of climatic behavior was "not only of theoretical significance, but of practical significance as well."[35]

Evans moved from beans to primarily study *Lolium temulentum* (darnel, or cockle ryegrass) in Australia. In his memoirs, Evans specifically notes that, to him, *Lolium* appeared a far more significant organism for study than even *Arabidopsis*, partly because *Lolium* directly competed with cereal crops but mostly because he felt "it provided the most sensitive and quantitative system for studying the induction of flowering by LD [long days]."[36] The core problem for Evans over much of his working life was the nature and cause of conditions that induced flowering, in particular the exact conditions and combinations of long days and short days that might initiate flowering. *Lolium* appealed to the budding phytotronist because of its measurability, its quantifiability, and its sensitivity. Evans sought to elucidate physiological principles. His "research on the physiology of flowering," he noted, turned into "a lifetime pre-occupation, central to which was my discovery that the grass *Lolium temulentum* required exposure to only one long day for its flowering to be induced."[37] In short, Evans's research career provides another comparison to Caltech's own story, notably in highlighting the tension between the phytotron as a general facility and every researcher's predilection toward specific plants.

In 1955, the young postdoctoral student clearly impressed the new Australian chief of Plant Industry almost as much Went's phytotron itself: Frankel insisted that Evans join his division. Evans moved to Canberra in 1956, and found the plans for an Australian phytotron already under way. According to Evans, on hearing of a phytotron, most established Australian plant scientists in CSIRO expressed considerable skepticism. At one of the first meetings Evans attended at Plant Industry, he witnessed rising anger and dismay among his new colleagues. As soon as Frankel outlined his idea of building a phytotron, he was met with a chorus of opposition. Evans believed that Frankel had purposively sought to bring out the most immediate opposition early on. Some criticism stemmed from the rather unprepared nature of Frankel's thoughts on the role of a phytotron in Plant Industry, while other comments reflected that many members of each specialty within CSIRO did not yet relate to a grand expensive facility in their midst. As Evans remembered, "several of the many agronomists in the Division spoke forcefully against the project, on the grounds that their field stations were already starved of funds. [Some] ecologists on the other hand, said they would prefer to spend the money on several transects across Australia's native vegetation. Biochemists didn't believe they needed more reproducible plant material. And so it went on."[38] Organizationally and institutionally, agronomists, biochemists, physiologists, and ecologists all existed simultaneously within CSIRO's Division of Plant Industry, but Frankel's phytotron threatened change because the facility would centralize ecological, biochemical, and physiological experiments. At the same time, self-interest seems to have played a role: the ecologists, imagining an enormous sum of money suddenly and unexpectedly available for Frankel to disperse, questioned why he wanted a phytotron, so they not unreasonably queried why the money might not be spent on their pet project rather than his!

Evans rapidly became indispensable to Frankel, particularly in the early design and testing phases at the cabinet level. The pair spent much of 1957 debating the merits of vertical versus horizontal air movement at various velocities in order to achieve a stable temperature gradient across the cabinet. Evans conducted "exhaustive tests with the wind machines" to persuade engineers that the air flow had to remain below 600 ft/min and flow downward from the top of the cabinet to the bottom.[39] As I highlighted in the discussion of the Caltech phytotron, here was the beginning of the transformation of biologists into technologists. Seeking to improve on several elements that Evans found lacking in Went's original phytotron, his goal for the Australian phytotron was that it possess greater climatic ranges held to greater precision. Moreover, anxious to avoid creating square environment profiles as trolleys of plants were moved from one environment to another as they were in Caltech, Evans and Frankel had CSIRO engineers design the freestanding autonomous climate cabinets where climates could be adjusted internally. As Evans recalled years later, his hope was that once he possessed "an excellent experimental system" he could "proceed directly to the question I wanted to resolve: in a single long day did leaves produce a stimulus that initiated flowering?"[40] To attack that question required an elaborate technological system.

Armed with a basic design by the beginning of March 1958, Frankel and Evans launched their campaign to fund their phytotron. Events moved extremely rapidly. The offices of the Prime Minister's Department heard the first overtures requesting a major percentage of the funds in March and by the end of April, the cabinet had "agreed that the cost of the project should be met solely by the government and any promises of financial assistance to date from outside bodies should not be taken up," and voted to approve the entire estimated cost of £500,000.[41] As the periodical literature reported afterward, the Australian government embraced the political economy of the proposed "multi-science institution" and willingly invested in a grand facility as an appeal to local Australian researchers as well as "overseas guest workers, particularly from the Colombo Plan area."[42]

Significantly, the Australian government rejected "outside" assistance, construing the facility as a national endeavor, while, at the same time, simultaneously anticipating that international scientists would use the new facility. Indeed, the irony of many national projects is that they require legitimacy, like the nations that built them, courtesy of international visitors. For national instruments, international recognition and reputation often moved governments into action. Evans later admitted that he and Frankel had sold the phytotron "as a national facility" yet "it quickly became an international facility with visiting scientists from many countries."[43] By the time the phytotron opened it had assumed the lofty purpose of helping to "Cherish the earth" (indeed, the quote eventually appeared on the phytotron building in Canberra). When Frankel handed over the running of the phytotron to Evans the year after it opened, Evans expanded the international aspect of the facility to "make it an interactive centre, indeed an agora for the plant sciences."[44]

At the same time that the scientists sought international cooperation and vindication, it was also clear that selling the phytotron to the Australian minister for Science and Foreign Affairs, R. G. Casey, took little more than noting how *practical* the instrument would be to Australian agriculture. Australians of the 1950s loved to quip that "Australia rides on the sheep's back," so a phytotron addressing the "requirements of sheep grazing" alone, the promotional material boldly stated, promised to benefit Australia to the tune of £50 million per year. Fully aware of easy and persuasive arguments for the phytotron based on economic and industrial utility, Clunies-Ross argued, "it is essential that Australia should have a facility of this sort if our plant scientists are to have the best opportunity of bringing about another great advance in the competitive efficiency and general development of Australia's agricultural industries." While Frankel noted how "Australia leads the world in many phases of plant research," two members of the Prime Minister's Department stressed that "since we are essentially dependent on the products of the soil, [Australia] cannot afford to lag in this kind of research."[45] Establishing a phytotron, the argument went, would ensure Australian leadership in such a vital area of research. Within the Australian government, pragmatic concerns held sway. The overtly applied aim of the technology occupied the mind of the first assistant secretary of the Treasury, H. J. Goodes, who prepared the Treasury minute brief for the cabinet meeting. He tellingly noted on the CSIRO submission: "unlike some pieces of equipment already installed or proposed the Phytotron does not rely upon academic prestige or keeping up with the international Joneses to justify its existence. Its justification is earthy: it will short circuit experimental plant breeding processes . . . and thus increase the output of new plants suitable for the country's agricultural industries."[46]

What Goodes no doubt obliquely referred to was Australia's new giant radio telescope, already taking shape in a sheep paddock near Parkes in New South Wales. The telescope's scientific architect, Edward Bowen, envisioned a giant instrument for basic astronomical research in the Southern Hemisphere. Bowen's giant radio telescope took shape within the CSIRO's Division of Radiophysics, the cost shared between the Australian government and two American foundations, the Carnegie and the Rockefeller.[47] It served as a model for the phytotron. In a booklet prepared in late 1957, Frankel made a formal submission to the CSIRO Executive in April 1958 very much modeled on Bowen's success. As Evans recalled, "the Treasury representative on the CSIRO Executive, Bert Goodes, had suggested that only half the funding should be sought from the Australian government and the rest from foundations. So Otto [Frankel] had arranged visits to several of these, in both the USA and the UK, with this in view, particularly as Goodes had warned him that Taffy Bowen, seeking to building the Parkes radio telescope, was already in the queue and that Taffy 'was reaching for the stars' i.e. he had a more attractive proposal."[48] In some quarters, it seemed that a phytotron was more akin to radio telescopes than cyclotrons. Reviewing a book, S. C.

Harland noted that "the phytotron is to botany and agriculture what the radio telescope is to astronomy."[49]

Direct comparisons between big projects went even further in Australia, when Ernest Titterton applied for over a half million pounds to buy a "tandem electrostatic generator" for low-energy nuclear research at the Australian National University. Recommending the project, the Australian Academy of Science directly referenced both Australia's new radio telescope and its new phytotron as "pure science" facilities, which taken together would raise the "scientific morale of the country."[50] Evans later recalled the process of "airing [him]self" to gain the necessary patronage, which evidently the CSIRO minister, Lord Casey also had to do. The implication is that "airing oneself" is the process of promising anything and everything that a practical, national government would want from their scientists but that is actually quite removed from the practice of science and the purpose of the facility. The phytotronists sought knowledge of basic plant physiology but promised food abundance. Likewise, the radio astronomers and nuclear physicists sought basic science of stars and atoms, but often advertised their telescope's and generator's usefulness for national defense: Titterton could "not claim" that "practical results" would come from an investment in his tandem electrostatic generator in the same breath as he promised it was a "vital research instrument" for "training personnel . . . into an era of nuclear power, nuclear traction and nuclear weapons."[51]

The history of many sciences during the Cold War has often revolved around the tension between scientists' demands for basic research and governments' expectations of applied results. CSIRO's own promotional materials argued that merely erecting a facility to address specific national needs or to bolster one's national standing distinctly undersold the real significance of the phytotron. According to Clunies-Ross, the phytotron would "make possible a kind of research of great potential which otherwise could not be attempted."[52] The real value lay in that which could not be valued because it only existed as potential. As another publicity piece by Frankel put it, "short term empirical approaches are sometimes fruitful, . . . but more often a basic understanding of the processes and reactions involved must precede a practical solution."[53] The information supplied to the government noted, for instance, that the instrument "could not reproduce climates" since they varied "hour by hour, day by day, year by year." Rather, "the salient and relevant features of any climate" could be achieved—imaged/imagined—such as the introduction of old plants into new habitats, or new plants into old habitats, or the tailoring of new varieties to anywhere's conditions.[54] Designed to mimic environmental conditions rather than duplicate or re-create them, the phytotron promised security and status for the agricultural and plant sciences by replicating the success of the physical sciences, which evoked continued dependence. From the specifics of any one experimental run, the phytotronists stressed control over analysis of "the complex field environment." Experimenters, the promotional pamphlet chided, could not ever know which "sudden heat wave or . . . erratic downpour, . . . single frost, or . . . unseasonable warm winter spell" had foiled them. Results, like climates, would be generated "unvaryingly, accurately, and reproducibly."[55] As a model production system, the phytotron simulated not one place but many. Like the nation, the phytotron occupied both meanings, delineating a distinct national maturity as well as evoking a sense of inclusion in the world.

While the Australian government would readily support a pragmatic, applied science scientific facility, Australian scientists proclaimed a far grander purpose for the phytotron. They sought to make it an instrument of experimental biology, in the sense that the chairman of CSIRO Frederick White expressed: "before one can understand an aspect of nature one must control it and make it susceptible to measurement."[56] In part, the case for the phytotron evoked precisely the goal of reductionist science: under the subheading "What is a phytotron? What can a Phytotron do?" The pictorial answer to those questions shows four boxes illustrating rain, cloud, sun, and snow—spring, autumn, summer, winter—transformed into a step function on a chart recorder. The image suggests that the seasons themselves can be transposed from the discrete to the continuous, from the unpredictable to the uniform, repeatable, and regular; the evocative "hand-of-man" controlling the mercurial weather. Thus overlaying a pastoral scene showing sheep being mustered is a three-part graph breaking down the availability of high- and low-quality feed and feed requirements by lambing and lactating ewes: the pasture might be messy, but the graphs are even and regular. Such representations, as Sharon Traweek observed, possess the same features as "the language of physics [which] is rich in negative images of change—deviance, decay, annihilation, fluctuation, instabilities in the beam—and centered on positive images of stability."[57] Agriculture is replete with ruination, star-

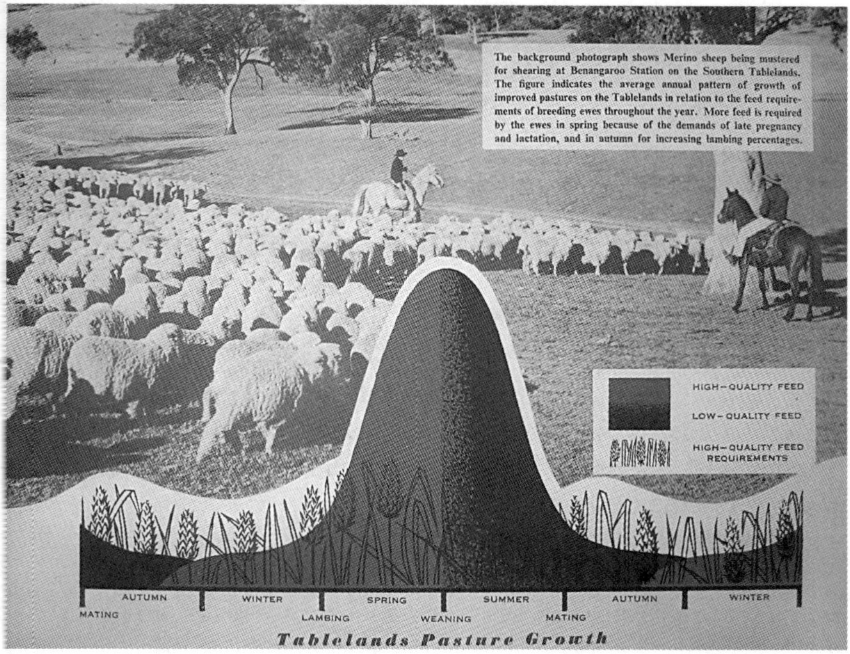

Figure 4.2. Picture of sheep mustering overlaid with twin graphs of seasonal feed requirements of sheep. From "What Is Needed for an Australian Phytotron?" March 28, 1958. NAA A4940, C2060. Attached to a letter from CSIRO Chairman Ian Clunies-Ross to minister for Science and Foreign Affairs R. G. Casey. Image courtesy of the National Archives of Australia.

vation, and hunger and so the image of phytotronic science was regular, repeatable, and secure.

Figure 4.2 suggests more, I think, than just an image of agricultural stability and predictability through the climate-controlled science available in a phytotron. In the regional context of Australia and Southeast Asia, agricultural stability meant political stability. Many political elites believed that Southeast Asia would become a bulwark against communist expansion if it could be supplied with repetitively and exactly adapted varieties to specific climates. Moreover, the creation of centralized facilities to supply a controlled experimental environment also served to tie together regional expertise as well as to display Australia's regional leadership. From the outset, it was the CSIRO Chairman Clunies-Ross's ambition that "guest workers would come from State Departments of Agriculture, the Universities, and from the countries of South-East Asia" to work at "a national facility."[58] As early as 1958, the ability of the phytotron to attract overseas scientific visitors, Cabinet-level Notes assured its readers, would in turn contribute to Australia's scientific reputation as a first-world nation working at the leading edge of research.[59] Australia's scientific and political values would accompany the particular plants developed in the generic phytotron back to their specific native environments, effectively tying together their political, cultural, and scientific worlds. Plants and values would be transplanted via "guest workers from South and Southeast Asia who, it is hoped," Clunies-Ross said, "will use the Australian phytotron for the study of problems of their own in close collaboration with their Australian colleagues." Such an experience "would serve as a training ground for the use of phytotron facilities which are bound to spread in their own countries."[60] Both in Australia and in the United States, national research facilities have often been touted as much for their pedagogical value as for their research potential.[61] Moreover, through the phytotron these guests would learn about Australian values and plants, replicating not only a system of knowledge but also a social system.

At least in Australia, greater international cooperation took place largely through the Colombo Plan. The genesis of the Colombo Plan lay in Australia's realignment from its traditional status first as a colony and then (after 1901) as a Dominion of the United Kingdom, to something more akin to independent nationhood after the Second World War. Australia and India assumed much of the mantle for the containment of communist expansion in their region. For example, Philip Charrier's stressed two concomitant strategies for the Indian containment of communism. First, both Britain and India stressed the creation of a "regional co-operation strategy" immediately after 1945, which served Britain to acknowledge and deflect nationalist charges against any continuing imperial overlordship, as well as local administrative efforts to establish India as a nation state. Second, in the wake of Indian independence and increasing communist activity throughout Southeast Asia, *both* Britain and India sought ways to appeal to the United States for economic assistance, uniting the two nations in a common cause.[62] Yet, economically burdened with both the reconstruction of Japan and the European Recovery Plan (the Marshall Plan), the United States had little money for anywhere else. The United

Kingdom acted to aid the Indian Ocean region economically, but its own economy could not support the development needed; indeed, early postwar British governments recognized that if the United Kingdom looked to maintain any place within Southeast Asia or India via economic reconstruction programs, it would have to appeal to the United States as well. On the other hand, Indian prime minister Nehru feared a crisis of respectability if India needed to appeal to the West for food, and believed that any such move would necessarily increase the appeal of communism.[63]

Daniel Oakman argued that the Colombo Plan demonstrated how an Australian "insular society" chose to face "its regional future" instead of retreating inward.[64] In the face of a rapidly evaporating British postwar Empire, Australian foreign policy became not singular or isolationist but committed to regional cooperation and alliances, invested in international participation and involvement, and welcoming of cultural and educational exchange. "Although grounded in Cold War politics, the Colombo Plan was one of the few post-war creations that achieved consistent, bipartisan support and allowed the humanitarian internationalist and the Australian nationalist, fearful of the outside world, to come together."[65] Australia's representative to the first Colombo summit, its new foreign minister Percy Spender told reporters before he departed for Colombo that it was now an "inescapable fact" of Australia's geographical proximity to Asia and the role it wanted to play. "Could not the old Commonwealth countries contribute part of their resources for the economic development of this area?" he challenged his fellow members.[66] Though Australia would contribute less financially than the United States, the United Kingdom, Canada, or Japan, Australia accepted disproportionately many more students than any participant other than the United States, well over six thousand from 1951 to 1964/65.[67] In fact, Oakman notes, it was the student scholarship and exchange programs that were among the most successful of all Colombo Plan schemes, and they threatened to overshadow the vastly more expensive economic developmental programs in numerous countries. Not coincidentally, also in 1964, the governments of India and Australia were deep into negotiations to build a phytotron.[68] In other words, the Australian phytotron as a solution to problems of biological knowledge was embedded in the practical solution to the problem of the realignment of the nation states in and around Southeast Asia.

MODERN DESIGN = MODERN SCIENCE

In that larger political context, the move from the uncertain field trial to repeatable experimental knowledge of the phytotron qua laboratory paralleled the maturation of the nation from imperial subject to regional power. Modernity assured regularity and certainty in political regions via the ordering and rationalizing of sciences. The scientific and national commitment to modernism was most visible, as in the Climatron that took shape at exactly the same moment, in the architectural form of the phytotron building. As we have seen, the Climatron at the Missouri Botanical Garden took a modernist shape under a Buckminster Fuller geodesic dome. Likewise, Frankel explicitly desired the "Australian phytotron to be novel and distinctive in both engineering design and architecture."[69] The facility's design reflected the distinct break that a controlled laboratory space cleaved from field research. A large central hallway dominated the interior. Along the glasshouse side, large twelve-foot doors sealed each of the fifteen controlled-environment rooms, and the electronic panel beside the door registered the rooms' individual settings. Smaller cabinet-level environments were also available, which combined the temperature, nutrient, and humidity variables of the rooms with light control. Evoking the angling roof of a factory, the doorways, windows, and ducts with dramatic white borders, modernist lines that strikingly mark the exterior of the phytotron. In particular, Evans recalled how the outer building's designer, the eminent Australian architect "Roy Grounds succeeded in giving the 'factory building,' as Otto referred to it in contrast to Roy's Academy dome, some architectural distinction in what he called the 'eyebrows' (i.e. the window surrounds) and the entrances."[70] Grounds had vaulted into the scientific world with his design for the Australian Academy of Science building, which is now known as the Shine Dome with its shallow arcaded concrete dome sheeted with copper described as an "unconventional, futuristic design."[71]

Architecturally, Grounds displayed an evocation of the future in the detailing of the phytotron building. Much recent work in the history of science and technology has demonstrated that the architectural design of a knowledge-producing space realizes much of the founder's implicit and explicit assumptions for the value and conduct of science. The research building for the International Rice Research Institute (IRRI) in the Philippines in the mid-1960s is a design distinctly parallel to

Grounds's futurist design of the Australian phytotron. The architect in that case, Ralph T. Walker, denied any necessity of a "Philippine style," completely deploying instead a "modern idiom." As Nick Cullather relates, the IRRI facility was "a sprawling one-story aluminum and glass structure that featured modular walls to encourage an egalitarian office culture. Air conditioning, tiles, plumbing, and upholstery conveyed 'the power and richness of American life,' and also a sense of permanence. Walker felt his buildings emblemized 'a new type of imperialism' based on 'specialized knowledge generously given to backwards peoples.'"[72] No doubt many held such sentiments, though not Frankel, as a new phytotron donated by the Australian government to the IRRI opened in 1974; it was probably seen as another specialized instrument given to the same backward peoples not least because the phytotron used air-conditioning as well.[73]

THE PHYTOTRON'S COMPUTER-LIKE NATURE

Great fanfare accompanied the opening of the Australian phytotron in 1962. With its modernist controlled climates, it was an imperial vision of a new science serving to centralize many of the biological sciences in a single experimental place. Speaking at the opening ceremony, the new CSIRO chief Frederick White celebrated that the creation of the phytotron now promised to provide "simultaneously for a great variety of experiments and satisfy a great multiplicity of interests in biology and agriculture." Under the leadership of the technologist biologist, in the new phytotron, biology was no longer a dominion in the world of science but a science grown to manhood, and was now equivalent to the physical science. The transition from periphery to center involved a change of scale, Frankel said. Biologists had made a giant leap by building a phytotron, an "enterprise which in all conscience is a big one. A big one even to physicists but an enormous one to us biologists and agricultural people."[74] With this grand facility, the new phytotron would itself become a new imperial center of plant science. The colonials, whether they were from other disciplines or other countries, were expected to travel to the center to work under proper conditions, the standards of which were established by the center.

The weekend-long seminar that followed the opening outlined the justifications for heralding the Australian phytotron as a new center of plant science. Went, for instance, acknowledged that the Canberra phytotron had now moved beyond the original limitations of his first Caltech phytotron. Significantly, he noted that the Australian phytotron had solved the feedback problem his phytotron had encountered, and could now isolate and control variable environmental conditions "independently of each other," effectively eliminating the last element of uncertainty from biological experiments. Later that evening, he confessed in his diary that he believed the Australian phytotron to be "very much patterned" on the Earhart Laboratory, and subsequently assumed the role of founder of a style of science clearly being replicated around the world.[75] A dominant feature of that pattern was the biologist as technologist, which starred as a central theme at the symposium accompanying the opening of the Australian phytotron. For Frankel, engineers had become vital to the phytotronists' ambitions of more complicated mechanically controlled environments. "Plant scientists have . . . welcomed advances in electronic engineering and lighting which make possible the growth of plants to normal maturity under artificially controlled climatic conditions."[76]

In the same vein, Evans's paper presented at the symposium evoked a long and broad discussion that "reemphasized the computer like nature of a phytotron." The ensuing discussion of Evans's paper flowered into an even greater rhapsodizing about the new potential of the technology: "if provided with specifications and means, engineers will produce practically any environmental complex: constant, continually changing according to a given program, and probably even fluctuating with complete randomness. However, phytotrons, much as computers, will provide reliable, intelligible answers only when presented with intelligent, logical questions. The quality of the answers will depend on the quality and completeness of the data on which they are based."[77] For Evans, the epistemology of the phytotron had become more akin to computers than to cyclotrons because the definition of the "environment" they anticipated working with had evolved from a static square waveform to one that required fluidity or even randomness.

In contrast, Went's own contribution pointedly announced with his usual rhetorical flourish that biology too now recognized "the principle of complementarity of Bohr," which he defined as meaning that "as one measures one parameter with greater and greater precision, one has to sacrifice the analysis of the others which have to be kept under less and less normal conditions."[78] Here Went was dealing explicitly with the emergence of feedback by appealing to the knowledge of the physicists. Indeed, Went's studies of testing tomato plants feedback seemed to signal the closer integration of physics and biology: "one cannot properly measure development under completely constant conditions, since the reaction system requires a circadian rhythm to react normally. Here we are dealing with the uncertainty principle in biology . . . even though the measuring of the system may not interfere with the measurements, the experimental setup interferes with the system."[79] In other words, there was also the gradual realization of the technologist biologists that the sort of idealized physics they replicated in their phytotrons was only a simple reduction of the emerging insights about the physical world and the process of measurement itself, especially at the quantum level. Publicly, Went appealed to physics as he sought to remake plant physiology and botany, ecology, and probably all of biology. Privately, however, Went noted in his diary that night that the integration of physicists into the world of the phytotronists hardly helped the cause of phytotrons or shone any light on the problems of growth and development in plants and animals. Immediately after Went at the symposium came not one but two talks by physicists, which "turned out to be so indefinite and complicated," Went seethed, "that practically no calculations can be made and are so imprecise as to be useless."[80]

And so, as the assembled expertise of plant science returned home from Australia, the imperial moment dawned. At the phytotron, the study of the whole organism came into view with a facility that controlled genes and environments. At the phytotron, disciplinary borders were crossed to equate biological science with the physical sciences: Frankel, Evans, and Went all claimed that biological knowledge would now be produced as securely as the physicist's knowledge. At the phytotron, national borders were crossed to forge new regional arrangements as the Cold War eroded older centers of power.

If biological knowledge was to be gained and legitimated through technological control, the phytotronists reasoned, then greater control, over both the range and precision of each environmental variable, was always the goal. In 1959, just as the phytotron was taking shape, Harry Highkin, who had been a graduate student during Evans's postdoctoral years at Caltech, toured Australia. Even after witnessing the expansion of the Earhart Laboratory, the sheer size and elaborateness of the technological complex in Australia shocked Highkin. Writing to the Australian physiologist Robert Robertson after his trip to thank him for his hospitality, Highkin swooned over the new phytotron: "So far as the CSIRO facility is concerned I think it is the best thing I have seen or heard of yet."[81] Like Highkin, Robertson was enamored with ever more elaborate technologies because they promised plant physiologists more exacting control of the environment. As Robertson noted a few years later, "Bigger and better phytotrons [since Went's] have now been designed and are being put into use with a wide range of conditions for experimental plants. They are expensive to establish and to run—on the standards of expense usually associated with biological research, though not expensive by physicists' standards!"[82] Robertson implied that while the technological size and expense of the phytotron began to equate biology and physics, only with comparable expensive facilities would the two fields' standards finally meet.

Privately, however, the single-minded pursuit of technological control, Highkin feared, dangerously moved a biological science of the environment away from organisms themselves. The creation of second-generation phytotrons made this fear real. As he flew back to California, Highkin wrote to Robertson confidentially about the troubling underside to the heretofore celebrated transformation of biologists into technologists. Simply, they had begun to worry more about technologies than organisms. Even in the midst of triumph, Highkin wrote: "it seems to me that the degree of control Les Ballard and Lloyd Evans are trying for has reached more than the point of diminishing returns. I had the feeling that both were more interested in growing thermometers than in growing plants."[83] This criticism of the entire endeavor, one of the most potent, still stung Evans even late in life. In his memoirs, he specifically denied that he was more interested in the engineering of controlled spaces than in experimenting with plants in those spaces. However, he admittedly saw the criticism as quite justified for "many phytotronists." Without naming names, Evans acknowledged that too many had "trapped themselves into a continuing preoccupation with improving the design of their units rather than using them for experiments."[84] The desire to match the physicists' standards of technological size and precision seemed in retrospect to have seduced the physiologists. At the height of technological modernism in the late 1950s, Highkin described to Robertson how Evans and Ballard had painstakingly sought to control every variable of their environmental systems. Yet "some of their measurements make no sense to me at least," Highkin confessed: "They measure twenty-five points in the boxes and spend time adjusting the airflow to give them exactly the same temperature at all twenty-five spots. And they've achieved that. However the minute that they put a plant into the box they will go nuts trying to get all twenty-five points to be the same. [I] don't think its worth the trouble. Frankly I don't think it's possible either. Our experience has been that most of the experiments people want to do don't require that much control."[85]

Highkin's criticisms hint at the deeper issues already undermining the imperial ambition of phytotronists as they sought to establish a grand experimental system to reveal the environmental causes of growth and development. As in Went's own experience in California, about which Highkin was no doubt fully aware, feedback had emerged as the next hurdle to further control of the environment. The Australian phytotronists deployed even more technology, but paradoxically it did not seem to help; rather it only further undermined their control. Even more dangerously, Highkin cautioned that the pursuit of more technology by plant scientists meant that they had begun to lose sight of the reality of botanical work on whole organisms. As phytotrons grew larger, in other words, technology became not only a form of biological knowledge but also nearly an addiction as biologists sought doggedly to establish comparative standards with the physical sciences. Furthermore, the phytotronists' associated ambition to make a new center for a regional power fostered demands for grander technologies of control. This was bad enough in Australia, but as we shall see in chapter 6, when the superpower United States built its national Biotron, the pursuit of the grandest technologies of control spiraled out of control, largely divorcing the entire project from the study of organisms altogether. Therein lay the emerging paradox facing the phytotronists.

THE POSTCOLONIAL PHYTOTRON

Once the Australian phytotron began hosting experiments and experimenters, unexpected revelations about the character of biological knowledge followed right behind, just as in the first one at Caltech. Unlike California's phytotron, the massive technological infrastructure of the Australian phytotron aimed to overcome feedback between the divergent environmental systems. With its multiscience facility replicating and controlling the gamut of plant environments, the Australian phytotron was expected to function as a centralized facility for universal knowledge about the biological environment. In fact, just a few years after it opened, comments by Frankel suggest that what he witnessed coming out of his phytotron was not universal but particular knowledge. Flying in the face of the expectations for modernist science and technology, Frankel noted that "much basic research, whether in Australia or overseas, has little meaning to agricultural practice until it is 'coded' for local environments." Moreover, he explicitly denied universality: "There can be no generally applicable blueprint for regional research."[86] This local coding is what historians call postcolonial knowledge.

Warwick Anderson and Vincanne Adams once said that using the term "postcolonial" was like throwing an "incendiary device."[87] The postcolonial signals the decentering of power and authority, and recognizes the agency of the local and the voiceless as much as the great and powerful. As the historian Itty Abraham has tellingly argued, "science and technology, both as desired forms of modern practice and as privileged instruments ensuring fundamental change are central to an understanding of the postcolonial condition" in the late twentieth century. The intertwined relationship between national science and technology, and science's own claims to universal knowledge and authority has been nicely illustrated in the case of India's atomic energy program and its atomic bomb.[88] In India, nuclear science was made local, Indian, postcolonial and divorced from its Western (imperial) origins. Abraham tellingly notes that the shape of nuclear reactors, the "most modern of objects," mirrors "the *lingams* found in countless Shiva temples."[89] As Abraham concluded, in a postcolonial state the past (but perhaps also the future) must be remade to serve ambitions of independence and security, appropriating other technologies and then coding them as local triumphs.

In the case of the Australian phytotron, as we have seen, the modernist design explicitly signaled Australia's central place in the region. Likewise, biological knowledge was expected to flow outward, from the center to the periphery, solidifying new regional geopolitical arrangements with the expectation that it was the modern Western countries that would solve the problems of less-developed nations. By the mid-1960s, many feared a Malthusian end for Spaceship Earth, realizing that the enclosed environment of the earth's biosphere was incapable of supporting exponential population growth, fears that culminated in Paul Ehrlich's significant exposé of 1968, *The Population Bomb*.[90] As a modernist and imperial laboratory, the Australian phytotron seemed ideally placed to address these issues: indeed, shortly after he had seen the Australian phytotron through its first year, Frankel gave an address titled "Internationalism in Agricultural Science," which laid out his convictions on the problems facing the world and the role of plant scientists. "We are all aware that there are two problems of overriding importance facing mankind [the] limitation of human populations, and the provision of adequate means of sustaining all mankind in comfort and dignity." Frankel advocated finding local solutions to global scientific and political issues. He proposed a greater role for agricultural scientists in asserting the reality and value of international links. "Half or more of the world seems incapable of winning the race for even minimal food supplies," wrote Frankel, and "above all is the shadow of dissention and war over the very regions in greatest need of social and economic development."[91] Agricultural aid and development given by Australia, Canada, New Zealand, and the United Kingdom became the first link of the Colombo Plan. On the eve of the Vietnam conflict, "The key to the political problem of South-east Asia is food," said the Australian parliamentarian Kim Beazley (Snr).[92]

Food and the fear of communism spreading through starving populations served to galvanize the search for universal solutions to feeding the world. In fact, the director of the Brookhaven National Laboratory, Alvin Weinberg, specifically addressed that very issue. In 1967, Weinberg anticipated that his own field of nuclear science would release vast storehouses of energy that people might use to confront "the overriding concerns of society in the next few generations [which] must be the question of peace and the questions of population." Weinberg invoked Thomas Malthus's conjecture that the consequences of uncontrolled population growth would rapidly overburden the available energy resources; the "imbalance between the energy available to man and the energy he requires" would inevitably lead to conflict. That logical outcome had not occurred, he told his readers, because big "science, at least in the West, [has] thus far forestalled the consequences of Malthus' dilemma."[93]

In other words, science had provided the West with answers that worked everywhere. That was the imperial science of the physicists, and, in the same way, it should have been the imperial science of the phytotronists, especially as Frankel, the director of another big science institution, was also directly concerned with the production of stable, viable, and reproducible plants to secure food supplies for the "less developed" parts of the world. *Instead,* Frankel not only stressed that facilities like phytotrons served as centers of expertise, authority, and food supply but also, surprisingly, claimed that the products of science required a "social responsibility" appropriate to peoples in other places: "No longer can we afford to drift along the currents of invention and discovery without examining the social implications for the individual, the nation and the world [especially] in the less developed countries across the Indian Ocean, whose development is perhaps the greatest problem of our time for the world at large, and for Australia as their neighbor in particular."[94] Remarkably, though speaking on the same theme, Weinberg and Frankel already held very different worldviews by the mid-1960s concerning the ability of scientific facts to move to other places. On the one hand, Weinberg's is an imperial vision of science: having succeeded in the West, it will succeed everywhere else. Science, to forestall Malthus, surely must now be sent to, perhaps even imposed on, other places in the world in order to provide universal peace and security. Weinberg's insights into science reflect an imperial mindset at work, namely, that "alongside and after military suppression, knowledge is the principal weapon used by imperialism in its attempts to control and silence the colonized. [The] values of empiricism—an emphasis on quantification and utility over qualitative concerns—are the same as those which structure and regulate capitalist imperialism."[95] On the other hand, Frankel strikes a very different tone. Science is a singular entity, but its "social implications" change from the frame of the individual to that of the state to that of the world. Indeed, Frankel noted that science's implications are acknowledged to be potentially quite different in less developed parts. Far from a direct imposition of the science of "the West" upon those "less developed regions," Frankel stresses an examination of the implications of discovery because clearly they will not be the same everywhere. The vision of producing species to be readily transplanted into any and every distinct environment to produce their maximum growth collapsed at the level of moving from the abstract laboratory world to the real world of farmers and food. In short, a postcolonial Frankel denied Weinberg's imperial vision of science's universal applicability.

This was all quite unexpected for a modernist facility. In fact, Frankel's phrase—the knowledge needed to be "coded"—was the inverse of Weinberg's expectation of how science travels, namely, that the underlying facts about the world are divorced from humans and move readily because they have usually been coded via mathematics into a universal script. Moreover, speaking as a director of a new national phytotron, no less than a facility designed to be geographically universal through the technological production of any climate, Frankel claimed the process actually went the other way around: basic research is devoid of meaning until it is understood locally, "coded" he said, for local practices, conditions, and understandings.

And so once again, technological control over the biological environment produced an unexpected conclusion about biological knowledge itself. Recall the experience of the Californian technologist biologists who discovered that the very act of entering a space to control it was also the mechanism by which corruption penetrated the facility of control. Or, that the attempt to more rigidly control one element of climate, like light, was also the mechanism by which another element like temperature fluctuated more radically and so the imposition of a control system led back to the instability that necessitated the creation of the control system in the first place, in other words, the discovery of feedback. The first phytotronists, like the cyberneticists, became increasingly aware of the role of feedback as they struggled to impose stricter control regimes on their experimental facilities. Feedback undermined hopes of a plant science able to individually control each environmental variable to identify the normal growth and development of any organism because every variable was cybernetic—that is, interconnected to every other variable.

Given Frankel's realization about the necessity to locally encode knowledge, I suggest that feedback, as the plant physiologists encountered it in their phytotrons, is postcolonial. Importantly, the postcolonial is not simply after the colonial, it is also constitutive of the colonial, the reconfiguration of established and now global systems of power through the deployment of technoscientific systems to legitimate new forms of governance and social control.[96] Analogously, phytotrons suggested that biological knowledge was postcolonial knowledge because it was the establishment of a universal facility that precipitated the postcolonial condition of recognizing the legitimate position of other systems of knowledge production and social organizations.

In the wake of Went's grand ambition at Caltech to create a near universal biological environment (a goal replicated in Australia), postcolonialism appears especially powerfully to aid our understanding of the historical transformation of biology in the Cold War. For literary theory, the "Enlightenment's universalising will to knowledge" fed "Orientalism's will to power."[97] This idea underwrote so much of the nineteenth and twentieth centuries. The ideals of the Enlightenment elevated science to a lofty position from which justice, equality, and fraternity would stem, true knowledge forming the basis of true governance and imperial rule. Especially in the nineteenth century, Europeans, convinced that they now possessed the knowledge to produce good for all, imposed their systems and ideas on other peoples. Historians of science can add to our understanding of the postcolonial condition by considering how scientists themselves have identified methodologies by which they could realize truly universal knowledge. The history of science can make a valuable addition to postcolonial narratives because the history of science claims insights into the social production of the scientific knowledge that was so often the basis of imperial power. Previous postcolonial work on the Cold War world confirms that no one state could serve as a model for political and social development. Likewise, my general conclusion holds that the case of phytotrons demonstrates that no one methodology could serve as a model of scientific progress, not physics or computing, not the molecular, phytotronic, or ecological, not the West or the East, not the Nation or the community. All were necessary, and all were related to all the others.

The parallels go deeper still. The case of the Australian phytotron resonates with postcolonial narratives of the social and conceptual reorganization of power in recent science and nations. A postcolonial historical study like Alice Bullard's on the changing practice of psychiatry in French colonial, and then independent, Senegal fulfills an important function of rescuing and asserting indigenous knowledge after independence, in contrast to the oppressive imperial system. As she summarizes the change after independence, "post-colonial transcultural psychiatry [valorized] local beliefs and practices [so that] culturally specific healing practices [could] inform and sometimes guide transcultural therapeutic interventions."[98] The resurgence of traditional healing, Bullard argues, challenged a "new orthodoxy [of the 1960s] geared towards 'universal scientific research and a universal language of psychiatry.'"[99] In psychiatry in Senegal and in the phytotron in Australia the common theme seems to be one of social diversification leading toward greater scientific authority, with the interaction of multidisciplinary practitioners paralleling the union of local and foreign medical knowledge. Significantly, similar claims toward universal knowledge mark the eve of postcolonial moments when people have recognized multiple legitimate sources of knowledge. Indeed, in the postcolonial understanding of natural and social order, local knowledge assumes as much (if not more) legitimacy than the universal knowledge produced by the imperial center.

Historians of Empire have already noted the evocation and defense of the "universal" on the eve of a postcolonial epistemological shift. According to Michael Worboys, there was no "Imperial science" between 1895 and 1940, the great age of Empire! Instead, "Universalism and the 'internationale of science'" became the watchwords of both the British Association for the Advancement of Science and the Imperial Conference for decades.[100] Worboys argues that tensions between the parts of the British Empire evinced themselves in such calls for unity and cooperation, and thus the positivist dream of universal science displayed the politics of colonial science. Postcolonial scholars have appreciated for some time now how only after the end of Empire do we realize "the irony of Empire": Britain's very claim to Empire "expose[d] the British perspective as partial and local in the very act of asserting its universality.'"[101] Likewise, umbrella ambitions were at the heart of the imaged phytotron in Frankel's reorganization of Plant Industry to permit disciplinary specialists to "inter-act," and especially in the Australian government's support for a "multi-science institution." Such social organization of science thus paralleled the new place of the Australian nation within the global community: its enrollment in the multilateral Colombo Plan, the ANZUS Treaty, and its passionate support for the United Unions.

In the postcolonial world, rather than just the French and British Empires as the legitimate political powers, we have Britain, Australia, India, France, and Senegal all as equally legitimate political entities; instead of simply British or French knowledge, we now have Indian, Australian, Senegalese, Algerian, and so on knowledges. It is this legitimate plurality that exposes any singular notion of "science," or a scientific method as having emerged from a local European center and imposed upon the world, consequently eroding science's claims to a transcendent truth. Likewise, the creation of independent nation-states in the postcolonial world politically delegitimized European political centrality and power. Menzies poignantly stumbled over the litany of valid scientific identities for the individual incarnations of the inclusive, universal instrument before him: as he told the crowd, "I declare the phytotron, the controlled environment, the CERES, I'll call it, I'm now told for short, whatever you call it, I declare it open."[102] In a postcolonial frame, one is not just like another: it *is* the other.

This chapter concludes that plant science in phytotrons encountered the postcolonial, but not purposefully, indeed reluctantly. It was an unexpected revelation when the creation of a national facility emphasizing basic science via the production of universal climates served, its director conceded, not to produce "imperial science," namely, universal, black-boxed knowledge, but rather postcolonial science, namely, gray knowledge that needed to be locally coded to possess meaning and usefulness. Went did not seek or anticipate feedback in the first phytotron, and it appears that Frankel was equally surprised to find that knowledge was postcolonial. In stark contrast then to the simultaneous rise of a deeply reductionist style of biology in molecules, the phytotronists, seeking environmental causes of growth and development exactly like the molecular biologists sought genetic causes, found a cybernetic and postcolonial style of biology.

Postcolonial narratives are "believed to demonstrate the fragility of 'grand narratives'[,] the erosion of transcendent authority[, and] the collapse of imperialistic explanations of the world."[103] The postcolonial lesson here is quite simple: a plant physiologist appropriated the epistemology of physics to build his science on firm foundations and to recreate the miracle, as he saw it, of the science of physics. He sought only to emulate, to reproduce, what he regarded as a superior epistemology to gain knowledge about the natural world. He thus firmly colonized life science. Under the banner of a colonial project, our botanist damned inferior local knowledge, botanical research conducted via traditional field-trials or in other noncontrolled ways. He, like generations of imperialists before him, constructed local knowledge and practice as the illegitimate "other." He stressed the authority and legitimacy of his own knowledge and practice via claims to universalism. But when he succeeded in building a scientific facility that universally controlled whole climates and could reproduce the environmental variables exactly to the standards demanded of the imperial science of physics, he did not merely replicate triumph but made the object of their adoration into a joke, because it is not that plants can be like physics but that physics can be plants. For an imperial physicist like Weinberg, of course, Western science possessed a privileged position of legitimate authority and had to save the less developed world from its own population. The postcolonial Frankel, in stark contrast, recognized that the social implications of scientific knowledge changed from frame to frame—science had to be "coded" for other countries and other parts of agriculture. Most importantly, as with so much postcolonial thought, such realizations stemmed from the active pursuit of the imperial vision of universality. You build a phytotron to replicate every environment, only to discover that its products need to be coded for every environment.

CHAPTER 5

THE TWIN PHYTOTRONS OF THE RESEARCH TRIANGLE BETWEEN DUKE AND NORTH CAROLINA STATE

> If we wish to preserve botany as an important part of biological science we must modernize our courses, our teaching and our research.
>
> — **Paul Kramer, Duke University, 1961**

CONCLUDING HIS year as president of the Botanical Society of America (BSA), the Duke University botanist Paul Kramer spoke at the annual dinner and impressed on his audience that botany faced a critical moment in its development. "The biological sciences are undergoing changes as far reaching as those occurring in our political and social systems," Kramer said, explicitly likening the solution to the problems of knowledge to the problems of the political order. While classical botanists increasingly railed against the intrusion of the "advance in chemistry and physics" into their subject, Kramer implored his audience to continue the modernization of botany. "It would be very unfortunate if we fail to make full use of new methods and new concepts to produce a more interesting and more productive science." Back in his student days,

Kramer had embraced "modern biology," notably during his senior year at Miami University when he had to teach the remainder of a course on plant physiology after the instructor fell ill. While his classmate "detested" so-called test-tube botany, Kramer said he preferred it to morphology or taxonomy.[1] Some twenty years later, the danger in delaying modernization lay in the very real threat that "botany will be taken over by biochemists, biophysicists and molecular biologists," Kramer prophesied, because they had effectively incorporated physics and chemistry into their practice while botany had not. In the era of modernity, it seemed clear to Kramer, such newer sciences aligned with social and political reality. Consequently, Kramer's departing message to the BSA membership in 1961 was that the old must give way to the new: "if we wish to preserve botany as an important part of biological science we must modernize our courses, our teaching and our research."[2]

Little was more modern in the early 1960s in the plant sciences than phytotrons. After working in the Caltech phytotron between 1956 and 1958 before moving to Duke University, Kramer was an early evangelist for controlled-environment plant science. Kramer formed a plan for a local phytotron as early as 1956, but did not secure the necessary funding until 1964, after Duke University had joined forces with another institution, North Carolina State University, to amass a war chest of nearly $4 million to build not one but two phytotrons within their so-called research triangle in the American Southeast.[3] As this chapter shows, the creation of the twin phytotrons in the research triangle reinforces the claim that with the widespread modernization of biology, technologist biologists built a plant science of controlled-environment biology at the same time that the political and social order faced the abyss as the Cold War climaxed around the Cuban Missile crisis. Scientists as much as their private and government backers hoped to contribute to major practical problems like overpopulation and economic vitality but also to make fundamental discoveries about growth and development.

Kramer was instrumental in this broad shift in plant science. He had occupied prominent roles in the broader American plant-science community and amassed impeccable institutional and disciplinary credentials, notably being elected vice-president of the Botanical Society of America in 1960, and its president the following year. Kramer had also served as the president of the American Society of Plant Physiologists, in fact, two years before Frits Went's presidency, and was elected to the American

National Academy of the Sciences in 1962. In other words, under the leadership of scientists like Kramer, Went, Pierre Chouard, Anton Lang, and Lloyd Evans, phytotrons became quintessential components of what leading plant scientists believed was the proper, modern, and altogether normal development of the study of the plant for at least the first three decades of the Cold War.

Where Kramer tilled the ground to plant a phytotron in the American Southeast, another member of the community, Henry Hellmers, husbanded the institution to maturity. Kramer recruited Hellmers for Duke in late 1964, specifically for the phytotron. Hellmers had gotten his PhD in plant physiology from the University of California at Berkeley immediately after the Second World War, and then secured a position as a Caltech research fellow between 1949 and 1955, where he had met Kramer. For the next decade, Hellmers worked for the United States Forestry service, before moving to Duke. Upon his appointment, Duke's dean assured Hellmers that "We take a great deal of pride in having developed one of the best departments [of Botany] in the country."[4] Disciplinarily, the addition of Hellmers to Duke's botany department promised "a large contribution to physiology and ecology" no doubt reinforcing the noted Duke ecologist Dwight Billing's interests.[5] In the memories of the Duke botanists, the historian Jessica Harland-Jacobs noted, Duke's phytotron helped their botany department maintain its strength "in ecology despite the shift from whole plant studies towards cellular and molecular biology."[6]

Unlike the California or Australian experience though, the host institutions of the twin phytotrons of the research triangle demanded innovative pedagogy and training alongside innovative research and greater recognition via larger research grants. In fact, much of the impetus to recruit Hellmers came from another axis of the drive to modernize biology, namely, that plant physiology, botany, ecology, and forestry needed eager recruits. Students stand as apt markers of any field's prestige. The recruitment and training of new disciples has always been part of science's reward structure as well as a key expression of the power of a scientific program, school, or institution;[7] the historian Robert Kohler argued that students should be placed equally alongside "authority and access to tools and craft knowledge."[8] It seems likely that the lack of a pedagogical program contributed to both the decline of plant physiology and the spectacular rise of molecular biology throughout the 1970s, as neither in Caltech nor in Canberra was the creation of a phytotron consciously connected to the recruitment and training of students. In contrast, Hellmers was explicitly expected to "begin a graduate program" once the phytotron became "operational" in the American Southeast.[9] In contrast, the desire to recruit new disciples and the expectations of their training shaped the twin phytotrons of the research triangle. Whereas Kohler only speculated that "surely recruiting and teaching students must also affect the definition of research problems and modes of academic practice,"[10] the historian of science Andrew Warwick pioneered study of the material culture of instruction. Working through a world of pen and paper, Warwick not only charted the twin familiar loci of institution and discipline, Cambridge and Mathematical Physics, but also explained pedagogy as culture. Uncovering a culture of coaching, of problems, of exams, of hard work, sport, teamwork, masculinity, and individual glory for the Senior Wrangler (and for the winner of the wooden spoon) reveals how scientific communities work through the problems of normal science by training new disciples.[11] In the case of the twin phytotrons, Kramer and Hellmers pressed for the flexible production of environmental conditions and a physical facility that would accommodate potential research programs of future students. They recognized that disciples as well as international visitors remain normative measures of the success of science.

The twin phytotrons weathered the dark 1980s far better than many phytotrons. Partly, a shift in research focus insulated the Southeast's facilities as plant physiology in phytotrons faded through the 1970s and environmental ecology gained momentum, interest, and disciples. Aided by efforts to draw students and visitors, the shift in research focus preserved support from patrons and institutions. They also benefited from being among the last men standing, as Hellmers later reflected, by "having the only functioning phytotron in the nation, the laboratory has been a drawing card for the botany department and the university. Some scientists that have worked here or visited the phytotron have used it as a model to develop some of their own controlled environment facilities. Studies at the phytotron have influenced environmental research throughout the country."[12] For Hellmers, the twin phytotrons continued to stand as "drawing cards," something like standards around which whole organism botanists might rally in the face of the molecular onslaught. Unlike many phytotrons, the twin phytotrons fought to preserve their funding and usefulness by transforming from facilities that only controlled the environment to study plants to facilities that controlled plants to study the environment. Indeed, the twin phytotrons were among the first facilities of the Age of Climate after they were refitted to experiment with variable concentrations of carbon dioxide.

"THE BEST TOBACCO MAKES THE BEST SMOKE"

The story of the twin phytotrons in the research triangle begins in the research group at North Carolina State devoted to research on just one plant, tobacco.[13] In May 1962, the Z. Smith Reynolds Foundation announced a grant of $750,000 to the North Carolina State University toward the construction of a phytotron. The foundation, an arm of the tobacco giant, R. J. Reynolds possessor of the ubiquitous Camel brand, became the third major private benefactor of phytotronics in the United States after the Earhart Foundation and the Campbell Soup Company, and the fourth globally after CSR sugar in Australia. Generous support from private corporations like Reynolds provided an alternative path toward large scientific facilities during the Cold War. Though the exact origins of North Carolina State's grant remain mysterious, evidently it was Charles Sprinkle, a research manager within R. J. Reynolds, who prompted the idea, and North Carolina State's dean of agriculture, H. Brooks James, brought the Reynolds Foundation together with the United States Department of Agriculture (USDA) in order to secure the initial grant.

Long a center of tobacco research, North Carolina State's tobacco scientists had identified various diseases such as black root rot and brown spot that appeared to attack tobacco in different ways in different environments, but also saw the counterintuitive situation in which pathogens were present yet did not attack plants. Likewise, pests like hornworms required climate-controlled studies both to evaluate their own environmental responses and to study the changing efficacy of pesticides under changing environmental conditions.[14] In short, the applied tobacco scientists of a state agricultural college sought to break open the multiple dependent relationships between plant, pest, and climate. While the initial grant was an extension of the relationship between a commercial enterprise and an academic institution, it seems that a phytotron held the same appeal for the tobacco industry as it did for a tomato soup company. After visiting James Bonner and the Caltech phytotron in mid-1961 and Went at the Missouri Botanical Garden in early 1962, J. A. Weybrew, the head of the North Carolina Tobacco Laboratory, and Kenneth Keller, his assistant director, championed a phytotron for tobacco research.[15] Unlike many appeals of scientists to large amounts of funds, Keller and Weybrew's formal proposal to the Reynolds Foundation emphasized the singular study of the growth and development of one organism, tobacco. "It is essential that North Carolina State College [have] a phytotron facility," the foundation proposal read, because only through a phytotron might tobacco scientists gain the "maximum opportunities . . . to realize major research achievements relating to tobacco production, breeding, genetics, disease, insect and quality investigations."[16] For the tobacco industry, the kinds of practical knowledge they desired and were willing to fund concerned the growth and development of the tobacco plant and its pathogens, each locked into a mutually dependent feedback cycle with the other. In contrast to Went at Caltech, Keller and Weybrew stressed to their tobacco patron the importance of pest and pathogen studies under controlled conditions.

As a second-generation facility, North Carolina State's botanists and foresters rehearsed the standard arguments of controlled-environment biology flawlessly: to meet the standards of science, biology required phytotrons because a range of environmental variables all variably fluctuated and rendered an experimental biology outside of controlled conditions impossible; the mixed ingredients of climate betrayed field research because environmental and genetic causes were "impossible to disentangle"; capricious climate emasculated the tobacco scientist who found himself "incapable of reproducing in a second growing season the conditions which existed the preceding year."[17] The impression offered the pursuit of control over genes as much as control over environments as the normal trajectory of an evolving science of plants. Once more, the advantages of phytotrons were displayed as obvious by their champions: First, a phytotron produced "experimental material grown under reproducible conditions." Second, the facility enabled scientists to "study the effects of variations in the principle environmental factors on plant growth." The Reynolds Foundation repeated the refrain from earlier phytotronists that the physicists and chemists had already "rigidly" controlled environmental factors in their experiments while only recently had biologists "recognize[d] the importance of the genotype and the environment in the growth of their plants and are becoming more aware of the important role of controlled environment facilities in obtaining basic information about plant growth."[18] Third, as all might derive pleasure from a cigarette, one major research facility would be available and useful to "agronomists, physiologists, geneticists, pathologists, entomologists, botanists, agricultural engineers and soil technologists." In short, almost everyone interested in biology above the cell needed a phytotron.[19]

Swayed, the Z. Smith Reynolds Foundation threw its considerable financial resources into support of a "modern research tool."[20] Almost immediately, news of the large Reynolds grant reinvigorated Paul Kramer's original idea for a phytotron at Duke University. During the first round of National Science Foundation (NSF) funding for phytotrons back in 1959, Kramer and Duke lost to Wisconsin's far grander Biotron. Still, as Kramer explained to his new provost in early 1962, Duke's original proposal, though not funded, had remained active at the NSF. Subsequently, after news of the Reynolds grant broke, Kramer learned that NSF administrators had once more begun to look favorably on a phytotron at Duke. Subsequently, Kramer and H. F. Robinson at North Carolina State College's Department of Genetics looked to expand North Carolina State's own phytotron via a connection with Duke. The idea was to create a far larger phytotron for the Southeast via seed money from Reynolds fertilized with substantial NSF funding. The plan was inspired: instead of two small phytotrons inadequately catering only to local needs, the combination of federal and private money might permit a phytotron of a "wider range of environmental conditions than either group would have."[21]

Kramer believed that botany and plant physiology needed large departments in big institutions, grand facilities, and enormous budgets to complete in the modern world. The university measured its success in access to newer technologies, but above all Duke prospered on the backs of increasing undergraduate numbers and government contracts, especially in the sciences. As Duke's president, J. Deryl Hart, informed his trustees in 1961, the physics department at Duke had brought in over $7 million in contracts and grants over the past twenty years. "Men having prestige based on scientific ability," he concluded, "can bring into the University large amounts of money to support their work and build the department." This money would come from "Government and Foundations" whose funding of university research had "increased rapidly and there are indications that these will be further increased."[22] Duke's botanists and physiologists not only appropriated various practices from physics and chemistry, including a desire for more controlled laboratory space, but also replicated the physical sciences' expansions via investments in large facilities by governments. At Duke, both the botany and zoology departments learned from the NSF campaign to fund a national Biotron that sciences could expect increasing amounts of funding for new staff, new research, and especially new facilities in this golden age of government and private philanthropy. Consequently, barely six months after the Z. Smith Reynolds Foundation grant had been awarded, the dean of agriculture for North Carolina State, H. Brooks James, found himself writing a conciliatory letter to the foundation's vice-president outlining the new joint arrangement with Duke.

The creation of the twin phytotrons in the Southeast was in part a struggle between botany and biology. At Caltech, plant physiology had already disappeared by the 1960s and biology came to be primarily molecular biology. In contrast, the botany department and the zoology department at Duke remained independent of a larger biology department until the 1960s, though they had begun contributing teaching resources to general introductory biology courses co-taught by all relevant departments. Although the botanists at Duke rejected incorporation with biology, they acknowledged that botany required a new focus to maintain its unique identity.[23] Whereas several members of the botany division certainly did not support Kramer's phytotronic vision, by modernizing botany through an explicit acknowledgment of the breakdown of "old and familiar boundaries" between scientific disciplines, the instrument insulated botany against the sort of takeover that Edward O. Wilson witnessed at Harvard.[24] In part, as much as in California, Canberra, or the Philippines, botany needed air-conditioning. Everywhere, air-conditioning signaled the transformation of old institutions into modern ones: outgoing Duke president Deryl Hart celebrated the air-conditioning of Duke's biology building, among many projects, as evidence of Duke's intention to "take its place among the leading universities of the world."[25] Even more important, botany required a facility that would support a new inclusive biology that focused its on "plants and animals at all levels, molecular, cellular, organismal, and community."[26] The technological control of the environment, in other words, was understood and advertised as a unifying methodology across the multiple levels of functioning life.

More than a dozen phytotrons had already been built worldwide: optimism and excitement ran high. Duke and North Carolina State botanists claimed their combined resources as "an instance where one plus one will yield three or four in terms of scientific productivity."[27] Duke's botany department impressed upon its provost its "real need for a relatively large amount of controlled space," preferring "one large, integrated facility rather than a collection of haphazardly assembled units." As Kramer noted, Duke's botanists chose a "medium sized phytotron on this campus rather than a larger one in the research triangle," but acknowledged the practical reality that it was better to concede to "this less desirable alternative" if it was the only way to secure a phytotron at all.[28] Perhaps as a concession to lingering resistance within Duke's larger biological community either from a newly dominant botany or, more likely, having to work with the state college on the wrong side of the tracks, the agreement between Duke and North Carolina State formalized the differing roles each phytotron unit would have, effectively maintaining some separation. Duke's phytotron "would concentrate on (a) ecological or environmental physiology, including analysis of the role of various combinations of environmental factors in controlling length of growth season, amount of growth, and distribution of plant species, and (b) study of effects of various environmental factors on specific physiological and biochemical processes." At the same time, North Carolina State's phytotron would both contrast and complement the Duke unit through investigations of "genetic studies. Population dynamics as influenced by environment, reactions of various progenies to extreme environments, and effects of environmental stress on biochemistry, growth processes, mineral nutrition, and the nature of disease and insect resistance."[29]

The agreement also established the Phytotron Board, a governance body composed of two people from each institution, and a chairman selected by agreement between the college presidents. Even before Duke's President Hart had signed the agreement, all parties had accepted Kramer as chairman of the Phytotron Board to begin the process of acquiring funding from the National Science Foundation. A cooperative arrangement possessed, however, a very serious drawback for universities looking to establish themselves in the Cold War world: as Kramer noted to his provost, "neither institution will obtain the prestige which would come from having a phytotron on its own campus."[30]

WHO WILL BUY MY SWEET RED ROSES / TWO FOR 4 MILLION

In 1963, the year after the Australian phytotron opened, Duke and North Carolina State Universities submitted a joint proposal to the NSF to build a two-unit phytotron in the North Carolina research triangle. Duke asked for $1.8 million, and North Carolina State requested an additional $1.3 million to add to the $750,000 from the Reynolds Foundation. To secure that sort of funding for a pair of controlled environmental laboratories, notably in addition to the millions already being allocated for the national Biotron in Wisconsin, required the support of a vast array of patrons, and, consequently, those patrons' demands and expectations also shaped the twin phytotrons.

Locally, Duke's geneticists were among the early supporters of Kramer's phytotron. Established geneticists as much as plant physiologists and botanists of the 1960s fully expected a science of genes to be accompanied by a science of environments. The head of Duke's genetics department wrote in support of funding for a phytotron, for example, stating that controlled environments were considered necessary for "studying genotype–environment interactions." His support conveys the idea that well after the emergence of biophysics and molecular biology, many genetics departments accepted that new environmental control facilities would aid the development of biological research: "the possibility of measuring various genetic components of variance under different controlled environmental conditions may lead to an entirely new approach in breeding techniques."[31] If plant scientists kept their focus firmly on living organisms, Duke's classical geneticists also looked no lower than the cell. No matter how low their organism of choice, from monkeys, to rats, to flies, they remained broadly within the phytotronists' claimed realm of new biological knowledge, the world of living organisms. Duke's anticipated program involved an array of organisms including corn, tobacco, drosophila, and mice.[32]

Regionally, Duke, and eventually North Carolina State, also possessed a useful ally in the United States Forestry Service, which had taken on board many aspects of the phytotronists' argument for environmentally controlled facilities.[33] They saw a clear need for studies on the "effects of environmental factors on the yearly cycle of growth—initiation of height growth, stem elongation, time of early wood formation, time of late wood formation, cessation of height and diameter growth."[34] The Forestry

Service was doubly useful to Kramer and Hellmers because, as someone told Louis Levin at the NSF, there was very little chance of the Forestry Service obtaining its own phytotron and yet their own research clearly foresaw a need and use for a facility in the American Southeast. Indeed, the Forestry Service seemed to have the most stalwart justifications for a phytotron of almost anyone in the period.

But it was at the national level that Duke and North Carolina found the patronage apparatus substantial enough to fund their grand visions for southeastern biology. Significantly, to evaluate proposals for research programs and new instrument developments, the NSF assembled various committees of experts and professionals from among the scientists whose work they funded. When Kramer proposed twin phytotrons for NSF consideration, the membership of the foundation's own committee structure were his trump cards, particularly the several phytotronists that sat on the NSF's Divisional Committee for Biological and Medical Sciences (BMS). Within the BMS, two phytotronists and one declared ally held the chairmanship for four successive years: Kenneth Thimann from Yale, who belonged to the committee between 1958 and 1963, and was chairman from 1961 to 1963, was replaced in 1963–64 by Rene Dubos, the author of a screed in *BioScience* supporting phytotrons, and then Kramer himself took over in 1964–65. In other words, from 1961 to 1965, exactly the moment when grand plans existed for both the Wisconsin Biotron and the twin phytotrons at Duke and North Carolina State, phytotronists held prominent, leadership positions on an influential NSF committee charged with evaluating proposals for grand new controlled facilities.[35]

Duke and North Carolina State sought NSF funding via a common argument: "maximum use in graduate student training."[36] From agricultural engineering to botany, crop science, forestry, genetics, soil science, and plant pathology, North Carolina State boasted nearly 200 professional staff and nearly 250 graduate students. North Carolina State claimed to have already accepted expressions of interest in a phytotron from scientists at Emory, the University of Florida, and the Tuskegee Institute.[37] In furthering the democratization of education, they argued, "the availability of a phytotron unit should be a strong stimulus to graduate training in plant sciences in the Southeast and should attract an increasing number of students to do research in population dynamics as influenced by environment, genetics, reactions of progenies to environmental extremes, effect of environmental stress on the biochemistry, growth process, mineral nutrition and disease and insect resistance of plants."[38] In fact, according to the preliminary outline Kramer forwarded to the NSF program director Jack Spencer in early 1964, research was a less important criterion for the NSF than education. A long and unending Cold War was now a reality for Americans and, consequently, their scientific and technological institutions geared up for the long haul by emphasizing the recruitment and training of future scientists and engineers to continue the fight. By the 1960s, Kramer was playing the educational card to win, noting that the establishment of various types of cooperative arrangements between institutions would realize larger "regional facilities" of greater power. As Kramer suggested, the two-unit phytotrons offered a "wider range of plant science specialties than if it were on one campus. The "exchange of ideas," Kramer argued, would make both institutions more scientifically productive, not through competing phytotrons but through complementary instruments, staffs, and students.[39]

At the same time, the established rhetoric of basic science underpinned the recruitment and training of students. The chancellor of North Carolina State, John Caldwell, and the director of Research at its Agricultural Experiment Station, R. L. Lovvorn, both emphasized in their letters of support of the phytotron granted access to the "whys of plant behavior." Went, as well as Otto Frankel and Lloyd Evans in Canberra, maintained that the phytotron could extend the boundaries of botanical and agricultural science beyond mere application into an unexplored region of basic science, the "whys" of things, not just the "hows" of things. Research was not divorced from an educational mission but was, in fact, intimately connected with it. Students were not merely the by-product of research, but its heart and soul. A phytotron, Caldwell said, "is a facility which will not only encourage basic research studies but will also serve as an inducement to maintain and enhance our academic standard of excellence. It is mandatory as we plan for the future that we continue to provide our scientists and graduate students with an atmosphere conducive to stimulating ever-increasing desire for seeking answers to the unknown, and the tools."[40] Unlike at Caltech where plant physiology disappeared as an undergraduate subject around 1970, at Duke, Kramer noted that undergraduate enrollment remained high in botany's introductory physiology course throughout the 1960s, with laboratory space and assistants for all. With plenty of students, Duke's botanists, foresters, and ecologists all "urged the need for another whole plant physiologist."[41]

WHERE TO PUT THE COMPUTER?

Kramer and Hellmers labored to improve the basic design elements to perfect the technological system of the phytotron. Given how closely both Kramer and Hellmers were associated with Went, it is no surprise that the research triangle's twin phytotrons took the lessons of the Earhart Laboratory especially to heart. Of course, no scientific instrument or facility is the final word but its creation and use is part of the process of discovery itself. For Kramer and Hellmers the first phytotron was a marvel, but it was far from perfect. Researchers complained constantly, especially about the unduly rigorous decontamination procedures. Went's procedures did not really defend against bacteria or pests but only restricted the "nominal degree of accessibility which any research worker must have with his plants."[42] Besides, technologies of material and biological containment had become common by the mid-1960s: one unnamed North Carolina State scientist suggested just borrowing from the space program where air curtains had successfully kept out pollutants and disease.

In addition, Kramer returned to Durham after attending the dedication of the Canberra phytotron in 1961 bewildered by the Australian decision to have many separate refrigeration and control units, nearly one for each growth chamber and greenhouse, many with flexible conditions. Flexible conditions permitted the technological production of something more like natural nature, but also inflated the necessary technological systems. The underlying issue, already faced by Harry Highkin at Caltech, was that any new facility required a firm statement on the character of the experimental environment itself for plant science. In other words, Kramer and Hellmers had to decide whether the transition between climates was to be abrupt or smooth. Back in the first phytotron, the climate of every room was kept constant and it was the plants that were moved. A decade later, second-generation phytotrons could instead achieve "smooth changes in temperature (sine waves) instead of abrupt changes (square waves)."[43] Plants would stay put, and control systems would change climatic conditions, and, moreover, change them in any way desired, abruptly, slowly, sharply, or constantly. For plant scientists, the specification and measurement of the environment encompassed not only its static physical characteristics but also its dynamic changes over time. Folke Skoog at Wisconsin was committing the Biotron, as we shall see in chapter 6, to flexibly smooth temperature gradients on the basis that nature did not move plants from one fixed temperature to another between daytime and nighttime. In the research triangle, Kramer had initially adopted Went's fixed day and night room temperature conditions, but by the late 1950s all phytotronists agreed that both individual rooms and cabinets should be equipped to accommodate any variable setting of various environmental conditions.

Consequently, the rooms and cabinets at Duke and North Carolina State were built to accommodate ranges of environments. Beyond flexible temperature ranges, the remainder of the space inside later-generation phytotrons presented a wealth of possibilities. Kramer wanted a "room for rain and fog," one for studies of "monochromatic light," another for "high light intensity of at least 10,000 foot candles," and even a room for "time-lapse photographic studies of growth phenomena."[44] At the same time, not everything was changed from the first phytotron: Kramer's idea for rooms for photoperiod work were "arranged as at Earhart" with six dark rooms and six artificially lighted rooms.[45] One element of the design of phytotrons remained especially sacrosanct. "Where shall we position a digital computer?" North Carolina State's W. E. Splinter asked.[46] Indeed, more powerful computing was now necessary to regulate and monitor all the expanded environmental systems.

All these decisions placed considerable extra burdens on the technologies of climate control to both produce a greater array of environments and maintain them constant within ever-closer tolerances. Moreover, they created an enormous financial burden. Duke's initial $1.8 million grant from the NSF did not come close to the initial estimates. In fact, even with its Reynolds grant, North Carolina State had to cast around for possible donors to make up the emerging budgetary shortfall. North Carolina State took the obvious tack of writing to every major tobacco company, which "seemed appropriate" simply because their phytotron planned to "devote approximately one-half of its research facilities for use in solving tobacco problems."[47] It seems that no other tobacco company took an interest, though the Reynolds Foundation did cover some of the shortfall. Desperate to cut costs, Duke would later rue a decision made around February 1965 to "forget about CO_2." In the revised schematic outline of the research triangle's phytotrons, the architect F. Carter Williams laid aside plans for a central CO_2 system, opting instead for a fitting in each chamber to "receive CO_2 piping from storage bottles." Kramer, it seems, crossed out the entire paragraph, presumably in an effort to eliminate optional substantial costs.[48] It was only in the 1980s that the Duke phytotron finally gained an external CO_2 greenhouse for, by then, the all-important measure of climate change.

It requires the support of a great many people to create a facility, and these supporters demand space dedicated to their own interests in exchange, so large scientific facilities become necessarily heterogeneous. For example, Dwight Billings became a key supporter of Kramer's phytotron after he was assured that the facility would include "low temperature and high light intensity" rooms to work on alpine and arctic plants. Suddenly, those rooms appeared in preliminary designs, and the botany of plants in extremely cold conditions demanded "more provision for controlling root temp independently of air temperature."[49] Having received his PhD from Duke in 1936, Billings returned to Duke as an associate professor in 1952, and edited the journal *Ecology* from 1951 to 1956. Billings flirted with the military-industrial complex, serving as a member of the Advisory Committee on Ecology for the United States Atomic Energy Commission (AEC) from 1955 to 1958, and then for a short time as a member of the Special Task Force (Ecology) at the Nevada test site. Subsequently, Billings shifted his interest away from AEC ecology work toward NSF-funded work with a $50,000 grant for research on "factors determining upper and lower altitudinal limits of alpine and subalpine plants." It was with this new research project in mind that Kramer's application to the NSF for a phytotron assured the NSF that "if a more controlled environment becomes available, . . . Dr. Billings will intensify his research in the physiological ecology of geographic races."[50]

Moreover, Billings had rendered and visualized the "environment" in his famous "holocoenotic environmental complex" diagram, which portrayed the "sum of all external forces and substances affecting the growth, structure, and reproduction of that plant." In 1952, Billings defined "the environment of a plant as sum of all external forces and substances affecting the growth, structure, and reproduction of that plant," including "heat, light, water, [and] elements." Among its notable claims was the display of the interconnectedness of all the forces at work;

indeed, the diagram took on a weblike structure. Like Kramer's call for the modernization of botany via physics and chemistry, Billings also broadened plant science by explicitly noting that "other plants" were one of only fifteen distinguishable factors defining the environment of a "plant."[51] Billings's conception of the environment was influential, not least in theses written from research performed in the phytotron: Duke's students frequently cited Billings's work on "the need to recognize the existence of variability in populations when conducting physiological and ecological experiments."[52]

PEDAGOGY AND THE PHYTOTRON

The twin phytotrons officially opened on May 10, 1968. Hero and founder Frits Went spoke at the dinner at Duke's faculty club, one of his last public appearances until the controlled environments conference at Wisconsin in 1979. No sooner were their cabinets operating than the triumvirate of Kramer, Hellmers, and the new director of North Carolina State's unit, Robert Jack Downs, began to beat the bushes for projects to bring both *dignitas* and denarii to the Southeast's newest facilities for biology. They advertised that several technological features would lure potential researchers to North Carolina's research triangle. Unlike the fixed-size rooms of other phytotrons, both units' chambers could be flexibly resized to accommodate various projects. As the result of considerable engineering effort, the growth chambers in both phytotrons possessed a high degree of temperature uniformity from wall to wall as well as from ceiling to floor, via a bank of lamps on one wall facing a reflective surface and a rapid downward flow of air. In addition, the growth chambers were illuminated by intense lights housed in their own air-conditioned subsystem, a feature shared by the Biotron at Wisconsin and the French phytotron at Gif-sur-Yvette outside Paris. As the trio noted, "keeping the lamp compartment at a fixed temperature aids in maintaining a constant level of illumination." They celebrated that their designs for individual light caps permitted access to the lightbulbs, initially for cleaning, but also, as Kramer, Hellmers, and Downs speculated, "if it ever becomes desirable to use a new kind of light source."[53] The future was expected to bring new sources of light and the research triangle's phytotrons were prepared.

From the outset, the steady and continuous work of the training of

Figure 5.1. Duke University graduate student, Patrick Tesha with Paul Kramer in the Duke phytotron. From John, "A Phytotron Is Good for Plants," 5.

graduate students occupied a substantial portion of the Duke and North Carolina State phytotrons. Consider, for example, a young man named Patrick Tesha, who, in 1970, completed a master of science degree in Duke's Department of Forestry via work completed in the new phytotron. It seemed obvious to him that the phytotron was a useful instrument for his study of pine seedlings because only in that facility could specific environmental factors be varied while all others were held constant. Like many others, Tesha cited Billings as his source for "the need to recognize the existence of variability in populations when conducting physiological and ecological experiments."[54] To new plant physiologists, the variability of organisms within any population was a scientific fact but part of their training involved cautioning students that they needed to control for the variability to produce repeatable results. In short, Billings's students clearly learned the same lessons as Went's back at Caltech as they conducted the "Bean Test."

Kramer supervised Tesha's thesis topic—the effects of thermoperiod on the growth and morphology of three tropical pines. The topic sat firmly within by-then traditional phytotronic work. Tesha referenced Went's original conclusion about the combinations of day and night temperatures that would account for maximum growth. Noting that trees seemed especially sensitive to thermoperiod, Tesha concluded that pines grown quickly might address the pressing issue of large-scale afforestation in tropical climates. He worked on three species of pine, *Pinus caribaea, Pinus leiophylla,* and *Pinus oocarpa* collected from Central America. In the phytotron, the seedlings grew in several discrete combinations of day and night temperatures, ranging from 20°C to 32°C during the day and 11°C to 23°C at night. To evaluate the range of growth, Tesha measured the diameter of the root collar after ten months and also the dry weight of the plants. Once more, as in Jean Paul Nitsch's early experiments on gherkins, Tesha's experiment revealed that at a constant daytime temperature of 20°C combined with increasing nighttime temperatures, *Pinus oocarpa* at 11°C was nearly half the size of the same plant at 23°C. At a daytime temperature of 32°C, however, *Pinus oocarpa* maximized its growth at a nighttime temperature of 17°C, at 414 mm, compared to 387 mm at the next increment of 32°C/23°C (daytime/nighttime).[55] Tesha's discussion of the growth rates of the tropical pines, true to his physiologist roots, gave great weight to the unfavorable balance between photosynthesis and respiration at higher temperatures, which caused plant growth to level off and then decrease above a nighttime temperature of 23°C.[56]

Alongside students' theses that served to continually demonstrate the necessity of controlled-environment work in botany and plant physiology was a growing array of research projects. Of course, students were one marker of success, but so were visitors. Downs, for example, cited numbers of visitors, especially from foreign countries as a significant marker of the importance of his phytotron. International visitors from twenty countries were as much a marker of the success of the phytotron as the range of studies—"biochemistry . . . pathogenecity."[57] At the same time, efforts to continually improve the systems of environmental control counted as notable successes, such as throughout 1972 when Downs told his patrons how he had worked "for better, more reliable environment control systems," "temperature fail-safes," better relative humidity control and CO_2 control, and "automatic watering systems." With new red and far-red light sources alongside both step and continuous temperature change programs available in several rooms, Downs considered the North Carolina State phytotron now fully equipped to answer the important question of "whether 'normal' plants can be grown in artificial environments."[58] More familiarly, highlighting numbers of researchers and research papers produced in the phytotron signaled its impact in the scientific world. Because the phytotron did not constrain research into particular organisms, Downs could celebrate a breadth of research via the "28 different plant species" used in phytotronic experiments that year alone. Finally, he noted that a number of projects would contribute toward student theses, and thus contribute to the growth and development of graduate students, plants, and science.[59] Once those projects were completed and, more important, published, however, the whole process would have to continue. The excavation of genes and environments throughout the entire and evolving biological kingdoms was only just beginning. Looking to the future Downs concluded that "people need more information about effect of environmental factors on physiological processes."[60]

A number of standard projects applied for space in the twin phytotrons in the early 1970s. One assistant agronomist from the state of Hawaii, Duane Bartholomew, used several rooms for eight months studying the temperature effects of CO_2 uptake on pineapples. While the pineapples in Hawaii were highly concentrated geographically, and thus, Bartholomew argued, "probably [of] the same genotype," the "wide variety of environments" in Hawaii required studies of the plant responses.[61] For Downs and the Phytotron Board, Bartholomew's project represented a near standard experiment in the phytotron. The experiment used a specific genotype, exposed it to several temperature ranges, and collected specific parameters of growth performance (leaf elongation, dry mass of plants, consumption of water, CO_2 consumed, etc.). Moreover, as with most phytotronic experiments, Bartholomew's research used some novel pieces of equipment, notably a "leaf porometer" for measuring the stromatal aperture, which permits leaves to take in CO_2 and respire O_2, and some kind of "pineapple leaf chamber," which Bartholomew had evidently constructed himself.[62] With comments on research like these, it was clear that the obsession with 'growing thermometers' gradually took hold in North Carolina State as well. Writing to Kramer in mid-1973 while at a conference opening New Zealand's new phytotron at Palmerston North, Hellmers inquired about Duke's search for a postdoc. He threw his lot in with Tom Wilkinson because "he is pretty equipment oriented."[63] As NSF funding began being directed elsewhere after 1973, Hellmers hoped that future grants would "include the electronic tech."[64]

PATRONAGE EVAPORATES

By 1972, the twin phytotrons of the Southeast were four years old, and Wisconsin's Biotron had finally opened. The year before, Caltech had razed the world's first phytotron to make way for a new developmental biology building, yet news from Japan that same year announced the creation of a national biotron center at a cost of over $6 million.[65] That year, Nitsch unexpectedly died in Paris, and the following year Kramer lost a crucial ally at Duke when Billings replaced Went at the Nevada Desert Research Institute. According to the historian Paul Edwards, the United States government only became concerned with the environment around 1973, when an African drought caused famine, the Peruvian anchovy fisheries collapsed, and the Soviet wheat harvest failed, although the Environmental Protection Agency had been founded in 1970.[66] None of these events, nor the later National Climate Program Act, which authorized $50 million annually, altered the course of phytotronists' research, or indeed halted their institutional and disciplinary decline over the next twenty years. Even at Duke and North Carolina State where ecology was a research focus, those inside their phytotron remained fixated not on addressing or shifting research toward issues of climate change but on perfecting the technological systems of environmental control.

No patron had been as stalwart in support or as generous in largesse as the NSF. But around 1973, grants both for facility development and for operating costs dried up as the NSF experienced a sea change in it priorities. At first, Kramer, Hellmers, and Downs began to notice that the NSF favored the Biotron over the twin phytotrons in their annual appropriations. As Hellmers queried, "how did the Biotron come off so well?" He noted that the Biotron received twice the supporting grants of the twin phytotrons (nearly $200,000 per year), even though "we sure turn out a lot more research and papers." For Hellmers it brought into focus an outstanding issue that both the Biotron and the twin phytotrons had faced since they first accepted NSF funding. For reasons that remained a mystery to the phytotronists, the NSF disliked any system of visiting scientists, even at national institutions whose existence seemingly required a flow of casual visitors and researchers through their doors. Exasperated, Hellmers exclaimed, "I have never understood why NSF refuses to support visiting scientists. Once people have used the phytotron they get enthused, otherwise they tend to think of phytotron research as expensive. [A colleague] kept telling me how he was trying to encourage people to use the Phyto. I know he suggested it to [someone] but Calif. was pretty far to commute."[67]

In the mid-1970s, the grand agencies and programs like NASA or the large phytotrons once so feted by federal patrons gathered together, united in their ghettoization. Hellmers entertained the Bio-Environmental Systems Study Group from NASA in mid-1976, for example. In the wake of canceled manned-moon landings, NASA began deep-space probe missions (Pioneer I and II and Voyager I and II), but also continued to explore long-term space travel, an idea that necessitated "completely closed system environments," after launching Skylab in May 1973.[68] For well over a decade, NASA engineers had largely ignored environmental concerns beyond the immediate aim of keeping a man alive for a mission of a week to ten days, traveling, in effect, in the front of a Volkswagen.[69] With the prospect of living in space, however, physiological, environmental, and nutritional comfort and maintenance became a topic of serious investigation. Presumably contemplating the next manned space station, NASA's group toured Duke's phytotron before sitting down for an intensive seminar with Duke's botanists, physiologists, and graduate students. The phytotronists offered their considerable knowledge of controlled environment systems. As Duke's phytotronists notably asked about growing plants in space: "How much of plant growth is for aesthetic & psychological purposes and how much for serious nutrition?" At the end of the day, the phytotronists and the NASA group developed systems diagrams describing the feedback nature of organisms in living environments. Evocative of the diagrams of Harold and Eugene Odum, the phytotronists and the NASA group constructed circuit diagrams for the "inputs" of atmosphere and water flowing into the black processors of "Animals," "Humans," and "Plants" and their outputs having to be processed before flowing around the diagram to become inputs once more. Odum's circuits were "aggregated systems diagrams," showing the flow of energy and material into, within, and out of an ecological arena like the rainforest biome. These types of biological circuit diagrams evoke the deep connections between biology and physics, not only in their thinking but also in their actual inscriptions about nature: "storage reservoirs were represented as point-round-bottom symbols, external sources by circles, photosynthetic producers by bullet-shaped symbols and consumers by hexagons."[70] Likewise, NASA's imagined fu-

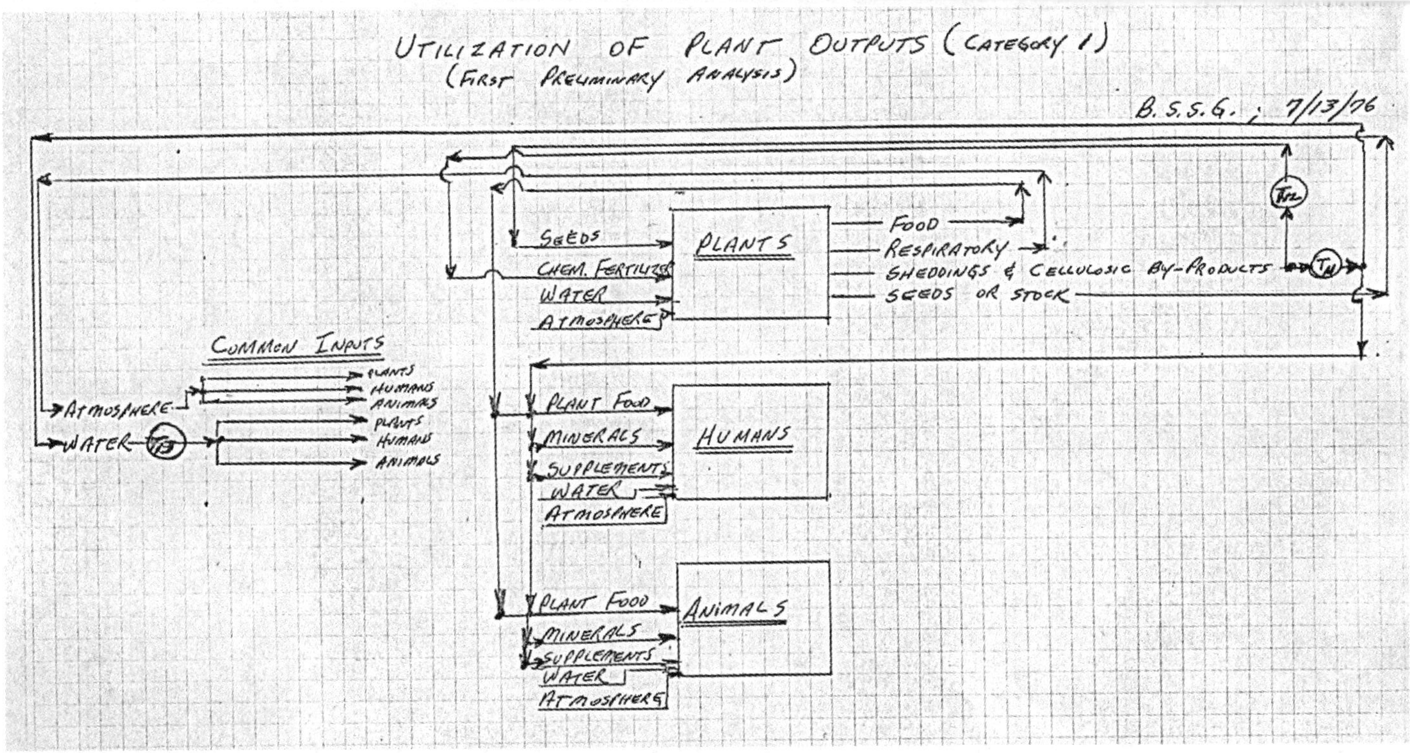

Figure 5.2. Phytotronists' circuit diagram "Utilization of Plant Outputs," July 13, 1976. Attached to Judith Thomas to Hellmers, August 2, 1976. Paul J. Kramer papers, box 11, file "Phytotrons, 1972–76." Duke University Archives.

ture space station was closed-system ecology, physiology, and electronics: the phytotron expanded to multiple organisms. The lines of flow crossed at junctions, or like circuits, intersected other components of the system, or like circuit diagrams, inscribed a little arch to show where they did not link at all.[71] In short, physics and biology met once again in another effort to specify and then measure the space "environment" for human habitation.

In contrast to phytotrons, these closed environments aimed to establish a controllable balance not an exactingly repeatable condition. Evolving plants or animals enclosed in a long-term, deep-space mission would fundamentally disturb the balance of the system. Therefore, NASA's discussion with the phytotronists centered on questions such as was there "any connection between taxonomic relationships and physiological processes?" The phytotronists answered that "all vascular plants have the same general metabolic systems" such as light respiration, but those similarities were less important than the search for "minor differences in a process." Clearly concerned about long-term stability of the environmental system, the NASA group asked, "What is the most 'fixed' physiological feature of related species?" to which the phytotronists answered, it "depends on conditions which limits" those features.[72] The NASA group not unreasonably sought the features of living controlled environment biology to economically and efficiently build a closed living environment with a small number of discrete but general systems. In contrast, the phytotronists sought the minor, specific differences between organisms, for only those gave the cause of an individual's particular growth and development. The NASA engineers wanted a standard black-box system, workable to process the general physiological processes of living organisms to sustain life at many levels, while the scientists in the phytotron sought only each organism's individual growth patterns. In fact, the large-scale test of a controlled environment to support an entire ecosystem including humans and aimed at long-term space travel had to wait until the late 1980s when Ed Bass spent $155 million building the 3.14-acre ecological laboratory called the Biosphere 2 outside Tuscon, Arizona. In its initial experiment, "Mission One," which took place between 1991 and 1993, over a thousand separate species of plants, animals, microorganisms, and eight humans were enclosed in the structure. The biospherians declared themselves the phytotronists' heirs because, as their messianic leader noted, "Biosphere 2 is the cyclotron of the life sciences."[73]

RAGE AGAINST THE MACHINE

The changes in structure and priorities, not to mention senior managers, of federal patronage bodies in the mid-1970s struck at the heart of all phytotrons. After two decades of strenuously arguing that the core mission of the NSF was the support of basic science, in the mid-1970s, substantial funds were redirected toward applied research. The twin phytotrons jumped at the new funding opportunity, immediately submitting substantial grants and even setting aside office space for the anticipated postdocs.[74] They did not yet know it, but the switch from a basic science institution to an applied science institution only served to further undermine an already weak support structure. The argument of Kramer, Hellmers, and Downs over the previous decade had stressed the unique basic science nature of phytotron facilities; could all that now be readily abandoned for applied projects ready to go? This quick inversion of priorities served, in fact, only to give substantial ammunition to the NSF office most opposed to phytotrons in the late 1970s. Within a few years, one man managed to almost entirely sever NSF support from the giant biological instruments the NSF had themselves built. His name was William Sievers.

By mid-1975 both the twin phytotrons of the research triangle and the Wisconsin Biotron suddenly found the NSF resistant to either further expansion of the facilities or, more dangerously, continued support of research projects. According to the records preserved for the Biotron in Wisconsin, the program director William Sievers of the NSF's Biomedical Sciences Division threatened to withdraw NSF funding unless the Biotron secured substantial alternative funding within a year. Sievers began directing larger portions of the new Biological Research Resources budgets not to institutions but toward large collections of plants and animals clearly supporting the rise of systematic biology.[75] Sievers favored systematic biology and its program to understand evolutionary history and processes. Much the same message went to Kramer and Hellmers at Duke, though evidently tinged with rather more of a threat. While Theodore Tibbitts in Wisconsin complained of a shortage of NSF funding, Kramer thought the Biotron was well-rewarded for far fewer users. "I really do not know why the Biotron got so much money in view of their limited usage," Kramer noted to Hellmers. "I suppose they expect the new assistant director to find more users."[76] From the perspective of the major patron of phytotrons, the NSF, it was time to rationalize the foundation's support for these expensive instruments. All three facilities were still comparatively new, but increasingly expensive to run and maintain. One solution the NSF advanced was for the twin phytotrons in North Carolina to more fully cooperate with the Biotron. Hellmers thought the suggestion reasonable, but unworkable. "Possibly we could guide some animal research their way if we hear of any" was his only alternative.[77]

The 1970s were tough times for many scientists: NASA was severely downsized, even physicists could no longer find employment, and the scientific-military-industrial complex came under sustained social and political attack.[78] The outlook for the phytotronists looked equally bleak. Where the gilded promise of the 1950s became the grand modernist engineering schemes of the 1960s, new funds to support such facilities evaporated in the 1970s, particularly long-promised maintenance and research funds that fell far short of expectations and needs. As the phytotronists' demands on the NSF became more insistent and desperate through the 1970s, they faced a new seemingly implacable enemy in the form of Sievers. Kramer met with Eloise Clark, NSF director of the BMS Division in June 1975 over "the hard line taken by Sievers and by the fact that he showed no interest in the scientific work in the phytotron." Kramer's comments about the meeting clearly indicate his fear that the NSF was on the verge of severing support for phytotronic work.[79] A few days later, at a meeting in Wisconsin, the biotronists and Kramer pleaded their case yet again for continued NSF funding. They expected the number of users of both facilities to increase, they said, while other sources of funding were being continually sought. It is hard to recognize in retrospect their justification for expecting that a great influx of new users would flow into phytotrons in the mid-1970s, but Clark, for one, shared their expectations. Phytotronists "were ahead of our time when we built the phytotron and the Biotron," she noted. Under the assumption that history would catch up to the facilities, their role in training new graduate students, especially from other countries, strongly justified continuing the funding of the phytotron.[80]

Kramer cited the "numerous papers . . . published in journals with national or even international circulation" as evidence of the importance of the phytotron.[81] Such pleas to NSF management betray the frustration faced by those Cold War scientists who embraced the mantra of "basic science": neither the number of papers nor their international circulation counted for much if the facility could not be justified on pragmatic grounds of medical discoveries, new technologies, or, lastly, student training. All three facilities now faced a hostile patron. When Kramer saw Sievers in Wisconsin the day after Clark's meeting, Kramer noted that Sievers appreciated Duke's prompt accounting, whereas "the Biotron data were unsatisfactory." Kramer also heard that "Director Williamson at NCSU had threatened to drop all NSF support if their management plans had to be changed in order to obtain it."[82] Sievers did not pull NSF funding.

Whether Sievers merely retreated or Clark successfully overruled her subordinate NSF administrator, we do not know. Certainly, Kramer believed that Clark had diplomatically moved around Sievers to continue to fund the twin phytotron and the Biotron.[83] In any event, the respite won in 1976 for phytotron funding was short-lived. If Sievers had been bypassed, he had taken steps to reassert his authority over phytotron funding, and moved further toward halting NSF funding for phytotronic work completely. As far as Kramer and the Phytotron Board of the twin phytotrons knew, Sievers was not receiving instructions to curtail funding from his superiors, but possessed some ulterior reason, which Kramer was never to learn. Kramer even officially submitted a complaint to the NSF in late 1977 to challenge Sievers: the complaint accused Sievers of "policy decisions without or contrary to the advice of scientists in the field of environmental biology." Sievers insisted that academic users be charged for using space in the phytotrons, even though, the complaint noted, "no other NSF facility is required to charge." Sievers complained that outside users of the phytotrons were "not financed by NSF," even though he had specifically conditioned further NSF support on gaining external funding. All these may have been real or imagined issues, but Kramer held the opinion that Sievers simply made "capricious charges in policy and procedure." Sievers seemed particularly dismissive of the cooperative arrangement between the Duke and North Carolina State phytotrons. "This is probably the first time a program director has ever actively discouraged cooperation and it shows very bad judgment," Kramer concluded.[84]

A year later the phytotron still stood, but so did Sievers. He had successfully dismantled the formal cooperative arrangement between Duke and North Carolina State, now only an "informal information exchange." Hellmers had reached retirement age, and Duke had elevated Boyd Strain to the post of director of the phytotron, placing extra burdens on him to get more users and funding for the facility. As Duke's chief academic officer wrote to Sievers in October 1978, "these signs suggest new awareness on the part of the Botany faculty of their 'corporate' stake in the new maintenance and effective use of this superb facility."[85] Until 1978, no expression resembling "a corporate stake" had appeared in any phytotronist's records that have been preserved, even when actual corporations like Z. Smith Reynolds and Campbell Soup had funded phytotrons. One might thus conclude that Sievers's victory was itself the sign of a larger shift in the policies of support for American science. Certainly, the historian of the NSF, Toby Appel, has pointed out that over the 1970s the AEC and the National Institutes of Health had steadily withdrawn support for plant science, which left only the NSF as a major federal source of funds. However, even as the NSF recognized its unique role and diverted major funds toward the plant sciences in the late 1970s, the subject of controlled environments barely registered in its "selected research areas," namely, plant genetics and metabolic biology.[86]

In short, the postwar elitism of scientists and the priority of basic science were discarded in favor of bureaucratic accountability, administrative control, and above all patents. As good as their word, North Carolina State received no direct NSF funding after 1977. They continued to accept external NSF-funded projects and thus the foundation did continue to indirectly contribute to the upkeep of North Carolina's phytotron. Meanwhile, plaintiffs, concessions, and restructuring brought only temporary relief to Duke's phytotron: Sievers still insisted that NSF funding for the facility decline to zero by 1981.[87] The decline of the late 1970s is evident in the compiled statistics from Duke's phytotron in the early 1980s. The proportion of space used in the facility fell from a high of 95 percent in 1973 to only 33 percent in 1978, and remained below 60 percent in 1977 through 1982. Publications over the same period remained fairly constant, however, averaging twenty per year, but the numbers of graduate students stayed below twenty per year, down from the high of forty-three in 1973. In short, over a mere five-year period, the numbers of two of the three measures of the phytotron were halved.[88] No facility, especially a largely taxpayer–funded one, could hope to survive unchanged.

As the new director of Duke's phytotron, Boyd Strain faced the future squarely. He admitted, even in 1979, that one possible solution to the financial problems confronting Duke's phytotron was simply to close the unit. Instead, the phytotron remained active though it shifted its primary research agenda away from plant physiology itself. Gaining new support from the forest industry and the USDA, the twin phytotrons moved fully into experimental ecology and environmental research. It was here that the NSF, notably with Sievers's support, funded a major upgrade of the phytotron: nearly a half million dollars toward "CO_2 enrichment research."[89] Strain emphasized to the Duke administration that its phytotron placed "Duke University 'on the map' when environmental research centers are considered."[90] Thus a subtle but important shift in the use of phytotrons and the meaning of environment had occurred, namely, that until the 1970s, phytotrons controlled the environment to study plants, now they would control plants to study the environment. This shifting meaning allowed the twin phytotrons to survive through the dark 1980s, and emerge in the 1990s as once more significant research centers as humanity realized it knew precious little about the biological effects of global climate change.

In many ways, the emphasis on environmental research for the phytotron took the facility full circle. Back in his presidential speech of 1961, Kramer had sounded the call to address the pressing ecological issues, almost a decade before the influential book of Paul Ehrlich, *The Population Bomb*. Kramer said that "most of the serious biological problems of the next several decades will result from our rapidly increasing population and from the effects of man's disturbance of his environment." Kramer's speech came at the moment of the Cold War focus on the space race and the missile gap—issues that Kramer already sensed were more political distractions than scientific and technical challenges: "solving the problems of overpopulation and air and water pollution may be less glamorous than putting a man on the moon, but it is much more important to the welfare and perhaps even the survival of man."[91] Thirty years later, John Allen and seven fellow biospherians sealed themselves in the Biosphere 2 to test ecological and technological systems to help man survive by in fact venturing out beyond the moon because the earth was becoming uninhabitable. They barely survived the two years and twenty-two minutes locked inside. As expected, Duke transitioned from botany and plant physiology over to environmental science and ecology; it may be that it will turn out to be the most important change in role for any early phytotron.

CHAPTER 6

BIG BIOLOGY IN THE BIOTRON

> Facilities like the Biotron are important for the stature and uniqueness of UW-Madison. Such units contribute to UW-Madison being viewed as a GREAT university, not just an AVERAGE university.
>
> **—Minutes of Ad Hoc Committee on Campus Plant Growth Facilities**

TECHNOLOGICAL MODERNISM reached its high-water mark in the decade surrounding 1960 as Sputnik streaked into space, nuclear power reactors opened, and computers starting forecasting the weather and elections.[1] Certainly, there was no shortage of confidence in complex environmental technology as a means to understand and control the biological world following Frits Went's California phytotron, the rapid construction of the French and Australian phytotrons, and substantial progress on similar facilities in at least a dozen other places. Flush with opportunity, the Botanical Society of America (BSA) initiated a campaign to establish something akin to a national phytotron in 1957. Looking to replicate other similar "national" scientific facilities such as Brookhaven National Laboratories for nuclear research and the National Radio Astronomy Observatory at Greenbank, the BSA received a mandate and funding from the American National Science Foundation (NSF), which had become convinced that a large, climate-controlled facility would serve its own goals as well as the needs of the wider biological community. For a week in mid-May 1958, the familiar and important names of

American plant science, including Frits Went, Sterling Hendricks, and Paul Kramer, held informational meetings at Yale, Duke, Madison, and Caltech, attracting some fifty researchers from forty-three institutions. It was a remarkable breadth of participation, not only because it conveys the extent to which the larger biological community looked favorably on a new national facility for controlled climate studies but also because it illustrates that it seemed to many that controlled environment facilities appeared set to accommodate a large swath of experimental work across the life sciences. Yet, because they wished to propose a figure that would easily fit within the NSF facilities budget after 1959, the assembled botanists and plant physiologists lowballed the enthusiasm for a biotron and too narrowly defined its potential usefulness.[2] Consequently, the BSA committee concluded that a "$1 million biotron" was the next step for plant science.

In fact, the NSF was already looking toward much grander laboratories for biology. As the BSA proposal cycled back to the NSF, the idea of a biotron was extended to include the environmental control of not only plants but also animals. The NSF initially considered building two trons sharing a single engineering plant and administrative space: one tron would be the "phyto" (plant) side and the other tron the "zoo" (animal) side. By early 1960, however, the two trons became unified into a single facility: in a marvelous and inevitable etymological and technological marriage, the "Biotron" emerged from the uniting of the botanical phytotron and the zoological "zootron" based on the winning proposal of the University of Wisconsin-Madison.[3]

To advocates of controlled-environment biology, the creation of a dual facility completed the transformation of the experimental life sciences. Once a distinct disciplinary and epistemological divide separated the study of the plant and animal kingdoms. Also solid were even the smaller boundaries between various classes of organisms be they viruses (virology), mosses (bryology), or lakes (limnology) for example. Such boundaries mattered distinctly less now that all experimental practice across the life science would be standardized regardless of organism via the creation of controlled environmental facilities. The new fundamental epistemological distinction, in other words, was now over questions about how to experimentally study genes and environments, not about differences between plants and animals. Regardless of the choice of organism, the Biotron completed the quest to make controlled-environment facilities centers of the experimental study of life to accompany the wondrous breakthroughs occurring in genetics. In addition, and perhaps not altogether coincidentally, the Biotron might also raise the lowly plant sciences to the level of the more prestigious animal sciences through the ingenious maneuver of insisting that an equivalent expectation of experimental control of the environment extend across all of biology.

The Biotron was planned and created to complete what might be termed the *tronning* of biology. Whereas in the 1950s plant physiologists were working in phytotrons, the 1960s now promised biologists working in Biotrons. One plant biologist, Jack Myers, "claimed that 1/3 to 1/2 of all problems in Biology at the U. of Texas could or should be investigated in a biotron."[4] Under the heady sway of powerful technological visions of astronauts going into orbit and nuclear submarines crossing under the arctic, it seemed that in the future all would be technologists in trons: once the Biotron was built, said Donald Griffin, the discoverer of echolocation, all that remained to complete biologists' experimental control over the natural world would be a "cycletron and a marinetron" for water biology.[5]

This chapter describes the first two decades of the Biotron. In one way, it is a classic Cold War era "big science" story. Big science has been characterized by a marked concentration of resources, the creation of specialized workforces, and the attachment of political significance to scientific projects.[6] Alvin Weinberg, the director of the Brookhaven National Laboratories, famously said that cyclotrons, those archetypical big science projects, were directly comparable to the pyramids.[7] For Weinberg, big science was a higher calling and a display of the ability to embark on great projects that might highlight a nation's greatness. As the size, cost, and vision of the Biotron expanded constantly over the 1960s to eventually nearly $5 million, it was always easily equated to the era's other big science projects: drawing on the ready-made comparison, when Went and Hendricks advocated the national Biotron in *Science*, they noted how its prototype "the phytotron has been generally accepted as an experimental tool, comparable to telescopes [and] particle accelerators."[8]

But the Biotron was big science in a very specific way. It was another example of a new class of laboratories that emerged during the Cold War, the "national" laboratories. Australia's national phytotron emerged within the scientific arm of its government. In contrast, in the United States the NSF largely underwrote the national labs on the recommen-

Figure 6.1. The Biotron at the University of Wisconsin-Madison. From University of Wisconsin-Madison archives.

dations of consortiums of universities led by a technocratic elite of scientists. Prominent "national" laboratories built in the United States in the 1950s and 1960s included the Brookhaven Cosmotron, the Greenbank radio telescope, and the Madison Biotron.[9] However, like the radio astronomers, most phytotronists pointedly separated themselves from the usual object of big science, the military-industrial complex. Remarkably, the phytotronists even rejected an offer from the Oak Ridge Institute of Nuclear Studies to accommodate a phytotron alongside their cyclotron in 1956.[10] The factories at Oak Ridge had produced all the fissionable uranium used in the Hiroshima atomic bomb in 1945, and afterward continued nuclear studies with the particle accelerator known as a cyclotron. Postwar, the Oak Ridge Institute of Nuclear Studies itself was a nonprofit organization of southern universities that offered regional participation in the atomic energy education and training activities in Oak Ridge prepared by the Atomic Energy Commission (AEC).[11] Deft maneuvering by the Duke University physiologist Paul Kramer, among others, however, drew a new phytotron away from the AEC toward the sphere of the NSF instead. "Puzzled," Kramer wrote to the chief of the AEC's Biology Branch, Paul Pearson, to evaluate Oak Ridge's seriousness about funding and staffing a phytotron. Kramer readily expressed his doubts that "anyone of their staff [really] knows a phytotron from a cyclotron," and widely advertised the fact.[12] Of course, the resources and reach of the complex ensured that it remained a seductive temptress: in 1965, Anton Lang left Caltech for the AEC-funded Plant Research Laboratory at Michigan State University.

In whatever direction Oak Ridge may have taken phytotronics, under the auspices of the NSF, the national biological laboratory privileged basic science to maintain the illusion that science was isolated from politics.[13] Quite generally, it has become apparent that Cold War era scientists on both sides of the iron curtain privately acknowledged the gulf between any overt rationale of their new lavish patronage (namely, winning the Cold War) and their own community's standards of legitimation (namely, knowledge about the natural world, divorced from practical benefits).[14] Indeed, the insistence on basic science shaped all the national labs in the United States: in the physical sciences, Peter Westwick noted that the national labs justified "basic science" both as a training tool for future scientists and "as a means to keep experienced scientists on tap for emergencies."[15] For the biological sciences, Toby Appel argued that Alan Waterman, the first director of the NSF and primarily responsible for the foundation's support of the Biotron, largely embraced Vannevar Bush's conception of basic science, namely, that the United States had spent too little replenishing the stock of knowledge from which applications and industries grew.[16] Bush's influential view was that basic science necessarily led to new practical application, and industries offered the ready-made argument to gain federal funding for grander projects. The American National Academy of Sciences Panel for the Plant Sciences, for example, readily accepted and advertised how "fundamental science and practical applications are closely interrelated."[17]

The project of the phytotronic era to specify and measure the "environment" in biology reached a climax in the creation of the Biotron, as the commitment to technological modernism shifted the study of nature toward laboratories that were fully divorced from nature and that were big, national, and basic. It took over a decade to build the Biotron because its creators labored to exactly produce, control, and measure each environmental variable to levels far beyond the tolerances deemed acceptable in previous phytotrons, all to befit a "national" facility. While the emergence of systems of control in Pasadena, Canberra, Paris, Raleigh, and elsewhere had unexpectedly revealed the pursuit of biological knowledge about the environment via technology to be laden with consequences, namely, feedback and chaos, as the 1960s dawned, plant biologists still trusted that those challenges would be met and mastered by the next grander phytotron. Yet, far from resolving the earlier unexpected consequences, the Biotron actually further opened up remarkable insights into the concept of the biological environment. A debate erupted, for example, over the shape of nature: was the environment "square" as phytotrons had constructed it, with lights and temperatures that changed almost instantaneously? Or was climate rather more a sine wave, or some sort of cycle? Or maybe, just maybe, the weather (to use the third term of this flexible category) was actually chaotic?

A plant science fixated on producing controlled climates to define and measure the environment climaxed with the Biotron by the late 1970s. Subsequently, through the 1980s, the Biotron and phytotronics generally declined even as a new burgeoning environmental science came to prominence. By the end of the Cold War, the assured modernism of biologists as technologists had been overtaken by the hype of the gene jockeys, and earlier enthusiasm for nationalist modernist technoscience replaced with the bubbles of venture capitalism underwriting biotechnology. As the AEC and National Institutes of Health (NIH) withdrew all funding for plant science, and NSF support for controlled environments evaporated, it spelled the end of the Age of Biology. In contrast to the Age of Biology's emphasis on the parallel sciences of genes and environments, the next age, the Age of Biotech, rested assuredly on the utopian promise of a genetic medical toolkit for mankind. Founded on much "hope and hype" about the potential of unlocking DNA and our genes,[18] the science of biology has become merely a "gigantic whirring biotechnology machine," said the eminent microbiologist and biophysicist Carl Woese.[19] Where once biologists believed they had rescued their science from the military-industrial complex and sought to underwrite the Age of Biology with basic science, the medico-pharma-industrial complex now dominates all in the Age of Biotech.

With the notable exception of Duke University, as we saw in the previous chapter, many technologist biologists in phytotrons also missed a pressing new area of research in the 1980s—that is, the growing awareness of climate change, which slowly returned funding and interest toward the subject of the biological effects of the environment. Though the Biotron was both the apex and the apotheosis of biological trons and the experimental pursuit of whole organisms in whole environments, the story of trons does not end with the Biotron. In fact, in the fine debates over the units of measurement for light sources, illumination patterns, wavelengths, and intensities, past-generation phytotronists anticipated much of what is now required for a science of climate change, specifically, the ability to specify the phenomenon in question and measure it. Though it is too early to tell the next chapter in the story of phytotrons, controlled environmental systems have returned in new guises and with new missions. New facilities have also been built, and, as this book concludes, they are inevitably called Ecotrons.

A NATIONAL LABORATORY IN WISCONSIN

Like many universities, the University of Wisconsin-Madison (hereafter, Madison) expanded, courtesy of new, deep federal coffers, to become a significant institution of science in Cold War America.[20] In a frenzy of construction, Madison built a number of facilities at the intersection of biology and technology, the Biotron among them. By 1963, Madison celebrated several new buildings, including a cancer research laboratory ($2.8 million), a primate laboratory ($1.9 million), a molecular biology and biophysics laboratory ($2.2 million), and the most expensive, the new Biotron ($4.2 million).[21] The price tags and purposes of each reveal the emerging fault lines in biology: cancer remained a separate entity, readily funded though the NIH, which did not "make grants for research with plants," except for some work for a "Cancer Research Institute."[22] At the same time, molecular biology was by then independent and growing rapidly, as were primate studies, which under Harry Harlow sadistically studied the comparative emotional psychology of rhesus monkeys.[23] With its focus on basic science and plants, the Biotron occupied a unique space in what was already the predominantly biomedical landscape of biology. As historian the Jean Paul Gaudillière argued, the American National Institutes of Health, pharmaceutical industry, and medicine all became vital agents in changing the emphasis and conduct of biological research, and while the "biomedical complex was not a replica of the military industrial complex" it still clearly bore a close resemblance.[24] A series of institutional decisions at Madison and disciplinary shifts throughout Cold War era biology saw the Biotron sit at a distance from the emerging landscape of the biomedical complex.

The Biotron came to Wisconsin as a direct result of the campaign of the noted plant physiologist Folke Skoog. Skoog was part of the larger plant physiological and botanical community that built phytotrons everywhere, not least because Skoog's early career both overlapped and paralleled that of Frits Went. In fact, Skoog and Went worked on growth substances in the 1930s at Caltech. Skoog also cowrote several papers with Kenneth Thimann, himself the coauthor with Went of the book *Phytohormones*, noting that they used Went's "standard conditions" when testing the *Avena* coleoptiles.[25] Throughout, Skoog sought growth-inhibiting actions, including one early experiment using X-rays to affect both the growth substance and the plant itself.[26] It was all classic plant physiology, always demanding greater controlled conditions.

In 1959, Madison submitted two proposals to answer the Botanical Society of America's call under Skoog's direction. No doubt seeking broader support via a wider range of biologists, Madison's submission contained one proposal from botanists and another from zoologists. The two were linked, however, on the basis that botanists and zoologists "confronted . . . common problems," namely, "the response of the organism to a complex of fluctuating environmental variables."[27] The BSA committee joined them together, emphasizing that the botanical and zoological communities claimed similar technological needs, and, moreover, that both groups would benefit from the "interaction of botanist and zoologist in a common institution."[28] Institutionally, local plant physiologists and zoologists quickly embraced the rhetoric and the reality of uniting agriculture, medicine, botany, and zoology through the environmental control of plants and animals.[29] Intellectually, these institutional arrangements merely reflected, as the Duke ecologist Henry Oosting noted in his 1956 textbook, an emerging consensus that while "There is [a] concentration on plant ecology or animal ecology as though the two were separate or distinct when, actually, they are often indistinguishable."[30] Similarly, the Illinois physiologist Ladd Prosser declared, "we all are primarily biologists, and only secondarily are divided by our materials."[31]

A national facility had wide-ranging support from across the biological sciences. Madison's NSF proposal specifically cited a large "conference" of interested researchers and intersecting disciplines, including agronomy, bacteriology, biochemistry, botany, entomology, genetics, horticulture, plant pathology, and soil research all interested in climate-controlled experimentation.[32] Under the banner of technological modernity, the Biotron would unite practitioners from across biology, not as we might imagine across the scale from the molecular to the ecological but across the divide that had long separated plant and animal science. Appealing to the NSF, Skoog highlighted a number of local biologists who were already committed to "studies of the influence of environment on higher plants and/or animals."[33] The lone voice of descent came from Knut Schmidt-Nielsen, the preeminent camel physiologist from Duke, who railed against any national Biotron because of "the travel involved" and because "zoologists and botanists could not work together."[34]

In fact, since disciplinary battles were practically absent, the reality of Cold War era science was more the mundane process of committee formation and decision making. Committees effectively made the Biotron—it was a committee from the Botanical Society of America that sold the idea to the National Science Foundation—while later the Biotron Committee administered the facility for the University of Wisconsin-Madison. Skoog chaired Madison's Biotron Committee. Costlier facilities became a reality on many university campuses precisely because they possessed mechanisms for control—specifically, greater oversight and accountability for the era's mundane administrators supplied via committees. In the apt phrase of the historian W. Patrick McCray's, Cold War era science was "science by committee." A practical solution to the problem of an overt concentration of scientific and material resources, leadership by committee expressed the value and power of democratic processes.[35] Among the more notable were the National Academy of Sciences BEAR Committee on the biological effects of radiation, the Physical Science Study Committee and the later Biological Sciences Study Committee, which effectively rewrote American science education, and the Presidential Science Advisory Committee, which gave unprecedented access to the executive branch of American government with its creation under Eisenhower in 1957 until being disbanded by Richard Nixon.[36] All fundamentally shaped science policy and education through the critical decades of the Cold War: one is tempted to suggest that "committeetrons," or big complicated technoscientific meetings, best characterize the era.

The Biotron Committee took the many expressions of interest from the wider biological community as strong evidence that the Biotron would be a valuable addition to the scientific laboratories of Wisconsin. They especially gathered proposals for animal studies in controlled or semicontrolled environments that drew on the established practices of botanical research in phytotrons, noting how both shared "obvious and universal" variables like temperature and humidity.[37] One expression of interest, for example, was from a scientist in the microbiology section of the Department of Veterinary Science. Citing promising results from early experiments under limited controlled conditions, the scientist sought to test Newcastle disease infection rates at various temperatures (34°F to 90°F) in chickens.[38] Similarly, a Dr. Bird, for example, submitted a formal "scientific method" proposal to study respiratory diseases in turkeys. Bird's "hypothesis" held that "undefined environmental factors so influence the turkey and its respiratory pathogens that the result may be transitory . . . or a severe or fatal infection." Bird's experimental regime "measured and recorded . . . selected environmental factors" to determine the "correlation [between] physiological activity of the turkeys and their infection history." Bird anticipated that if specific environmental conditions proved casually detrimental to the turkey, then "their critical levels" could be guarded against via "sensing and warning devices."[39]

Grand visions swirled in those first days: some noted that "special accommodations for burrowing or aquatic forms" would have to be provided; others that a larger space to accommodate both bigger and more numerous animals might be needed; a few insisted that space had to be available for interacting plants and animals if it was to be a truly unified facility.[40] Armed with absolute confidence in technology and with the federal government as patron, no conception of a national facility was too grandiose. For Prosser, one of the original members of the BSA committee, the Biotron promised to contribute to research in "controlled experimental ecology," "community ecology," and "experimental evolution."[41] In fact, only one avenue of research was declared off limits: man. From the outset, "medical interest in human responses to the environment . . . laid . . . outside the scope of this committee."[42] In other words, at exactly the moment that the attack on human diseases and hormones became the centerpiece of the new biomedicine of radioisotopes, controlled-environment biologists declared humans outside the boundaries of their newest facility.[43]

Joining plants and animals, the Biotron Committee also sought to join basic and applied research. The committee explicitly proposed a facility of "both pure and applied research."[44] However, it was Madison's inclusion of basic research that drew the attention of the NSF to the university's proposal: "We have had a number of requests which emanated from institutions where they essentially want to do applied research, and particularly from agricultural schools, and this is fine, but this is not our particular mission and so we were very much impressed by the basic research component here."[45] The NSF did not want one of its premier national scientific facilities to be distinctly second-class. What the NSF did not want were projects like those of the entomologist R. L. Metcalfe of the Californian Citrus Research Station, for example, who only pointed out that "the application to ecology is perfectly obvious." Undoubtedly it was, as his Hawaiian fruit fly project already possessed five climate-controlled cabinets that readily predicted "the possibility of this organism existing in any particular climate in the United States," and looked to expand the research in any new Biotron.[46] It was a straightforward, applied, and small-scale type of work. It was also not what the Biotron Committee was looking for. Instead, a Biotron more suited the work of C. M. Williams from Harvard's Biological Laboratories, who proposed, "animal studies and environmental investigation surely must include biological rhythms, ecological problems, . . . bringing with them the necessity for proper instrumentation to track these happening, behavioral problems and problems of development."[47] Williams flattered the Biotron Committee that its facility supplied the methodological tools to do basic science. In the case of biological rhythms, Williams concluded, the "whole study of photoperiodism in insects discovered about twenty years ago and rediscovered in recent years has really clarified a lot of things that were hitherto thoroughly opaque"; specific causes would replace speculation about the nature of growth and development.[48]

At the high point of technological modernism, basic science equaled funding. As science budgets expanded in the wake of Sputnik, was it any wonder that Wisconsin's administration encouraged its biological scientists to pursue ever-larger funding opportunities for their Biotron facility? When Louis Levin from the National Science Foundation spoke directly to the 1959 Biotron Conference, the rhetoric and reality of a biological laboratory devoted to basic research entirely meshed with the NSF's purpose in American science. According to Levin, the NSF existed "to help science wherever we can, basic science particularly."[49] Shortly thereafter, not only did Frits Went at the Missouri Botanical Garden get several hundred thousand dollars for his Climatron, but the NSF also awarded some $1.5 million toward the Biotron, in what was until then "the largest single facilities grant that the Foundation had made" in its ten-year history, Madison's dean Willard celebrated.[50]

The next few years witnessed a dance between the emerging design of the facility and frequent requests to many foundations and agencies for financial support. Willard remained committed to the strategy that had landed Madison the NSF grant, which was to publicize that the Biotron's "purpose was to obtain fundamental information on growth processes in all types of organisms."[51] Harold Senn, subsequently appointed the director of the Biotron, insisted that this fundamental information would come primarily from the development of increasingly exacting technologies to replicate and control the biological environment. Senn represented, in short, the apex of that generation of technologist biologists: when the Caltech plant physiologist Harry Highkin left Australia in 1959, he mentioned to his host that there was a phytotronist in Canada who was "more interested in growing light bulbs than thermometers"; that phytotronist was Harold Senn.[52]

With the support of Willard, Senn chose to concentrate fundraising efforts on the Rockefeller and Ford Foundations, as well as further NSF support, and above all the NIH. No doubt Senn and Willard were aware that private foundations had funded Went's first phytotron, the North Carolina State phytotron, and half the giant Australian radio telescope. Over the next three years, however, all wells came up dry. The funding of big science was a political game and revealed the new contours of American science. For example, though hoping to "find 'pay dirt'" with the NIH in mid-1961, Willard encouraged Senn to return from a trip to Washington via New York to visit George Harrar at the Rockefeller Foundation.[53] At the Rockefeller, Harrar listened to Senn's by then familiar argument that Wisconsin could build an "inadequate biotron" with the NSF grant ($1.5 million), or could hold out for a larger grant from either the NIH or the Rockefeller to gain the proposed $4.5 million.[54] Disappointingly, however, Harrar could make no commitments, but at least spoke sagely about the dilemmas facing the funding of big science. While government research contracts through the United States Department of Agriculture (USDA) and the United States Forest Service might defray "annual operating costs," and while the new National Aeronautics and Space Administration (NASA) looked promising, Harrar cautioned Senn against making "commitments with military or semi-military organizations."[55] At least according to an officer of the Rockefeller Foundation, even the consciously civilian space agency still resembled the military-industrial complex and thus required equal caution. Moreover, if, as Senn reported back to Willard, Harrar was right in believing the Biotron to be "an international facility to which foreign scientists might freely come to work," any connection with the complex or its allies might stifle freedom to work.[56] Subsequently, after much wrangling, Senn and Willard never did formally apply to NASA because the new space agency could not guarantee funding for basic research, unlike the private foundations. In other words, Cold War universities and foundations were painfully aware of the choice they faced. On the one hand, being beholden to the military-industrial complex dangerously undermined a scientific facility's ability to attract or employ foreign scientists. On the other, the Rockefeller's own agents saw clearly that the government patron was the only realistic source for the astronomical sums of money a national biological laboratory demanded.

Of course, parts of the military-industrial complex were very interested in the types of controlled environment research proposed for the new Biotron. Senn accepted an invitation from the United States Air Force (USAF) 6570th Aerospace Medical Research Laboratories in early 1962, for instance, to be the banquet speaker at a symposium on Biologistics for Space Systems. Charged with manned missions into space, the USAF looked to those "many workers," like Senn, "in the biological sciences who are conducting research which would apply to the various biological subsystems necessary to maintain man in the aerospace environment." The Air Force sought out the major groups of the Cold War scientific world, "universities, industry, and Government," to create the "maximum coordination of effort in developing the ultimate closed ecological system."[57] The Air Force appealed to established patterns of gathering and utilizing scientific manpower, and the incident became one more example of the military-industrial-academic complex at work. It seems likely that the military knew that the Biotron Committee had proposed "biological studies to be conducted in these chambers may include basic studies of the response of animals and plants to changes in composition of the major gases of the atmosphere," noting that "these studies have practical implications for submarine, space travel and concentrated food production."[58] Indeed, we know from their sponsorship of work on the Algatron that the Air Force's mention of the "ultimate closed ecological environment" had both outer space and underwater in mind. William J. Oswald and Clarence G. Golueke from the Sanitary Engineering Department at the University of California Berkeley built a scale model of the Algatron, now even more forgotten than phytotrons, under a research contract from the Air Force Cambridge Research Laboratories by the mid-1960s. The Algatron was an audacious attempt at a closed ecological system of living and growing algae to provide for oxygen generation/carbon dioxide absorption as well as "microbiological waste conversion" for "humans sealed within an isolated capsule," on its way to the Moon, Mars, or even "indefinitely long periods of time" among the stars.[59]

A skilled temptress in the 1960s, the Air Force appealed to many areas of the life sciences. Assuaging any concerns over secrecy, it pointedly declared the entire proceedings "UNCLASSIFIED," and looked to advertise the entire performance. Senn had flirted with both NASA and the Air Force symposium, and by having him speak, the Air Force had undoubtedly winked back. Still, no invitation was forthcoming. Talk of "space habitats" disappeared after about 1962, presumably as NASA became solely responsible for manned space flight. Perhaps Senn considered attempting to make the talks he had heard on "algal gas exchange systems" and "nutritional support from bioregenerating systems" relevant to the Biotron's basic science mission; we do not know. And, it must be said, many noticed that the military were excellent patrons: in the same era, even the Marxist biologist Richard Lewontin remembered being on Office of Naval Research and AEC contracts for fifteen years, and that their "program officers" never "intervened in any way except to remind me annually to send in my renewal application for money that had already been put aside."[60]

BIOLOGICAL CLOCKS, COMPUTERS, AND CONTROL

It would be four years before Senn, the Madison administration, and the local Biotron Committee would gather the remaining $3.5 million beyond the NSF's original grant. Finally, in January 1963, the Ford Foundation donated $1.7 million just before the long-sought-after NIH would contribute an additional $1 million. Still short, Senn successfully pleaded that the NSF increase its original grant by $300,000, followed a few months later by the state of Wisconsin itself funding the remaining $300,000. Defending the cost of the Biotron, René Dubos asked readers of the journal *Bioscience* to "compare" biology "with the luxury and modernity of the physical and chemical equipment!"[61] Biology became luxurious and modern over the next decade as the Biotron grew as a complex structure of negotiations between design elements, technological gadgets, and demands from scientists and patrons alike. In time, the Biotron would host a wealth of special facilities beyond the more standard requirements of climate-controlled facilities for experimental areas, from dew rooms to electrostatic fields, and from plant nurseries to a wind tunnel, the facility grew in size and expense every time a new stakeholder appeared.[62] And at every step, the facility grew with the expectation that either the NSF would decide to cover the escalating costs or that some other organization would assist. "When more money was needed" for a radio astronomy telescope and the Arizona Observatory (Kitt Peak), the Biotron Committee observed back in 1961, the NSF simply added "many millions of dollars."[63] It was certainly grand to pursue big science in the Age of Biology.

Big science required scientific management. In that capacity, Senn displayed a near obsessive attention to every technical detail of construction and administration. No parameter of the complex lay outside Senn's oversight; in late 1962, Senn mentioned to Paul Kramer that he had just finished the "user specifications" *on a room-by-room basis.*[64] Heroically, Senn shouldered the entire edifice of the Biotron operation, from the minute environment gradients of each room to the exact job details of every employee. He appears to have simply worked longer and harder as more and more rooms came into operation. As rooms gradually became operational from about 1966 onward, Senn discovered that in controlled experiments, even a short absence of the desired environment could ruin work. Each room demanded constant monitoring of the mechanical and electrical systems. Operating the Biotron around the clock placed a huge burden on both manpower and finances. Students, naturally, were sought as cheaper "state funded" labor, but students presented their own problems with demands to attend class and a predisposition to "telephone an hour or two before their duty period and say that they cannot come." Senn certainly did his duty to the Biotron: as he said to a colleague, "Usually it has fallen to the lot of the Director to fill in these gaps. . . . On many occasions I worked a full normal day and followed it by a full night duty period."[65]

As much in modern biology as in national security, computers promised both better control and an eternal watchman. As in both the Earhart Laboratory and the Climatron before it, Madison's Biotron Committee advocated that "a common control panel (supervisory data center) for both units of the Biotron seemed desirable so that a single engineer could supervise it. This could be given a central location and be open to visitors."[66] Once more, the computer itself not only became a detailed record keeper for the multitude of variables and data that the facility generated, but was itself a symbol of the larger goals of the biological trons. The style of these cabinet-sized machines on centralized display, like a new car in the showroom window, celebrated the values of Cold War control.

Senn's designs for the Biotron had long included a large computer. Originally priced at several hundred thousand dollars in the early 1960s, by the early 1970s, the price had already fallen by an order of magnitude. Then again, the day-to-day data recording of a dozen environmental variables across some sixty individual rooms required, a decade later, much larger computing resources. With the building completed by early 1971, Senn could finally spend the small amount of remaining Ford funds on a computer system. Not only would the Biotron itself be an experimental space for climate-controlled experimentation, but the systems of "equipment design and biological evaluation of controlled environment techniques" themselves became the experiment.[67] Compartmentalized, controlled, and flexible, the Biotron itself became an object of study by the early 1970s: in a nice postmodern moment, Senn supported a grad-

Figure 6.2. The Biotron computer. From University of Wisconsin-Madison archives.

uate student from Wisconsin's mechanical engineering department working on control theory. As Senn outlined, the "objective would be the development of suitable algorithms that will permit optimum computer control of our heating, cooling, humidification and dehumidification

Figure 6.3. The new breed of biologists at work. From University of Wisconsin-Madison archives.

processes."[68] The student probably got to use the Biotron's new computer, a PDP-8/E-AA, complete with "teletype control," "console," and processor. Senn felt that the needs of the Biotron and the budget would stretch just far enough to spend an extra $3,000 on an extra 4 K of memory, taking the machine's core memory to a whopping 8K.[69]

New computing power made it possible to create new experimental environments. The computer no longer just controlled individual variables, it "optimized" sets of conditions. Once again, the Biotron's creators saw it as the apex of controlled-environment facilities especially in its ability to produce varied environmental shapes. As we saw in chapter 2, in Went's first phytotron, nature was effectively a square-wave form: plants would be moved from a day-temperature room to a night-temperature room creating a nature that was a square-wave, and many of several critiques noted that no actual environment was square. The creators of the Biotron looked to more realistically define the biological "environment" through the creation of the very technological systems bywhich plant scientists might measure it. The historian of science Robert Kohler offered an important insight about science when he explained how "modern laboratories reshape and transform natural objects to suit the needs of experiment."[70] In the case of the Biotron, where the study of life was an exercise in the technological control of the environment, the ambition was to move beyond static conditions to the "controlled replication of that natural environment to which living systems are adaptively adjusted," because as someone said to the BSA committee, "square wave cycles of environmental parameters simply do not exist in nature."[71]

The commentator (now unknown) might have been referring to a moment earlier that same year when Caltech plant physiologist Harry Highkin was challenged at a Cold Spring Harbor symposium over exactly the issue of the static controlled environments of the rooms of the Pasadena phytotron and their square-wave patterns of climate change.[72] Highkin's challenger would no doubt have been pleased to hear the damnation of square-wave nature. Though rendered neutral in the official planning documents, Went's diary explicitly revealed that one of the members of the Botanical Society, Colin Pittendrigh, later known as the "father of the biological clock" for his fundamental discoveries about circadian rhythms, had also argued for controlled-climate experimental environments with "changing cycles."[73] Not coincidentally, Pittendrigh wrote to his friend Skoog during the planning of the Biotron to stress "*every* biologist wants more control than he currently has." Control, Pittendrigh insisted, could be gained in many far less costly ways, and usually locally. But better control had to be not just more control, rather it "*must* be that it makes possible, by virtue of the *range* of diverse environments simultaneously possible, a new type of experimental approach that is otherwise impossible."[74]

Pittendrigh explicitly questioned Went's model, which had forced movement of plant trucks from one condition to another and so "restricts one to highly artificial 'square wave' conditions and commits one to serious loss of control while the material is in the atrium at 'changing time.'"[75] Instead, Pittendrigh opened up a thorny practical and metaphysical problem, not to mention an engineering headache, for the technologist biologists for the better part of the next decade: should the environments of living organisms be best represented by square wave versus cycles, either stable or essentially chaotic programs? Could the random yet cyclical "environment" be re-created through square-wave technological systems? Pittendrigh, Skoog, and Went all offered differing organizations of the laboratory space that answered, implicitly or explicitly, these epistemological questions about the nature of the environment, and thus the nature of the living organism.

The definition of the experimental biological environment took shape in response to these questions. By May 1964, for example, Senn was deep in decision making concerning the range of temperature and humidity requirements of the Biotron's fifty rooms. It had already been decided that some rooms' temperatures would stretch from −25°C to +50°C, with a humidity ranging from 1 percent to 100 percent, whereas other rooms might only require a temperature range of +10 to +45°C, and humidity covering 5 percent to 100 percent. For Dexter Lazenby at the firm Hygrodynamics, with whom Senn had entered contract negotiations, the range was not his concern. Lazenby's charge was to provide precise monitoring of the range and a supply of the "highest possible accuracy" sensors (near ±1 percent), preferably with the ability to electrically record environmental data and even be equipped with a "transmitter" "for control and/or data logging."[76] Later that same month, Senn sought design information about Xenon lights from Adrian Godschalx. Again, the lights were not so much the issue as ensuring that any "fixture that will provide both suitable support for the xenon lamp and facilities for passing air through the fixture for cooling."[77]

Godschalx's commission was a microcosm of the broader task of designing closed environments once more. Beyond holding the lamp and permitting air to pass, the socket needed to accept either 10 kW or 20 kW lamps, be enclosed in a housing so that the cool air surrounding the lamp was "separated from the room air," and then the housing itself "designed to withstand the pressure differentials between the room and the inside of the fixture."[78] At its extremes (a +50°C lamp adjoining a −25°C room), Godschalx's housing might have to withstand pressure differentials of about a quarter of an atmosphere, or roughly a mountain 2,000 meters high. In addition, the greater exchange of air required for stability also required greater heating, cooling, and humidity control, which placed pressure on other systems. Moreover, it was necessary to identify these feedback pathways for all of the Biotron's systems, including cooling, heating, lighting, humidity, sound, air, and nutrients. Each system affected each other system. Even smell was examined: though smell might be easily controlled via more air, the quality of the air required increased monitoring for consistency, especially in the absence of exchange with the outside air.

Figure 6.4. Harold Senn deep in the complex of the Biotron. From University of Wisconsin-Madison archives.

THE BIOTRON AND ITS PROBLEMS, 1965–1975

The phytotronic community assembled to dedicate the newest and most luxurious facility yet built to control the biological environment. At the Biotron's opening, Arie Haagen-Smit spoke about his tortuous journey to botany, from being an undergraduate student of Frits Went's father in Utrecht to becoming the famed vanquisher of smog. The audience also heard from George Beadle, recently a Nobel Prize recipient—another thread from the Biotron back to the original phytotron at Caltech, which Beadle, we will recall, had had to approve as his first action at the institute. The original Pasadena phytotron had just been demolished, but the national Biotron took up its mantle.

A day of celebration, toasting, and speeches could nevertheless not hide the deep problems facing the newest, and grandest, tron. Harve Carlson, the director of biology and medicine at the NSF visited Madison shortly after the dedication principally to "discuss the Biotron and its problems." By the time it officially opened, the Biotron had cost nearly $5 million and ongoing construction remained a thorny issue, but the bigger problem was the "failure to fully utilize the facility with an adequate number of good projects."[79] Senn suggested a six-month moratorium on overhead charges to attract some business, but the dean of the Graduate School, Robert Bock, refused, citing the example of Madison's computing center. Consequently, Senn offered his resignation. Though Wisconsin's president refused it, Senn confessed to Bock that he had often felt himself a "major stumbling block" to the Biotron's success. Carlson had probably been candid with Senn, but certainly told Kramer that he considered Senn's "perfectionist attitude" to be in fact "discouraging users."[80] Senn himself realized what a large amount of effort he had put into the detailed construction of the Biotron, while leaving the promotion and research side to its fate, presumably convinced that the Biotron would blossom once the facility was completed. Senn was hardly alone in detrimentally privileging the technology of facilities over the study of life; when Joshua Lederberg computerized his exobiology laboratory the members of his group too got "bogged down" in the design, construction, and debugging of their instrument.[81] As phytotronists and exobiologists both learned, it is difficult to live in a house that is constantly being altered and changed.

Under financial and administrative pressure, the Biotron's leaders fell back on an appeal to basic science to maintain the university's support.

In mid-1976, for example, Bock offered the familiar appeal: "In the current projects, 1/3 are of very basic character asking fundamental questions with no applied end use as a direct motivation. . . . The University of Wisconsin-Madison has taken a leadership role in encouraging USDA to adopt NSF-like research grant policies for science basic to agriculture and nutrition."[82] By the mid-1970s, however, the conception of basic science so idealized by the 1950s no longer held much cultural capital. Not only that, the NSF itself challenged Bock's continued assertions that the Biotron even hosted any important basic research. To the NSF it seemed that the Biotron had, in fact, a "heavy emphasis on what can be called applied research in agriculture and medicine."[83] Consequently, the nemesis of controlled-environment biology, William Sievers, the program director of the Biological Research Resources Program of the NSF, threatened to withdraw NSF support unless a new director was quickly appointed and a "real effort [made] to seek other federal support."[84]

Throughout its long construction, Senn and the power brokers of Wisconsin remained convinced that the Biotron would always attract more than enough research projects able and willing to pay the hefty overhead charges. Those hopes were soon dashed. From the Biotron records, it seemed that only minor research projects flowed through the facility, and few possessed their own substantial funding. Most were small scale and local, and the Biotron Committee frequently awarded grants from local funds to offset space charges to even permit research. As soon as a few chambers were complete, the Biotron began accepting projects. However, between 1968 and 1971, only thirty-five research projects occupied Biotron space and paid for the privilege. And of those, only one spent more than $5,000, the Ralston Purina Company. Three others spent more than $3,000. The remaining 90 percent of Biotron business went to projects with budgets from only a few hundred dollars to, more often, about $2,000. All told, the combined budgets of all research projects over four years to 1971 did not even clear $100,000, while the yearly maintenance bill already topped $500,000.[85]

Back in 1959, the complex's founders had stressed that the facility would be expensive to build and run: "planning this sort of machine involved a very intricate planning of needs for very precise control against the economics that saw greatly increased costs as you go to finer and finer limits of control."[86] Confident of a continuous stream of federally funded projects along with substantial NSF, NIH, and private foundation support, the Biotron became a white elephant, as no new patrons or new users came forward. From the list of researchers we know that the University of Wisconsin, via NIH grants, remained a big source of paying projects, as did the NSF and other elements of the federal government, including NASA, the public health service, and even the Bureau of Reclamation. Only five private companies sponsored research, alongside the same number of foundations, including McGraw Wildlife and the American Meat Institute.

Nor did the expansion of extended controlled-environment research to the animal kingdom yield many paying projects. Such as they were, they certainly appeared to fulfill the planners' expectations of applying similar methods for experimental control over the biological environment, whether plant or animal. In 1968, one of the earliest experimental runs in the new Biotron, for example, looked at temperature stress on turkeys exposed to bird flu. For D. R. Anderson and B. C. Easterday from Wisconsin's veterinary science department, the important part of the experiment rested in the comparison between the controlled fluctuating temperatures of the Biotron's turkeys versus the "equal control groups" held outside the facility. Only via the comparison, the researchers noted, could "temperature stress [be] used to determine whether clinical disease may be precipitated by such stress and whether the virologic and serological responses may be influenced."[87] As the pair of scientists explained to the Biotron Committee that approved the use of the facilities, the idea of temperature stressing turkeys with influenza paralleled the idea that in cold weather humans seem to get the flu. A month later, turkeys did indeed succumb to avian influenza more readily under near freezing conditions, but suffered no appreciable effects under hot conditions.

Among the experiments conducted in the early days of the Biotron, a revealing one is the story of a researcher and his experiment on leaf movement. It is interesting because of the controversy it generated around what it meant to be a technologist biologist, but also because the primary investigator would end up becoming the director of the facility in the mid-1970s, unbeknownst to the actors at the time.

In an article in the 1970 issue of the journal *Plant Physiology*, readers might have considered the study titled "Circadian Rhythm of Leaves of *Phaseolus angularis* Plants Grown in a Controlled Carbon Dioxide and Humidity Environment" to be of interest. The sponsor of the work, Theodore Tibbitts, was from Madison's horticulture department, and the primary author was a PhD student named D. K. Alford. Alford neatly outlined a series of observations of daily leaf movements and speculated on their environmental cause. Alford noted that several previous studies of leaf movement had used controlled conditions, especially uniform light and temperature, but mainly those controls had been limited. Taking advantage of the "facilities at the Biotron," Alford argued, permitted an investigation of other potential climatic factors. "Perhaps fluctuations of . . . CO_2, relative humidity, nutrition, or water availability initiate and maintain leaf movement," he hypothesized.[88] In good scientific form, Alford then described the experimental procedure: taking adzuki beans grown from seeds, and placing them into the Biotron's controlled environment. In fact the experiment was doubly controlled because the plants were grown in controlled plexiglass chambers inside controlled Biotron rooms. The various climate factors were constantly monitored, per standard Biotron operation, and the plants themselves photographed at six-minute intervals via a camera attached to the clear plexiglass chamber. Over the course of fourteen days, the temperature and light intensity of the chambers was held constant, at 24°C ± 0.5°C and 600 ft-c ± 25 ft-c respectively, but Alford gradually increased the relative humidity while simultaneously decreasing the carbon dioxide levels in the chamber.

After two weeks, the experiment concluded, "there were no apparent correlations of movement response to the environmental factors." In other words, the experiment was a failure, producing no further insight into the cause of leaf movement. We might ask, why then was the article published at all? The answer hinges on the new patronage networks available by the middle of the Cold War era. Let us chart this particular experiment back to its proposal. Before Alford did his PhD work, Tibbitts had himself been working on leaf movement under controlled conditions for several years before Alford's experiment and its negative result. It is probable, though not certain, that Alford's financial support stemmed from a grant Tibbitts had received from the American National Aeronautics Space Administration. NASA, for its own reasons, granted $58,000 for the study of leaf movement. As Tibbitts had done using the earlier grant, Alford and Tibbitts continued to develop, as their article's abstract noted, an experimental "a system . . . for growing plants for extended periods while collecting data with time lapse photography."[89] Tibbitts ingratiated himself with one of the biggest of the big sciences in the Cold

War period, becoming part of the space race. One significant part of NASA's mission was to conduct scientific experiments in space, especially those that possessed no seeming military or industrial purpose, in other words, basic science. There Tibbitts's leaf movement seemed to strike just the right chord.

If a negative experimental result requires some explanation, a NASA-funded study of leaf movement surely needs substantially more. In fact, Tibbitts's own proposal to NASA reveals that his circadian experiments were a collaborative effort with T. Hoshizaki of the Space Biology Laboratory at UCLA, and C. Bentley at Wisconsin's geophysics department. In many respects, of course, the Biotron itself is the silent fourth partner in this somewhat odd endeavor. What is clear is that Tibbitts's proposal clearly registered with NASA. Given the success of the proposal, and Alford's subsequent work, NASA evidently sought "base-line data for leaf rhythm experiments that are being developed for space flight; information on plant regulating effects from . . . magnetic fields, atmospheric pressure, atmospheric ionization and cosmic radiation [that] fluctuate significantly in the course of space flights." Perhaps more important, Tibbitts noted that developing "a photo-electric sensor" "for more effective leaf monitoring . . . may have a significant use in a flight experiment."[90] For NASA, the controlled-environment setup may have been less important than the new photoelectric sensor drive unit that Tibbitts illustrated in the grant proposal. We also observed that Alford appropriated two of Tibbitts's three illustrations for his article in *Plant Physiology* directly from the original NASA proposal. Tibbitts's experimental apparatus could be automatically deployed during a space mission to conclusively demonstrate that the environment caused leaf movement.

Nor is that the end of this particular story. Tibbitts's NASA grant of 1968 was, in fact, an extension of an even earlier grant awarded by the Biotron Committee in early 1966. Two years before he received the NASA grant, Tibbitts had specifically invoked his "development of a leaf movement experimental system at North American Aviation Company" in, the Biotron Committee noted, his brief application. For Tibbitts, the "leaf movement . . . response provides a distinct and easily measured reaction upon the intact plant that can be monitored for continuous periods without disturbing its growth."[91] Tibbitts's original application sought 181 days of a Biotron chamber merely for its constant conditions, and requested $3,000.

Back in 1966, Tibbitts's original proposal impressed neither Senn, Skoog, nor Burris. After a dismissive review by Skoog and Burris, the Biotron Committee "rejected" Tibbitts's proposal "on the grounds that it does not appear that leaf movement is an appropriate phenomenon to use in the study of diurnal rhythms in space environments."[92] Tibbitts remained committed to work in leaf movements in the Biotron and appealed the decision to the chairman of Madison's Space Science Committee, Robert Alberty. Tibbitts's appeal stressed "the basic science information that can be obtained from [the] study," not to mention reiterating how, "as an outgrowth of my work here at North American Aviation," he had been "asked to act as a biological monitor at Goddard Space Flight Center during the forthcoming biosatellite flight."[93] A month later, Tibbitts received permission and some funding to conduct his experiment in the Biotron.

The explanation of the Biotron Committee's turnaround is also the explanation of the nature of the Biotron as a scientific instrument complex. Skoog believed that Tibbitts's rewritten proposal now planned at least "feasible" experiments. Still, Skoog stressed in his response to Senn, "the ultimate purpose . . . of using bean plants to test the presence or absence of diurnal rhythms in space must be considered as nonsense."[94] Skoog fundamentally denied that leaf movement would occur outside of gravity. For Skoog to give his permission it was necessary for Tibbitts to narrowly confine his research to the development of an experimental setup. Even then, however, Skoog did not consider Tibbitts's experiment a proper use of the Biotron's "expensive facilities." Tibbitts evidently did not require "programmed conditions," merely constant conditions, which by then were readily available in any "commercial chamber." In other words, Skoog understood the Biotron as an instrument that provided a controlled changing environment against which the organism might be tested. His comment about the general availability of simply controlled chambers bespoke a botanical world in which Went's initial vision of a general use of his phytotron was becoming a reality. Rather, in a surprisingly frank admission, Skoog acceded to Tibbitts's proposed work simply because, first, "no more suitable projects are on hand," and second, "if government funds are to be wasted by decree, we are in as good a position to do it as anyone else."[95] Skoog's admission is essentially a pragmatic scientist's vision of the military-industrial complex. Skoog believed NASA was simply wasting money, but if governments carelessly and extravagantly expended funds, then scientists too had as much moral right to benefit as anyone else. In short, the moral economy of science was reshaped in the 1970s by scientists' perception of the 1960s as a profligate decade.

THE NEXT FRONTIER

The Biotron is still there, right along University Avenue, across from the gymnasium, halfway to the towering medical school. The geographic imagery reflects the plight of the facility in contrast to biomedicine and molecular biology, who grew to dominate modern biology far more than nuclear physics ever dominated physics. According to Toby Appel, although there was new technology and vast sums of new money entering biology, it never permitted the appearance of an instrument so unique or expensive throughout the 1960s that it truly functioned as an accelerator or a telescope.[96] In other words, it was not that the phytotronists asked for too much, it was that they asked for *so little* in comparison to their physicist cousins. But there was another difference as well. To a greater extent than its counterparts in Caltech and Australia, from the mid-1960s until the 1990s, corporate-sponsored work remained almost entirely absent from the Biotron. It hosted few projects from biotechnology, or medicine, or agriculture even as the gene jockeys mounted their biotechnology steeds in the late 1970s and as the state withdrew from funding basic science.

The Biotron went into marked decline throughout the 1980s, but never closed. In the late 1980s, Tibbitts was told that a weakness of the Biotron's staff was in its "not trying to help and to educate scientists in using the facility well." Tibbitts, perhaps displaying a moment of frustration, noted "if they [visiting scientists] don't know enough to ask, we shouldn't ~~waste~~ use our valuable time."[97] Two year later still, in 1989, Tibbitts not only had to educate the scientists, he also had to explain the notion of the Biotron to each new incoming administrator. When John Wiley was appointed dean of the Graduate School in mid-1989, Tibbitts made a note for their first meeting: "Biotron what is it"?[98] During their meeting later that day, Wiley made the suggestion, presumably he thought helpfully, to "encourage [Tibbitts] to talk with ________ at Synchrotron over priorities for Biotron use."[99] Exactly forty years after the first phytotron, people still confused the trons of physics and biology.

The Biotron continued to be used for plant research to a far greater extent than for animal science, and encompassed many of the old topics stretching back as far as Went's first phytotron, including pea physiology and floral induction. In a major project stretching from 1984 to 1993 Tibbitts himself worked on "space potatoes" (to which Andy Weir's stranded Martian astronaut probably owed his life) while others looked at "trees, insects, and CO_2" and "CO_2 & tree growth."[100] The potato had been chosen by NASA as one of eight potential plants to be incorporated into evolving conceptions of "bioregenerative life support systems for space bases" under the Controlled Ecological Life Support System (CELSS) program. Potatoes were nutritious, possessed a high yield for their weight and size, but also had several advantages, not least being "palatable and acceptable to most people."[101] As Mary Roach has detailed, since humans started going into space, food has been a constant complaint of astronauts, some of whom preferred starvation for two weeks over anything NASA served up.[102] Tibbitts's team experimented with over twenty cultivars of potato, beginning with the white potato (*Solanum tuberosum* L.), which they found would supply a single person's nutritional requirements in an area of twenty square meters. While they confirmed that certain potato cultivars like the Kennebec required periods of darkness (more than eight hours), many others grew continuously in longer light periods as long as the temperature did not rise about 24°C. At the same time, potatoes were also found to grow at the highest rates in elevated carbon dioxide levels. In short, they concluded, "adequate information is now known to utilize potatoes effectively in space and this crop has distinct advantages that encourage its inclusion in space life support systems to provide food, supply oxygen, purify water, and removal of excess carbon dioxide."[103] This was the mature science of complex life at work.

Consequently, by the 1990s in the Biotron research was conducted essentially in three areas: "space biology, potato biotechnology, and global warming." As in the twin phytotrons, carbon dioxide emerged as a central research topic, attached to environmental research. In addition, however, the animal rooms housed transgenic mice for the entire year, alongside monkeys, aging rats, and hibernating squirrels because back in the mid-1980s, with genetically engineered plants receiving wide attention and entering the testing phase of development, the United States imposed "unrealistically rigid containment requirements" on the company's testing. Here was exactly a requirement the Biotron could fulfill.

New projects needed bioisolation, rather than environmental control, but they paid for the space just the same. Under the legal necessity of containment, not the scientific fixation on control, perhaps the dream of a biology stretching across the entire spectrum from molecules to elephants might finally be realized. While neither NASA nor global climate change realistically offered sufficient funding opportunities to support the costs of running the Biotron, the likely accelerating pace of environmental research, Tibbitts believed, might attract those long-sought-after "Federal agencies" "interested in subsidizing the Biotron."[104] In short, as the physicists at the superconducting super-collider also painfully learned, after the fall of the Berlin Wall, big sciences faltered if they continued doing business only in the traditional Cold War way of appealing to governments for the funding of basic science. Climate change research, in contrast, represented a radical departure from the Cold War era model by necessitating government support of science on a global scale.

CODA II

THE PASSING OF THE AGE OF BIOLOGY

> Progress in crop ecology may even be inhibited by the distractions which phytotrons, like computers, offer to the young research worker and the result is that research stations tend to acquire new hardware faster than they produce new ideas.
>
> —**J. L. Monteith, 1971**

WRITING TO his longtime friend and colleague David Keck from the steamship *Liberté* in May 1952, the famed American botanist William Hiesey offered the subject "The Golden Age in American Botany" to solve Keck's evident block in finding a topic for an upcoming talk. Hiesey told Keck that botany's golden age "is the age that is now ahead of us." According to Hiesey, "U.S. botanists" had passed the "rugged exploratory stages," "inherited rich collections of material, facilities and libraries," and now had "new tools available for further real advances." Though confessing that the "French wine" on board ship had worked him up, Hiesey believed that "for the first time," botanists were in a "strong position to evolve a mature, well-balanced, integrated development of the plant sciences embodying all the results of the efforts from different special fields."[1] Hiesey knew what he was talking about. The very next year, he went out to California and got to work in Frits Went's phytotron, subsequently both coauthoring articles with Went and contributing an entire chapter to Went's foundational book describing the first phytotron. Phytotrons established, scientists like Hiesey said, equivalent experimental control over environments and genes but also established biology as methodologically equivalent to physics.

This book has been about that optimistic and modernist golden age, more or less from 1945 until the mid-1970s. The American National Academy of Sciences called it "The Age of Biology," in which the dramatic discoveries in genetics were paired with the exciting technological advancements in controlled environments and computers. In the Age of Biology many were prepared to pay handsomely to build an experimental science of whole plants in real environments, in Paris and Pasadena as well as in Canberra and the Carolinas. As in the case of most technoscience of the era, costs quickly escalated: while Frits Went's phytotron in Pasadena cost hundreds of thousands of dollars, Folke Skoog's and Howard Senn's Biotron in Wisconsin cost millions, and Pierre Chouard's *le grand phytotron* outside Paris, tens of millions. But such expensive creations were entirely justified, the technologist biologists maintained, because they always aimed to more exactingly specify and measure the biological "environment."

The golden age of biology lasted about three decades. As Harry Highkin feared back in 1959, what became painfully true by the 1970s was that a rhetoric of technology had overtaken the technologist biologists: they came to experiment more on their technology than on their organisms. The rise of technologist biologists and phytotrons coincided with the emergence of troubling questions about modern science's clear obsession with technology. Technocrats and technoscience generally fell under widespread suspicion for waste and pollution, and for being warmongers, as students and hippies connected the Vietnam conflict and the military-industrial complex in their protests. Well before the National Science Foundation (NSF) reoriented its funding priorities away from basic plant science, a broad cultural sea change had already taken place that undermined public confidence in and political support for any massive infrastructure of science and especially the large national laboratories. As the historian Matthew Wisnioski concluded, "by the end of the 1960s, disparate criticisms had coalesced into an interrogation of the technological foundations of modernity. . . . This ideological debate about the nature and control of technology offered a host of concepts—the mega-machine, the technostructure, the technological society—that provided commonality among the countercultural, environmental, civil rights, and antiwar movements."[2] The crisis fully eroded the modernist technological optimism of the 1950s and 1960s, and cyclotrons as much as phytotrons suffered scathing criticisms.[3]

For nearly three decades, the claim that greater experimental control led to better science appeared an effective rallying call for the phytotron. As it was during much of the era, the *rhetoric of technology* became more important than the technology itself.[4] The slow erosion of confidence in simply more and better technology emerged alongside the expansion of the phytotrons, and was ultimately the cause of the general backlash against the *tronning* of biology. By the 1970s, in other words, the technological hubris of phytotrons had become their greatest weakness not their greatest strength. In July 1971, that backlash was keenly felt as the environmental physics pioneer John Lennox Monteith attacked phytotron studies in his introductory remarks at the opening of a growth chamber facility at Glasshouse Crops Research Institute in Littlehampton, England. Instead of revealing the range of relationships between crop and field, Monteith suggested that "experiments in growth rooms or cabinets provide only limited information about plant-weather relations because the physical environment is unrealistically uniform both in time and space."[5] Going against nearly a century of plant physiologists who stressed the primacy of the laboratory over the field, Monteith openly advocated field studies in the face of a conference organized to open a new phytotron. For Monteith, "the availability of extensive and horizontally homogenous stands" in the field allowed both sampling over a significantly reduced population and the dismissal of "edge effects," undercutting one of the core economic arguments for phytotrons since the 1950s.[6]

Moreover, Monteith argued that the enormous sums lavished on phytotrons in the past decade would have been more profitably deployed in "well-designed field experiments linked to complementary programmes of physiological work in the laboratory." Worse still, the phytotron qua laboratory had suffocated thinking in and about the field. Field experiments, Monteith suggested, needed to be better planned and more nuanced but they were not naturally inferior to phytotrons. Indeed, the phytotron merely reversed the imbalance between field and lab, and Monteith thought that each needed the other. Monteith feared that the technologist biologists had gone too far by smothering plant science research programs with technology, and creating a generation of scientists too busy tinkering with circuits instead of tilling plants: "Progress in crop ecology may even be inhibited by the distractions which phytotrons, like computers, offer to the young research worker and the result is that research stations tend to acquire new hardware faster than they produce new ideas," Monteith argued.[7] Still, aware that he was treading on some professional toes, he also disavowed any technological obsession at the new Littlehampton phytotron, or with its founding researcher, Warren Wilson.

At the same time, Monteith observed that the act of specifying and measuring the biological environment had begun to move from the employment of large brute-force technological systems to the widespread use of models and a greater emphasis on computers and statistics. He may have questioned younger people's growing obsessions with computers, but he also recognized that the ability of computers to process larger statistical samples had allowed those young research workers once again to move into the "construction of crop environment 'models'."[8] The old analogy of the phytotron likened to the cyclotron, however, no longer held in the new construction of models. Now, Monteith noted, the frameworks stemmed mostly from "financial or electrical analogies." In models that likened a plant to a business, "a balance sheet is drawn up to show how the income of heat, water or carbon to a plant or to a stand of vegetation is related to the expenditure in terms of convection, evaporation, respiration." On the other hand, in models that likened a plant to a circuit, "fluxes of heat, water vapour and carbon dioxide behave like currents and differences of temperature and concentration are the corresponding potential gradients."[9] Under both analogies, the era of big technical systems was being replaced by an emerging obsession with big data: Monteith saw "young mathematically minded ecologists" being especially overt about utilizing the methods of other statistical sciences. Instead of the experimental certainty laboriously sought after during the Age of Biology, young people now used new technologies to "reach general conclusions about crop behaviour from a limited amount of information," Monteith lamented.[10]

About a decade later, the Age of Biology had firmly ended, certainly at least as biologists like Hiesey, Went, and even Monteith understood its methods and purposes. It ended because the same technological obsession over control and precision that had served to create generations of biological trons also came to stand for what many, including NSF director William Sievers, saw as the problem with giant technoscientific facilities of environmental control. The directors of the remaining large environmentally controlled facilities, Theodore Kozlowski and Robert Downs, continued to press the National Science Foundation for more funds for Madison's Biotron and Duke's phytotron but they met increasing resistance and skepticism in regard to the entire endeavor of phytotronics. It seemed stunning to them that somewhere in the wider biological community, Downs said, were scientists who did not even consider their emphasis on the definition, control, and measurement of "light" as an important biological category: without providing a reference, he commented in a paper from 1980 that "illuminance . . . is periodically denounced for having no biological significance."[11] Such a dismissal was little short of heresy to a phytotronist like Downs, but it still seems incredible that parts of the biological community could be so resistant to the efforts of the technologist biologists that they might deny a role for light in the study of plants.

Changes in funding priorities undermined the maintenance of American phytotrons but also halted any new construction of phytotronic facilities for a generation. Unable to sway patrons back to the study of the controlled environment, the phytotronists' modernist facilities declined at the same time that the Age of Biology was replaced by the Age of Biotech. Still, the story of phytotrons is not yet completed. While this book has focused on the rise of phytotrons after 1945 and mentioned the general decline of most phytotronic facilities by the end of the 1970s, the science never completely disappeared. The science of controlled environments was utterly overshadowed by the rise of genomics and biotechnology through the 1980s and 1990s, when controlled-environment science shrank to represent less than a few percent of biological work measured by the number of papers published and the number of PhDs produced.[12] Where once the National Academy of Sciences equated controlled environments and new ideas on genetics, by the 1990s there was a near-total asymmetry of biological thought and practice with the overwhelming preponderance of genetic manipulation, the rise of biotechnology companies, and the creation of the Human Genome Project. Since the year 2000, however, studies in controlled environments and phytotrons have once more become the approach of significant numbers of new PhD dissertations. The story of controlled-environment biology and even phytotrons, then, is far from complete. Future work might begin back in North Carolina, where phytotronists once helped define the biological environment and laid much of the groundwork for measuring that environment, but also where some plant physiologists changed course to study environmental ecology as concern over the environment mounted through the late 1970s and 1980s. It was one of only a few early phytotrons able to contribute at the beginning to new understandings of the biological effects of climate change. Indeed, the growth of controlled-environment topics in the life sciences suggests that research priorities shifted by the 1990s and into the new millennium. It may well be, as this book concludes, the opening of a new Age of Climate.

CONCLUSION

THE NEW AGE OF CLIMATE

> Climate change has become an existential crisis for the human species.
>
> **—Naomi Klein, 2014**

A LACK of wonder in the novel *The Martian* disturbed the movie critic Dan Kois. "Space, to Mark Watney and the book's other characters," Kois wrote in *Slate* magazine, "isn't vast or unknowable or terrifying or awe-inspiring. Space is merely a series of problems to be solved—different from the problems one faces on Earth due to transmission delay and lack of oxygen, but nonetheless solvable with some math and a little elbow grease."[1] Kois was disappointed that Weir's image of Mars did not have any ineffable quality to be wondered at; perhaps he wanted some sort of all-powerful Force sweeping unexplained through everything. Instead, the astronauts of *The Martian* acted like that generation of optimistic modernist technologists that sent people into outer space and, to distinctly less fanfare, created the phytotrons this book has described. Awe existed for them all in slide rules and computers but mostly in realizing that once a problem is correctly identified and the parameters are known, if there is a solution to be found, a person can, with some math and a little hard work, solve any problem from a better mousetrap to global health. As readers of *The Martian* know, Weir's astronaut derived the solution to the large problem of leaving Mars alive by progressively identifying and solving smaller problems involving surgery, water, food, communication, heat, travel, and power. If any part could not be solved the overall result would be the astronaut staying on Mars dead, which is also incidentally the far simpler alternative requiring no math or effort.

In science, the real problem is to properly state the problem. For plant science in the middle of the twentieth century, the answer to any number of complex biological questions lay in the simultaneous creation of controlled environments for experiments because control over genes seemed assured. Consequently, the proper statement of the problem demanded specifying what exactly one meant by "the environment" in biology as well as what "control" meant to any meaning of that "environment." Further, it meant devising methods to measure the environment and its control. In short, plant scientists became technologist biologists as they sought total experimental control over the environment of the growing plant through air-conditioning and computing. In their evocatively named phytotrons, technologist biologists broke the biological environment into discrete technological systems. They identified the smaller components of the overarching problem of the biological environment including light intensity, light wavelength, light length, wind, humidity, nutrients, air content, root atmosphere, leaf atmosphere, and temperature: in their first laboratory of aluminum and glass in a land of milk and honey, they learned that 21°C was too cold for African violets, 26°C was too warm for gherkins, and that 25.5°C was just right for Idaho potatoes. By adding multiple controlled systems together, the technologist biologists said, plant science had solved the methodological problem of reproducible results and could properly move toward identifying the laws of growth and development in plants. Moreover, only once the science of the living plant was secure could the life sciences begin work at larger scales such as ecosystems or even the entire biosphere. Thus, for over three decades as space became the next frontier and atomic power promised electricity "too cheap to meter," so too did a cornucopian phytotronic era seem near at hand.

One goal of this book has been to recover a time when it was utterly commonplace throughout science and popular culture that the study and control of life at the level of the whole organism required both a science of genes and a science of environments. To drive home that point, I want to highlight how the idea formed the backbone of some of the most influential works of mid-century science fiction. Long before Weir's *Martian*, there was that famous technocratic modernist utopia of Aldous Huxley's 1932 novel *Brave New World*, in which the control of both genes and environments dictated the growth and development of every person and ensured "social stability." Early in Huxley's story a young technician challenges the director of the growth facility to explain why control over genes alone was not sufficient: the Director replied, "'Ass!' . . . 'Hasn't it occurred to you that an Epsilon embryo must have an Epsilon environment as well as an Epsilon heredity?'"[2] The world of Huxley's novel operated on the assumption that control over genes and environments meant that whole social classes were bred and raised from birth to be as interchangeable as the machines they operated; it was a process that created a society in which all people were honestly happy with their place literally from birth: "'Can't you see? Can't you *see*?' [The Director] raised a hand; his expression was solemn. 'Bokanovsky's Process is one of the major instruments of social stability!' . . . Standard men and women; in uniform batches. . . . 'Ninety-six identical twins working ninety-six identical machines!' The voice was almost tremulous with enthusiasm. 'You know where you are. For the first time in history.' [A problem] solved by standard Gammas, unvarying Deltas, uniform Epsilons. . . . The principle of mass production at last applied to biology."[3] Perhaps it was Huxley's world that first gave the philanthropist Harry Earhart the idea to fund Frits Went's controlled-environment laboratory at Caltech in 1946. Possibly it was a spur that convinced Robert Millikan that the problem of population, food, and social and political stability would be solved via a laboratory that studied the environments for any plant to grow and develop to its maximum size; there is no direct evidence either way. Regardless, standards of the genre of science fiction used environmental and genetic control as a central trope for generations: Robert A. Heinlein's *Stranger in a Strange Land* described the hero Smith as "a man by ancestry, [but] a Martian by environment."[4] Likewise, in Ira Levin's masterpiece of suspense writing, *The Boys from Brazil*, the infamous Nazi "doctor" Josef Mengele attempts to duplicate Adolf Hitler's genes and environments. It was not enough, Mengele convinces the Nazi comrade network, that a young Hitler possess the same genetics. The duplicate also had to suffer and experience the same forces at work on his development, notably the death of a petty-bureaucrat father around age ten. "You see, genes aren't the only factor in our ultimate development; I'm sure you know that."[5]

One suspects that what appealed to many of the most prominent science fiction writers as they forged their imagined worlds beyond the twentieth century was the imminent reality of genetic and environmental control even on the largest scales. In an episode of the iconic science fiction show *Star Trek: The Next Generation*, the crew of the Enterprise came into contact with the isolated "Genome Colony," a society sealed for generations inside a biosphere. "You see this is an engineered society," the colony's leader explains, "genetically engineered." But it is not only a genetically engineered society but also an environmentally engineered society: "For the most part we have achieved a fully integrated existence, not just among ourselves, but with our environment. We don't just live here, we're part of our environment. It is part of us. [W]e cannot separate from it without irreparably altering who and what we are."[6]

The true crescendo of the idea that all life is the product of genes and environments appears in Frank Herbert's saga *Dune*, which remains one of the true masterpieces of science fiction. Beginning with a dedication to "dry-land ecologists," *Dune* is the story of House Atreides coming to the desert planet Arrakis, known colloquially as "Dune," and their subsequent adaptation to the challenging environment. Throughout *Dune*, the narrative pivots on the idea that both environment and breeding shape the development of a plant or a person, but also shape a planet. The story line is driven by the demands of life on a planet possessing extremely limited water. In one scene, Herbert dwells on the ambitions of Arrakis's planetary ecologist to create a "Tansley effect," which is described as the process of creating enough plant cover to produce a system of water uptake and release. The details are convincing, as is always the case in the best of the genre, so much so that Herbert provided an explanatory appendix on "The Ecology of Dune." In Herbert's Cold War era imagination, the struggle to change the ecology of Arrakis parallels the larger social and political struggle of the galaxy. At the novel's dramatic conclusion, Paul "Muad'Dib" Atreides takes command of Dune and with it the Galactic Empire, announcing to his comrades that he is the product of ninety generations of "the proper combination of genes and environment to produce the one person their schemes required."[7]

Like the era's fiction, most life sciences built laboratories that controlled many or all parts of the environment to understand the facts of whole plants living in whole environments throughout the twentieth century, with phytotrons as the apex of those efforts. Phytotrons, as the cases in the preceding chapters have demonstrated, started as an optimistic and reductionist effort to understand the environment's role in growth and development as fully as the role of genes. Under the overarching assumption that every organism was the sum of its genes and its environments, plant biology was always both intellectually and practically a dual science of genes and environments. For a time, the phytotronists' unproblematically black-boxed genetics and readily used the stability of inbred plant lines to construct an experimental science of the biological environment, just as the geneticists' unproblematically black-boxed environments. The era's fiction played on the assumed fact that there was no conflict between the study of genes and environments. Likewise, although the founder of the first phytotron, Frits Went, later railed against the domination of molecular biology, he was featured in the *Encyclopedia Britannica Book of the Year for 1949* under the heading "Genetics" as he began a three-century-long study to test the viability of some 120 seed varieties.[8]

Yet a half century later even the word "phytotron" is largely forgotten. The Age of Biology did not become the Phytotronic Era once expected by Pierre Chouard. Though the 1980s saw an explosion of interest in both the environment and computers, the decade also saw many phytotrons shuttered or adapted to other purposes. At least one, Chouard's *le grand phytotron* at Gif-sur-Yvette outside Paris, was converted to serve biology's new master, relabeled as a laboratory for "biologie moléculaire et structurale."[9] The emergence of molecular biology has been rightly celebrated by the history of science, but the bright molecular vision has also created a historical blindness. The very success of recent molecular biology has served to erase a previous conception of biology that understanding the whole organism required an experimental science of environments accompanied by an experimental science of genes. Instead, the conceptualization and measurement of the biological environment shifted to become largely the purview of environmental science and ecology, where the study of the environment always took place at the level of whole ecosystems. However, in the wake of appreciating just how common the duality of genes and environments was throughout most of the twentieth century, the completeness of the erasure should give pause. Certainly, the public needs to keep in mind one lesson from the history of technology, namely, that the technologies of genetics may be seen to be best because they have triumphed, rather than triumphed because they are best. More important, however, without an awareness that biology was once the study of genes and environments it is difficult to understand the immediate and pressing threat that global climate change poses to

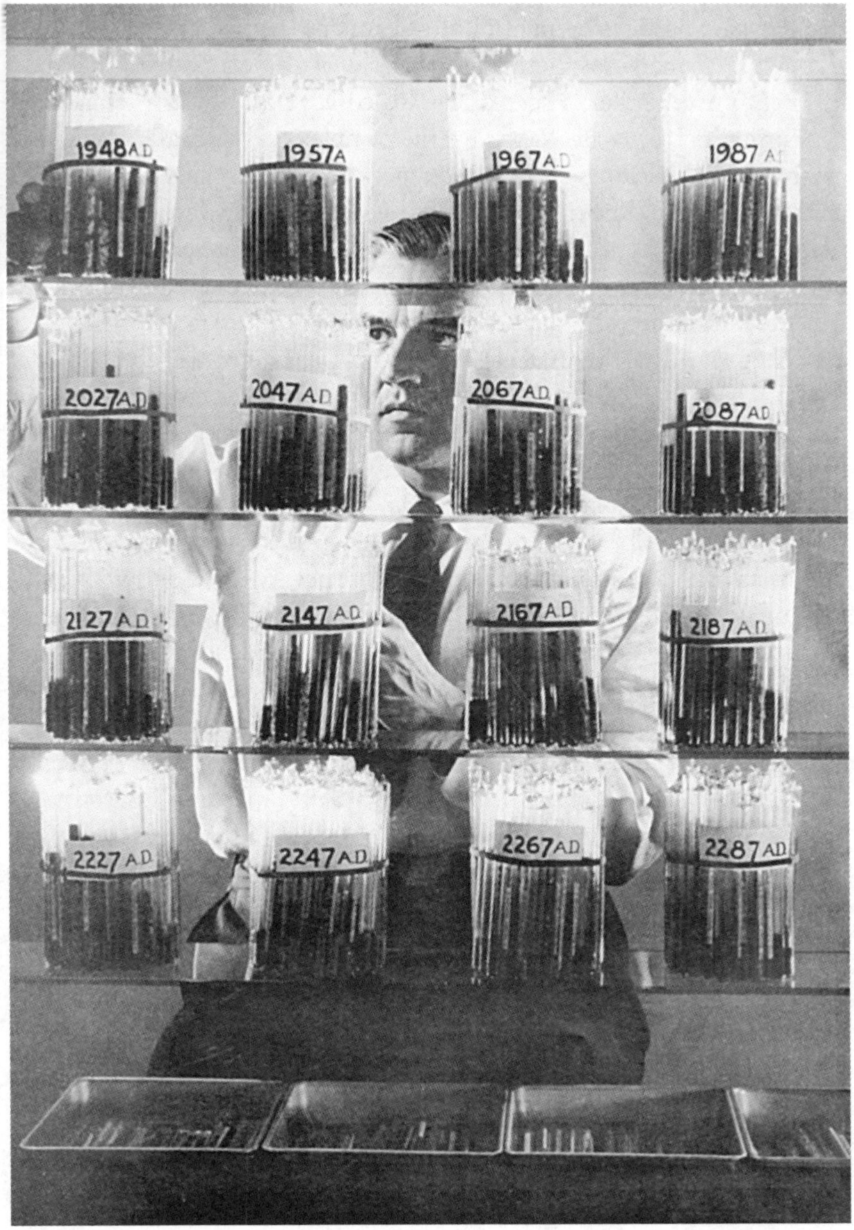

Figure C.1. Frits Went behind the "set-up for a 360-yr research on the genetics and longevity of seeds, begun at . . . Caltech. Twenty sets of dried seeds, each set containing 120 seed varieties, were packed in glass tubes to be opened at specified intervals between 1948 and 2307." *Encyclopedia Britannica Book of the Year for* 1949, under "Genetics."

the growth and development of all organisms, homo sapiens included, trapped in a small biosphere once evocatively termed "Spaceship Earth" by Buckminster Fuller or seen as the "Pale Blue Dot" by Carl Sagan.

Humanity slowly awoke to face the complex and difficult problem of establishing the causes and consequences of a rapidly changing climate around the same time that many facilities specifically devoted to the experimental control of climate went into decline or disappeared. The signs of climate change are everywhere: the glaciers of Greenland are retreating; the butterflies of England are advancing; and some Pacific Islands are disappearing. Dire expectations about the fate of the entire planet abound as ocean fish stocks collapse, deforestation accelerates, and atmospheric carbon dioxide levels rise. Already apparent, environmental shifts are changing the expected growth and development of major crops that underpin the global supply of food now and into the next century. As Lloyd Evans, the former director of the Australian phytotron and later president of the Australian Academy of Science, pointed out, with pressure on agriculture to raise crop yields the most significant challenges for future plant breeders and agronomists are "to adapt major crops and their agronomy to warmer climates, and possibly even more so to any abrupt change to the distinctly cold conditions which they could precipitate within a century, possibly accompanied by much drier tropics."[10] Compounding the problem, mounting urban expansion appropriates more and more arable land because of the tremendous pressure of an inexorably growing human population: between 1986 and 1998, the world population grew to six billion and by the end of 2011, to seven billion. All these people now consume more energy, primarily via burning more fossil fuels in their cars and over-air-conditioned houses.[11] Only the most deluded now disagree with the author and activist Naomi Klein that "climate change has become an existential crisis for the human species."[12]

What better time than now to recall a story about a forgotten biological science of the environment? Indeed, because history is not written for the past but for the present, I want to conclude this book with some provocative thoughts on the consequences of remembering the phytotronists' story. It is by confronting the assumptions of dominant narratives of biology via the recovery of the story of forgotten facilities and research on the biological effects of the environment that we might, to quote E. P. Thompson, "discover insights into social evils which we have yet to cure," most pressingly, climate change.[13]

First, recalling the story of phytotrons challenges the naturalness of the reductionist molecular approach that has come to dominate biology at the very moment humanity needed insight into the specification and measurement of the complex biological environment in the face of an impending climate change catastrophe. The decline of phytotrons through the 1970s and 1980s coincided with the climax of what Edward O. Wilson famously called "the molecular wars," which, after a long bitter struggle begun in the 1950s, resulted in the idea that life was solely the action of molecules. The molecular victory swept across many disciplines of biology from botany to zoology, erasing the environment from biological explanations as it conquered.[14] In triumph like Caesar after Gaul, the codiscoverer of the structure of DNA, James Watson declared, "if you can study life from the level of DNA, you have a real explanation for its processes." This was when he was at the height of his influence, speaking as the head of the pharaonic, Department of Energy–funded, Human Genome Project.[15] The Age of Biology became the Genomic Age in a flourish of hyperbole such as, "no matter how you slice it, the new science of DNA will transform everything [from] medical treatment [to] criminology [to] history. Ethics. Politics. And don't forget the economy."[16] If potentially patentable gene sequences could be exactly matched to diseases or maladies, they would be worth enormous fortunes. Thus, as historian Sally Hughes points out in her history of the first biotechnology company, Genentech, "molecular biology had acquired a patently utilitarian dimension."[17] In the popular mind, the molecular victory was perhaps no more firmly sealed than when Jacob Bronowski pronounced in *The Ascent of Man* that "Life today is controlled by a very few molecules—namely the four bases in DNA. They spell out the message for inheritance in every creature that we know, from a bacterium to an elephant," repeating Jacques Monod's famed phrasing.[18] DNA as an object, an illustration, and a metaphor pervaded Bronowski's widely seen television series being featured in a third of the episodes and compared to gothic arches, clocks, and perspective.[19]

E. O. Wilson credited the discovery of DNA's structure with having "injected into all of biology a new faith in reductionism."[20] Indeed, early molecular biology is now generally regarded as "the most aggressive phase of a [reductionist] tradition that has existed since Descartes declared that animals are only complex machines," argued the historians Peter Bowler and Iwan Morus.[21] That extreme reductionism generated protests from other biologists. Among the early voices to resist pronouncements like Watson's was a one-time colleague of Watson at Caltech in the mid-1950s, none other than Frits Went. Near the height of his influence, Went had ridiculed claims like Watson's "real explanation" as arrant nonsense: as he told a major symposium in 1962, "statements suggesting that if we knew all about DNA and RNA, we would understand the secrets of life, or that molecular biology holds the solution to the problems of biology can only be made by immature minds."[22] At the time, few paid any attention, but doubts about the extreme reductionism of early molecular biology slowly surfaced over the following decade, even among some molecular biologists. One of the field's pioneers, Erwin Chargaff, criticized the easy assumption that properties of molecules could be directly mapped onto organisms: he came to "dislike *E. coli* impersonating nature," he said, "the difference in talents is really too great."[23]

As the plant physiologists of the 1940s, 1950s, and 1960s acknowledged, none of this suggests that molecular biology has not made fantastic progress but rather only insists that a science of life without a corresponding study of environments under controlled conditions remains a job only half done. Early on, Jean Paul Nitsch, in presumably an attack on biochemistry, said that plant physiologists did not want a science of biology "dominated by chemicals out of a bottle,"[24] while later one past president of the American Society of Plant Physiologists rhetorically asked, "If the parts are taken away, what is left of the whole?"[25] Likewise, a number of leading biologists have urged the environment to be restored to prominence in biology. "Molecular biology's obsession with metaphysical reductionism," critiqued the eminent microbiologist and biophysicist Carl Woese in the new century, "stripped the organism from its environment . . . shredded it into parts to the extent that a sense of the whole—the whole cell, the whole multicellular organism, the biosphere—was effectively gone." "Knowing the parts of isolated entities is not enough," Woese said, because the whole remained invisible.[26] Plenty of voices agree, insisting that the understanding of life is incomplete without knowledge of the environment. For Stephen Jay Gould, "genes and environment interact in a non-additive way,"[27] while the maize geneticist and Nobel laureate Barbara McClintock speculated that the instructions for life came "from the entire cell, the organism, perhaps even from the environment."[28] Notably, after having worked in the Caltech phytotron for a number of years, the well-known plant physiologist and activist Arthur

Galston concluded, "even a solution of the molecular mode of action of phytochrome will leave unanswered many crucial questions," when he detailed the previous fifty years of photobiology.[29] Similarly, when Evans surveyed his thirty-year career, he noted the "great progress in plant physiology" that had come, yes, "from research at ever smaller scales of analysis, but when it comes to the performance of crops in the field," he highlighted, "it is important to set these findings in the context of the whole crop and of the whole plant breeding process, where they may have counterintuitive effects, as we have seen for leaf photosynthetic rate."[30]

In other words, while Evans and Galston both acknowledged the monumental triumphs of biology below the cell, they would not accept that all of biology might be reduced to research at ever-smaller scales, not least because feedback operated at the level of whole organisms in complex environments. In much the same vein, Paul Kramer, the founder of the Duke phytotron, noted in his last book with Boyer how "the current interest in research at the molecular level is very important, but should not obscure the fact that there is still need for research on water relations at the whole plant level."[31] Moreover, it seems that the dark years of the 1980s when the environment became invisible may have passed. As the plant physiologist William Laing told me, "it is only recently that the plant molecular biologists are realizing that accurate environmental control is critical to study the subtle effects of gene expression on plants and to unravel the combined small effects of multiple genes on plants."[32]

Second, the story of phytotrons reminds us of the likelihood that realistic responses to climate change will require answers to questions that revolve around defining and measuring the complex effects of changes in climate on biological organisms. As this book showed, where once the phytotronists were also deeply committed to reductionism and to modernity, they came to reluctantly accept that neither was sufficient as an explanatory regime. Actually, the story of the phytotronists is not a critique of reductionism per se, but of any system of knowledge dominated by only one explanatory framework. As the historian of science Andrew Pickering sagely observed about his closely related case, modernity's "ontological monotheism is not turning out to be a pretty sight."[33] Consequently, a greater concern is the continuing devotion of many parts of biology to reductionism. In small part, the sweeping adherence to reductionism saw a decline in other biological disciplines with alternative styles of biological thought including distinct conceptions of organisms and their measurement. While "botany," Erwin Chargaff lamented as early as the 1970s, was "all but banished" in universities, plant physiology also faded, with many in the American Society of Plant Physiology noting the decline in their membership and in the number of subscriptions to their journal.[34] The decline of their field was so serious by the 1980s that some plant physiologists even advocated changes to their flagship journal *Plant Physiology* because it "fails to attract the best papers in a number of emerging fields (plant molecular biology, cell biology)."[35] Even closely related sciences disappeared. For example, Barbara McClintock's biographer, Evelyn Fox Keller, admits to having to "learn about a science [cytogenetics] that, as a student of molecular biology in the early 1960s, [she] had never needed to study."[36]

But beyond the travails of disciplines is a larger concern—that is, to the extent that the science of life is considered primarily molecular and reductionist, those worldviews fundamentally shape how such questions can be addressed. The molecular victory means that it is now primarily the molecular that shapes the answers offered by the biologists who, as the historian of science Scott Gilbert noted, are being asked to be the final arbiters of many pressing questions of late industrial society, from race, gender, and sexuality, to ecology, health, and population, a mantle once worn by the physicists.[37] This lesson was forcefully made by the historian activists Naomi Oreskes and Eric Conway, who argued from the future that an emphasis on reductionism "made it difficult for scientists to articulate the threat posed by climate change."[38]

In some small way, recovering the story of phytotrons forces us to see that the environment as a significant concept of study for biological science still remains largely invisible even to those calling for radical solutions. Klein, for instance, noted the World Bank's conclusion that the world was "on track [to be] 4°C warmer," before ominously saying, "We don't know what a 4 degrees Celsius world would look like, but even the best-case scenario is likely to be calamitous."[39] Part of Klein's pessimism is linked to the erasure of alternative styles of science that might provide alternative nonreductionist solutions. It offers only cold comfort to realize that over the past century, plant scientists in controlled environments like phytotrons actually learned a tremendous amount about the wide-ranging effects not only of temperature but also of many environmental variables on growth and development. In 1950, even a lowly graduate student could have readily told Klein that a 4°C change in temperature will cause gherkin flowers to change sex, while a 6°C increase will yield no flowers at all. The future of life certainly hangs in that fine a balance, but apocalyptic pessimism is justified only if we remain in a genocentric world.

A NEW HOPE

The Genomic Age must become the Age of Climate. As this book showed, the technologist biologists using the first phytotrons in the 1950s and 1960s began to appreciate the difficulty of studying mutually dependent processes and systems. At the same moment that the molecular biologists became assured of reductionist explanations from genes, the plant scientists in phytotrons began doubting the possibility of reductionism from either genes or environments. Always seeking a greater range for each climatic variable and greater precision of control, the phytotronists unexpectedly discovered that their environmental systems possessed feedback, and that the whole amalgam of systems required to make whole environments was at least complex and perhaps even chaotic. Consequently, the controlled study of the environment transformed into those complex environmental and ecological sciences possessing understandings of systems, an appreciation that organisms are feedback loops of genes and environments, and that the watchword for modern capitalism, development, and biology, namely, "growth," was not governed by any single set of instructions but "is the integrated end product of many physiological processes."[40] In short, the old formula that had long guided the study of biology, "phenotype = genotype + environment," turned out to be some complex function p = *f*(g,e) even as some molecular biologists insisted that every organism's fate lay in its genes (i.e., phenotype = genotype) and the financial and scientific support for the further technological development of phytotrons evaporated.

Hope has come with new phytotrons at the forefront of research over the changing environment's impact on living things, taking up the challenge of engaging with the complex system of the biological environment. The Europeans have taken the lead in building several new phytotrons.[41] Also taking up the challenge of studying the whole plant and the whole environment, the University of Saskatchewan committed to a multimillion-dollar investment in a phytotron in 2011. In a similar example, when I visited the Plant Research Laboratory at Michigan State University in

Figure C.2. Room for growth in the Michigan State University Plant Research Laboratory growth chamber room. The structure on the right is a modern environmental control growth chamber unit. Author's photo.

2012, it had several rooms prepared for more banks of environmentally controlled cabinets, and its greenhouse manager Jim Klug told me the laboratory had no trouble finding funding and experimenters.

More promising still is a British facility that "is unique among controlled environmental facilities in that it attempts to construct, maintain and manipulate entire model ecosystems and simultaneously monitor population dynamics and ecosystem processes."[42] Inevitably, it is called the Ecotron. Situated at Silwood Park outside London, and attached to Imperial College, the Ecotron was built in 1991. The facility owes its existence to the United Kingdom's National Environmental Research Council's Center for Population Biology, which opened in 1989. The Ecotron contains sixteen, 8 cubic-meter "environmental chambers," each electronically monitored for ranges of environmental variables including light, water, air, air flow, CO_2, humidity, and temperature.[43] The facility has two distinct types of environmental chambers, eight on the upper level where sunlight is available, and eight on the lower level where light is fully supplied by technological means. Likewise, opened in 1998, the University of Joensuu in Finland now hosts research into plant root and soil systems via their dasotron. Both the British Ecotron and the Finnish dasotron are charged with "ecological and physiological research" on plants and ecosystems because of the "global human impact" on the environment.[44]

The creators of the Ecotron, the dasotron, and new phytotrons all acknowledge that understanding the global human impact on the environment is a complex problem. Defining and measuring the impact involves recognizing the multiple feedback loops at work. For example, the Ecotron explicitly studied eco-*systems* to understand complex relationships between the genetic, the environmental, and the ecological. It is the mutual interactions at all levels that are now part of the measure and meaning of "biology." New facilities like ecotrons continue the story of phytotrons. They evince the continuing development of biological thought and practice, notably that how to even work the problem of defining and measuring the complexity of the biological is still the problem. As the genetic, the environmental, and the ecological parts of biology thus unite, recent biology seems to hark back to a claim of Frank Herbert's planetary ecologist Pardot Kynes in seeing organisms and ecosystems as fundamentally systems: "The thing that the ecologically illiterate don't realize about an ecosystem is that it's a system. A system!" In short, a proper definition and measurement of the problem of the impact of climate change requires the study of weather systems, ocean systems, biological systems, and even human social systems. It is a tough problem not least because humans themselves have to make the necessary gestalt switch from seeing themselves as the most important being on this planet to accepting that homo sapiens are but one organism among many trapped and interconnected in a small biosphere.

Hope also stems from the reintroduction of the environment into the study of whole organisms. Of course in many parts of biology, the environment has always been a focus, especially at the level of the ecosystem, where environmental science and ecology have done tremendous work. However, sixty years after those early postwar plant and animal scientists working in controlled environmental conditions and readers of science fiction said that the controlled study of genes *and* environments was the only way toward proper study at the level of the whole organism, the environment component of the biologist's equation—even for Went's model organism, the tomato—remains a mystery. Among other results, that mystery has condemned generations of people to eating "plastic junk" tomatoes.[45] The tomato was one of the first organisms studied in Went's proto-phytotron, the Clarke greenhouses, at Caltech. It was the organism that led the Campbell Soup Company to fund an entire new wing of Caltech's phytotron in the late 1950s for the development of varieties for the hot and dry American South. In the early 1990s, the study of tomatoes became a now famous part of the molecular understanding of life when the "Flavr Savr" tomato became the first genetically engineered food approved in the United States. The Flavr Savr heralded acres of perfect looking but utterly flavorless tomatoes, consumers now remember.[46] As Barry Estabrook's exposé, *Tomatoland*, highlighted, the scientific and industrial organization of the tomato industry, its control and regulation in Florida by the conspiratorially named Florida Tomato Committee, and its gangs of pseudo-slave labor produced picture-perfect, readily packaged, and tasteless tomatoes year round. Pointedly, Estabrook was told, the large multimillion-dollar tomato industry in the United States is dominated by research from geneticists concentrated on eradicating diseases because "disease resistances tend to be simply inherited. For the most part, you are talking about single dominant genes that are fairly easy for geneticists and breeders to work with."[47] Disease was geneticists' overriding interest for years, and its successes illustrative of the reductionist molecular biological culture predominant in medicine and agribusiness.

Here may be exactly what a biological science of genes over environments produces, namely, disease-free but also taste-free fruit. In contrast, "breeding for something like increased yield or improved flavor involves multiple genes, so it is harder for researchers," noted Roger Chetelat, the director of the C. M. Rick Tomato Genetics Resource Center at the University of California, Davis.[48] In fact, by considering the environment, substantial progress has also been made on the mundane task of growing a decent-tasting tomato, notably by John Warner Scott, a professor of horticultural science, who released another tastier tomato called the "Tasti-Lee." As Scott stressed to Estabrook, genetics was certainly part of the story of the Tasti-Lee but "there is no easy way to breed for taste. It's not like there's one genetic marker that tomatoes must have to taste good." In fact, even if genetics produces a reliable product, the environment still plays havoc. According to Estabrook, "even if all else goes [according] to plan, a tomato can lose its taste if exposed to cold temperatures at any time between harvest and being eaten, after which point it can never recover it. Crop specialists even have a scientific term for this process: 'chilling injury.'" While the process has been labeled, the reasons for why "chilling reduced the fragrant volatile chemicals that are all-important in giving the fruit its distinctive flavor" remain "unknown." Indeed, without knowledge of the growing and developing environment, "years of effort by a plant breeder can be destroyed by a few days in a refrigerator."[49] The future, then, is that after a generation of perfect-looking tomatoes courtesy of only genetics, the addition of odd environments offers up strangely misshapen, but tasty fruit.

In conclusion, I have come to regard the story of phytotrons as important because without it the two great assumptions of molecular biology—its reductionism and its elimination of the environment as an experimental variable—go further unquestioned and appear natural to scientists, their patrons, and their historians. To address climate change or bring tang back to tomatoes requires research investigating the causes of plants' growth and development in the opaque and complex interactions between genes and environments. It is time to recall the work of Frits Went, Harold Senn, Jean Paul Nitsch, Paul Kramer, Robert Downs, and Lloyd Evans, who designed systems of experimentation on whole organisms in biological environments. Their hard-won but crucial insights into the mechanisms of feedback in complex organic and technological systems are more relevant than ever, especially to the existential threat of climate change. By remembering those efforts to define and measure an organism as the product of its genes and environments we may help sway social and political forces into action to address the now commonly accepted fate of an increase in global temperatures over the next fifty years.

Our age, scholar McKenzie Wark powerfully noted, is the Anthropocene where the rifts have become planetary, where there are no longer only state problems or realistic national solutions.[50] Wark points out urgently that "addressing the Anthropocene is not something to leave in the hands of those in charge, given just how badly the ruling class of our time has mishandled this end of pre-history, this firstly scientific and now belatedly cultural discovery that we live in a biosphere in a state of advanced metabolic rift."[51] To date, the oil industry has declared to its investors that they do "not see governments taking . . . steps" to address even rises of 2°C.[52] Needed solutions include imposing a harsher regulatory environment for carbon emissions, acknowledging that human life is no more or less valuable than any other life, and creating awareness of the interconnectedness of all parts of the biosphere. We need to think postcolonially: there is no "waste" opposite "goods," both are only material moved and changed. We need to remember systems such as the Algatron for algae waste management that lost out to the reductionist biomedical waste disposal system of the "fecal bag" for the Apollo program. Even at the embryonic stage in 1962 the Algatron was built with the understanding that humans and their waste existed within an ecological system as algae and its waste. It took remarkable insight to see that while people breathe oxygen and excrete urea and CO_2, the algae breathe CO_2, ingest urea, and excrete water and oxygen. A pair of sanitary engineers, William Oswald and Clarence Golueke, built their tron to mediate between the biological functions of each species, and that each species is connected via and to the waste of the other. They understood that the closed environment of a space capsule was itself just a microcosm of the closed system of the earth's biosphere: the space capsule was not just like earth, it was earth, an "ecological system," Oswald and Golueke said explicitly, that it was really "a miniature version of the grand scale terrestrial ecological system of which we are apart, in that the basic principles of the two systems are the same, only the size and variety of their constituents differ."[53]

To similar ends, I hope this book has fostered an appreciation of previous work investigating the relationships between genes and environments. Moreover, now rearmed with new phytotrons and ecotrons as well as powerful tools of genetic manipulation, we may yet come to understand the complex biological effects of climate change. It is beyond the purview of this book, which is a work of history, to judge whether this new role for trons will be more successful than the role they played in their previous incarnations, but as one organism in this biosphere, I can hope.

APPENDICES

APPENDIX I. CHEMICAL SYMBOLS

2,4-D	2,4-dichlorophenoxyacetic acid
CO_2	Carbon dioxide
C_2H_4	Ethylene
$C_6H_{10}O_5$	Glucose
$C_{10}H_9NO_2$	IAA, or indole-3-acetic acid
$C_{18}H_{32}O_5$	Auxin a, or Trihydroxy-monocarboxylic acid [imaginary]
H_2O	Water

APPENDIX II. PHYTOTRONIC UNITS OF ILLUMINATION

ft-c	Foot-candle, or "British standard candle, which was made of spermaceti, weighed 1/6 lb, measured 1 inch in diameter, and had a wick made to burn at the rate of 120 grains per hour
Lux = m-c	Meter-candle, or lux
hlx	Hectolux = 9.29 ft-c = 1.733 microeinsteins/m^2/sec
$\mu mol.s^{-1}m^{-2}$	Micromoles per second per square meter
nE	Nano-Einstein = 6.02×10^4 photons of light
$MJ.m^{-2}d^{-1}$	Megajoules per square meter per day

APPENDIX III. BOTANICAL TERMS

Arabidopsis thaliana	Thale cress
Avena	Oats
Bauhinia glabra	"Monkey ladder" or gut vine
Cattleya	Orchids
Cestrum nocturnum	Night-blooming jasmine
Cucumis anguria	Small gherkin
Drosophila melanogaster	Fruit fly
Dysphyma	Round-leave pigface (Cactus)
Eucalyptus globulus	Australian blue gum
Hevea brasiliensis	Rubber tree
Lycopersicon esculentum	Tomato
Lolium temulentum	Darnel, or cockle ryegrass
Nicotiana	Tobacco
Oenothera	Primrose; notably Lamarck's evening primrose (*Oenothera lamarckiana*)
Phaseolus angularis	Adzuki bean
Pinus oocarpa	Mexican yellow pine
Poa	Meadow grass; notably Kentucky bluegrass (*Poa pratensis*)
Saccharium officinale	Sugarcane
Saintpaulia	African violet

NOTES

ACKNOWLEDGMENTS

1. Harland, "Changing the Behaviour of Plants," 30.

PRELUDE

Epigraph: TRON: Legacy, directed by Joseph Kosinski, written by Edward Kitsis and Adam Horowitz (Disney, 2010).

1. Waisel, Eschel, and Kafkai, *Plant Roots,* 299–300. Finér et al., "The Joensuu Dasotrons; Inoue, et al., "The 'Assimitron.'"

2. Guerlac, *RADAR in World War II,* 420–21. Buderi, *The Invention that Changed the World,* 28.

3. "The Cosmotron: It Works," *Nucleonics* (July 1952), 34–35.

4. Proctor, "'-Logos,' 'Ismos,' and 'Ikos,'" 290, 291.

5. "The Eggatron."

6. In winter 2016, Australia faced a shortage of "free-range" eggs—"free-range" being defined as hens that spent substantial time outdoors—because unseasonably cold weather had reduced their laying. Even more appropos to the story of controlled-environment biology is that, as the *Sydney Morning Herald* noted, the egg supply from caged hens was unaffected because "hens in cages can be manipulated to lay all the time with artificial light." Liam Mannix, "Free-Range Egg Shortage Set to Last Two Months," *Sydney Morning Herald,* June 7, 2016. http://www.smh.com.au/business/freerange-egg-shortage-in-australia-set-to-last-two-months-20160607-gpdfog.html (accessed July 9, 2016).

7. Thanks to Kärin Nickelsen for introducing me to the Algatron. A full-scale model was produced under a research contract from the Air Force Cambridge Research Laboratories with William J. Oswald, Clarence G. Golueke, and Donald O. Horning around 1965. See Golueke and Oswald, "The Algatron"; Oswald, Golueke, and Horning, "Closed Ecological Systems," 45; Golueke, Oswald, and Gee, "A Study of Fundamental Factors," 1.

8. Oswald, Golueke, and Horning, "Closed Ecological Systems," 23.

9. Roach, *Packing for Mars,* 299. The mortifying process of using the fecal bag in space is best described in Russell Schweikart, "There Ain't No Graceful Way: Urination and Defecation in Zero-G," interview with Peter Warshall, 1976. From http://settlement.arc.nasa.gov/ (accessed June 30, 2016).

10. MacKenzie, *Knowing Machines,* 7. Looking at failure or just the roads not taken is one cure for the too easy assumption that successful technologies were somehow inevitable or natural. See McCray, *The Visioneers,* 19; Cowan, *More Work for Mother,* ch. 5.

11. Salisbury, Gitelson, and Lisovsky, "BIOS-3."

12. http://www.bbc.com/news/technology-30081453 (accessed November 18, 2014). I swear I did not put them up to it!

13. Eco, *The Name of the Rose.*

14. My thanks to Bruce Hunt for the background on the naming of elementary particles.

15. Darwin, "Use of the Termination -tron." My thanks to Gail Schmidt for this. One of the early researchers on mesotrons, Bruno Rossi, concluded his recollections by fondly remembering the field of early particle physics before "the big accelerators." He felt a "lingering nostalgia" for what he called "the age of innocence of experimental particle physics." Rossi, "The Decay of 'Mesotrons,'" 204. It was Robert Millikan's own textbook that introduced many students to mesotrons, see Kargon, *The Rise of Robert Millikan,* 158, and it was Robert Millikan who helped popularize the name *phytotron,* as we shall see in chapter 1. Important background is in Walker and Slack, "Who Named the –ON's?"

16. My thanks to Matt Wisnioski for several of these: for the mellotron, see http://en.wikipedia.org/wiki/Mellotron; for the Detectron, see the extensive historical file at http://national-radiation-instrument-catalog.com/new_page_8.htm. The Mu-Tron is at http://en.wikipedia.org/wiki/Mu-Tron; for Unitron, see advertisements in *Sky and Telescope* 14, no. 2 (1955). For the Accutron, see http://www.accutron214.com/Accutron History.htm, but for the Jumbotron, go to a sporting event.

17. The Metatron is "the Voice of God. But not the *voice* of God. An entity in its own right. Rather like a Presidential spokesman." Pratchett and Gaiman, *Good Omens,* 246. Alan Rickman played the Metatron in *Dogma,* directed and written by Kevin Smith (View Askew Productions, 1999). *Tron,* directed and written by Steven Lisberger (Disney, 1982). *Tron: Legacy,* directed by Joseph Kosinski, written by Edward Kitsis et.al. (Disney, 2010). *Sleeper,* directed by Woody Allen, written by Woody Allen and Marshall Brickman (Rollins-Joffe Productions, 1973). For *Voltron,* see http://www.nytimes.com/2010/06/11/arts/design/11keefe.html?ref=obituaries&_r=0; https://en.wikipedia.org/wiki/Gravitron.

18. For Morris, his Interrotron "creates greater distance *and* greater intimacy . . . it creates the *true first person,*" he said in an interview in 2004 with *FML Magazine,* "now when people make eye contact with me, it can be preserved on film." First used in his film *Fast, Cheap and Out of Control,* the Interrotron was named by Morris's wife, Julia Sheehan, "because it combined two important concepts—terror and interview." http://www.errolmorris.com/content/eyecontact/interrotron.html (accessed September 30, 2015). For a person seeking the truth behind the person sitting in front of him, it appears that the use of the -tron suffix is entirely natural, arising from his earliest exposure to cinema and comics. http://www.rogerebert.com/festivals-and-awards/errol-morris-megatron-son-of-interrotron (accessed September 30, 2015).

19. Advertising brochure for Phototron 2 found in "Phytotron vertical file," LuEster T. Mertz Library of the New York Botanical Garden, Bronx, New York. Quotes from Daniel S. Janik and Jeffery J. deMarco, "Engineering Testbed for Biological Water/Air Reclamation and Recycling," SAE Technical Paper Series, 901231, July 1990, p. 2. From "Phytotron vertical file."

20. www.phototron.com (accessed March 2014).

21. Janik and DeMarco, "Engineering Testbed," 5, 1. From "Phytotron vertical file."

INTRODUCTION

Epigraph: Went, *The Experimental Control of Plant Growth*, 2.

1. Weir, *The Martian*, 21.

2. Kingsbury, *Hybrid*; Phillips and Kingsland, *New Perspectives on the History of Life Sciences.*

3. Alberda, "On the Use of Plants," 591.

4. Weir, *The Martian*, 150, 71.

5. von Wettstein, *The Phytotron in Stockholm*, 3.

6. Pickering, *The Cybernetic Brain*, 7, 15. But the entirety of chapters 1 and 2 has deeply shaped my thinking about phytotrons and their era.

7. Nitsch, "A Critical Comparison," 17.

8. That image continued even into later incarnations, such as at the Biotron Institute in Japan, for example, where "nondestructive digital image processing" obtained "plant growth information for feedback to adjust the environmental conditions during computer control of the growth processes," as one later phytotronist celebrated. Downs, "Phytotrons," 465. When I visited the Michigan State University Plant Research Laboratory, they were in the process of expanding their computer systems. Individual computer units attached to each growth chamber were being centrally linked to a hub in the facilities manager's office to permit researchers to remotely monitor, or change, the environmental conditions. My thanks to Jim Klug for taking the time to show me his facility.

9. Gilles Deleuze observed that modern "societies of control" possess social norms that are both controlled by and have created "computers." Deleuze, "Postscript on the Societies of Control," 6; Edwards, *The Closed World*; Akera, *Calculating a Natural World.*

10. Wettstein, *The Phytotron in Stockholm*, 5.

11. Murphy, *Sick Building Syndrome*, ch. 1. Sheller, *Aluminum Dreams*, ch. 4. First generation *Gemini* spacecraft possessed only the bare minimum of systems for sustaining life. Apollo craft were better, but in reality says the official National Aeronautics and Space Administration history, "quite a lot of minor inconvenience could be tolerated by a man on his way to the moon." Only with the planning for Skylab in the late 1960s did any integrated thinking about habitability in space take place. Compton and Benson, *Living and Working in Space*, ch. 7.

12. Anker, *From Bauhaus to Ecohouse*, 69.

13. Berkner, *The Scientific Age*, 15.

14. Le Corbusier, *Towards a New Architecture*, 3, 57, 95. For Le Corbusier in architectural context, see Anker, *From Bauhaus to Ecohouse*, and in artistic context see Gay, *Modernism*, 305.

15. Chouard, "Phytotronics," 5. For the French CNRS phytotron at Gif-sur-Yvette, see P. Chouard, "Introduction." For a brief biography of Chouard, see Champagnat, "Pierre Chouard."

16. Kevles and Geison, "The Experimental Life Sciences"; Kohler, *Lords of the Fly.* On the overlapping physical and biological electrical and fluid experiments on the cell, see Stadler, "Models, the Cell, and the Reformations." See also Allen, "Naturalists and Experimentalists." While "turn of the century plant taxonomists [and] plant morphologists reacted unfavorably to the newer science of genetics and its experi-

mental approach" (Smocovitis, "Botany and the Evolutionary Synthesis," 191), biology in general was overtaken by the drive toward experiment. In contrast to molecular biology, where "different experimental systems often centered around a particular technology" (Rheinberger, "Internationalism," 252), early twentieth-century botanists embraced an experimentalism that did not center on any novel technology like an electron microscope, ultracentrifuge, radioisotope, or cellular equivalence circuits. See Hagen, "Experimentalists and Naturalists." An example of an overt experimental program in plant genetics can be found in Smocovitis, "The 'Plant *Drosophila*'"; Harwood, "Comments on Experimentation."

17. Parolini, "Charting the History."

18. Lawrence, "The Glasshouse," 129.

19. American Greenhouse Manufacturing Company, *American Greenhouses* (AGMCC, 1928), 5.

20. Lawrence, "The Glasshouse," 129, 133. Putting such claims together with a similar remark from the world of genetics, namely, that the production of new mutations via the chemical colchicine permitted plant breeders like David Burpee and William Hoag to "not just pick up sports, [but to] manufactur[e] new plant[s] scientifically," implies that the grander ambition for many biological scientists was to make "biology" into what they understood as a science complete with experiments, planning, new technologies, and a systematization of both methodology and outcomes, as opposed to "old established catch-as-catch-can methods." Curry, "Accelerating Evolution," 252. I think that geneticists and phytotronists were each engaged in the "engineering life," following Curry's lead (p. 10).

21. Pickering, *The Mangle of Practice*, 104–9. Likewise, the early philosopher of biology, F. S. Bodenheimer, dwelled at length on a "series of experiments" made under "apparently 'equal' conditions" by E. Roubaud that showed how experimental practices undermined the "assumption of an equal temperature" because thermostats were turned off and on, and cultures removed at length from one environment for cleaning. See Bodenheimer, *Studies in Biology*, 66.

22. Coen, "Living Precisely," 498; Nickelsen, "Growth, Development and Regeneration."

23. Crocker, *Growth of Plants*, 286. As Crocker noted, one of the limitations of their first constant-condition greenhouses was that the "lighting made it difficult to air-condition the room." This feature was among those directly addressed in phytotrons. While Garner and Allard made the phenomenon of *photoperiodism* apparent in plant science, the experiments of Frits Went suggested the dominance of the night period rather than the day period in flowering. The original paper is Garner and Allard, "Effect of Relative Length of Day." See also Sage, *Pigment of the Imagination*, ch. 1. Roll-Hansen, *The Lysenko Effect*, 28–32.

24. Rheinberger, *An Epistemology of the Concrete*, 131.

25. O. H. Frankel, "Report on a Visit to the USA May 3–Aug 3, 1955, under the auspices of a grant from the Carnegie Corporation of New York." Copy in James Bonner papers, file "Australia," p. 34. Other early controlled environment work as well as some experimental ecology are mentioned in Craig, *Centennial History of the Carnegie Institution*.

26. von Wettstein, *The Phytotron in Stockholm*, 3. Costing 4.5 million krona, the Swedish phytotron was financed by the Rockefeller Foundation, the Cellulose Industry Foundation, and even by the companies that made the air-conditioning system and the computer.

27. Somerville, "The Twentieth Century Trajectory"; Weevers, *Fifty Years of Plant Physiology*; Galston, "Plant Photobiology"; Pennazio, "Mineral Nutrition of Plants." Although I see no evidence that the term *plant science* was debated or questioned, at least by the plant physiologists—the Society of Plant Physiologists' official history said that the society was created in 1923 "from a desire to give plant physiology identity and recognition as a distinct branch of plant science" (Hanson, *History of the American Society*, 57)—the botanists have certainly debated such labels (Smocovitis, "Disciplining Botany").

28. National Council for Research and Development, *A Phytotron in Israel*, 2.

29. Roll-Hansen, *The Lysenko Effect*, 31. Roll-Hansen quotes the famed Soviet geneticist Nikolai Vavilov, who himself quoted the noted British cotton breeder S. C. Harland at a 1935 meeting of the Lenin Academy of Agricultural Science to the effect that "vernalization is the third greatest achievement in world science" (173). Twenty years later, Harland would write in the *New Scientist*, "the phytotron is to botany and agriculture what the radio telescope is to astronomy" (Harland, "Changing the Behaviour of Plants," 30).

30. Sage, *Pigment of the Imagination*, 119.

31. Rudd-Jones, "Controlled Environment," 1.

32. Went, *The Experimental Control of Plant Growth*, 110.

33. Bonner, "Summary and Observations," 215.

34. de Chadarevian, "Laboratory Science," 39–40; Kingsland, "Frits Went's Atomic Age Greenhouse"; Schürch, "Der Schlüssel zum Pflanzenwachstum."

35. Went, "The Effects of Climate," 57.

36. Went, "The Response of Plants to Climate," 492, fig. 4.

37. Chouard and Nitsch, *The World of Plants*, 97, 103.

38. Handwritten notes, n.d. (ca. 1951). Frits Went papers, Record Group 3/2/6/1, box 11, folder 38.

39. Rasmussen, *Picture Control*; Creager, *Life Atomic*, 4, 236, 391.

40. Galison, *Image and Logic*, xviii.

41. Belasco, "Algae Burgers," 621.

42. Bonner, "Fundamental Plant Physiology Research," 73.

43. Evans, Wardlaw, and King, "Plants and Environment," 207. Also Harwood, "Did Mendelism Transform Plant Breeding?" 362.

44. von Wettstein, *The Phytotron in Stockholm*, 7.

45. Frankel, "The IRRI Phytotron." As N. C. Brady's foreword pointed out, rice is grown in markedly varied conditions from the cool conditions in Japan and India, drought conditions in Africa, in dryland conditions in Australia, and under three meters of water in Thailand and Vietnam. See Brady, "Foreword." The case of the IRRI phytotron seems almost entirely unknown, though its Australian background adds to several major works concerning the United States and the Green Revolution, including Cullather, *The Hungry World* and Perkins, *Geopolitics and the Green Revolution*.

46. See Pauly, *Controlling Life*, 199.

47. Appel, *Shaping Biology*, 1.

48. Evans, *Feeding the Ten Billion*, 222.

49. Note David Kaiser's important point that probably the vast majority of scientists and engineers in the military-industrial complex see their profession as a job rather than a calling. Kaiser, "The Postwar Suburbanization." The vast budgets of research and development of weapons systems under the military-industrial complex began, of course, with the $2 billion spent developing the atomic bomb. Rhodes, *The Making of the Atomic Bomb*. They continued with the dreams of atomic-built harbors and canals at a cost of around $770 million (see Kirsch, *Proving Grounds*, 6); and Sugar Grove, which was designed to be the world's biggest bug but even by 1962 was a huge $50 million boondoggle (see Bamford, *The Puzzle Palace*, 217–220). Cloud seeding for rainfall was a large applied science project of the 1950s, subsequently developed by the United States military under "Operation Popeye" to wash out North Vietnamese supply routes during monsoons, while "Operation Stormfury" aimed to steer or extinguish tropical cyclones, to the estimated cost of $150 million. See Edwards, *A Vast Machine*, 359; For the development of the ICBM, see Hughes, *Rescuing Prometheus*, ch. 3. Hughes later noted that even figures on the order of hundreds of millions of 1960s dollars are still an order of magnitude below the defense contracts given to aerospace industries: Lockheed alone was awarded over $10 *billion* between 1960 and 1967. Hughes, *Human-Built World*, 81.

Gore Vidal was undoubtedly correct when he declared that the aim of the military-industrial complex "was to continue pumping federal money into companies like Boeing and Lockheed and keep the Pentagon full of generals and admirals while filling the pork barrels of congressmen who annually gave the Pentagon whatever it asked for, *with* the proviso that key military installations and contracts be allocated to the home districts of senior congressmen" (Vidal, "The State of the Union," 274). Nor has the end of the Cold War halted such expensive development of scientific projects for the military. Ongoing, for example, is the $12 million per year research budget for the speculative Hafnium grenade, a "miniature [atomic] bomb. Explosive yield, 2 kilotons. Size, five-inch diameter." See Weinberger, *Imaginary Weapons*, 10. Likewise, the MIT Institute for Soldier Nanotechnologies, which was "kick-started" by a comparatively paltry $50 million from the United States Army Research Office in 2002. See Milburn, "Nanowarriors." Both pale before the $55*billion* contract awarded to Grumman in October 2015 to develop and build the next-generation B-3 bomber. "Cruise Control," *Economist*, January 23, 2016, 22.

50. Quote from Nelkin, *The University and Military Research*, 75. A significant debate about "how the Cold War shaped and altered the trajectories of science and existing technologies," has long emphasized the emergence of the military as a patron of science. Oreskes, "Introduction," 3.

51. Appel, *Shaping Biology*. All the essays in Oreskes and Krige, *Science and Technology in the Global Cold War* speak directly to the ongoing debate that began with Paul Forman, who saw science as corrupted by the complex in response to Daniel Kevles's earlier claims that physicists, at least, maintained their autonomy in the face of the demands of the national security state and funding beyond the dreams of Solomon.

Forman, "Behind Quantum Electronics" directly challenged the narrative of Kevles, *The Physicists.*

52. Rader, "Alexander Hollaender's Postwar Vision"; Creager, "Nuclear Energy"; Curry, "Accelerating Evolution," ch. 6.

53. Smith, *A Peril and a Hope,* 77; Nelkin and Lindee, *The DNA Mystique,* 143.

54. DuBridge, "Basic Research."

55. Alice Kimball Smith's early work on the immediate postwar scientists' movement notably highlighted the contravening of assumed norms of scientist's moral economy, especially openness: "Work on secret projects [like the atomic bomb] had violated one of the most cherished articles of [scientists'] unwritten code" (Smith, *A Peril and a Hope,* 82). The foundational work is in Forman, "Behind Quantum Electronics" and Roland, *The Military-Industrial Complex.* Much of the debate is summarized in Galison, "Ten Problems," 114.

56. Rasmussen, *Gene Jockeys,* 25.

57. Rasmussen, *Gene Jockeys,* 26. The rise of genetics as an anticommunist endeavor was also connected to the substantial revision of American biological education in the Cold War. See Rudolph, *Scientists in the Classroom,* 52.

58. Frits Went diary, May 17, 1959. Frits Went Papers.

59. Dennis, "Our Monsters, Ourselves," 57–58; Appel, *Shaping Biology*; Schedvin, *Shaping Science and Industry.* Picard, *La République des Savants.* Of course, these institutions were dwarfed in the American context by the verdant research support offered by the Atomic Energy Commission and the Office of Naval Research. See Creager, *Life Atomic,* 2, 152–53, 246; Sapolsky, *Science and the Navy.*

60. Roll-Hansen, *The Lysenko Effect.* See also Levins and Lewontin, "The Problem of Lysenkoism."

61. See Murneek and White, *Vernalization and Photoperiodism,* 8.

62. For an exemplary moment, evocative of the fear surrounding McCarthyism, when unsubstantiated rumors without attribution or sources are casually mentioned, see Berg and Singer, *George Beadle,* 195. On science in the McCarthy era, see Wang, *American Science.*

63. The Duke University phytotronist, Paul Kramer, later lamented, "the half-forgotten pioneers of the late 19th and early 20th century, such as Sachs, Dixon, Renner, Livingston, and Went," who had forged plant physiology. Paul Kramer, unpublished Autobiographical Notes, ca. 1988, Biographical Files—Paul J. Kramer Papers, p. 28. The most prominent place to find Went cited is the now dated Ruhland, *Encyclopedia of Plant Physiology,* esp. vol. 16. Significantly for the phytotronist community, both James Bonner and Anton Lang were coeditors of the multivolume encyclopedia.

64. Went, *The Experimental Control of Plant Growth,* 319. This corresponds to David Harvey's reading "that most 'modern' writers have recognized that the only secure thing about modernity is its insecurity." Harvey, *The Condition of Postmodernity,* 11.

65. Went, *The Experimental Control of Plant Growth,* 97, 319.

66. John Holloway to the director of the New Zealand Forest Service, July 29, 1955. Folder "Research Phytotron (Controlled Climate Facilities) 1955–1973," F 1 W3129 (Box 247) 41/7. Archives New Zealand.

67. Chouard, "Phytotronics," 4.

68. Hanson, *History of the American Society*, 191.

69. Appel, *Shaping Biology*, 5.

70. Nickelsen, *Explaining Photosynthesis*, 2. Placing physiology at the center of the story of twentieth-century biology has fallen to Nicolas Rasmussen ("The Forgotten Promise of Thiamin). A number of participant histories offer some starting points for a larger historical account, including Weevers, *Fifty Years of Plant Physiology*; Went, "Fifty Years of Plant Physiology"; Galston, "Plant Photobiology"; Bünning, "Fifty Years of Research"; Somerville, "The Twentieth Century"; Pennazio, "Mineral Nutrition of Plants." See also the useful official history of the American plant physiologists in Hanson, *History of the American Society*. While Kremer, "Physiology," and Geison, "Toward a History of American Physiology" represent major synthetic studies of physiology, Kremer's article barely mentions plant physiology, which Geison, in favor of more medical physiology, does not mention at all.

71. Only in Appel, *Shaping Biology*, ch. 7, are phytotrons and biotrons shown as part of the general story of modern biology. Phytotrons are not mentioned in Allen, *Life Science in the Twentieth Century*; Mayr, *The Growth of Biological Thought*; Bowler and Pickstone, *Cambridge History of Science*; or Olmstead and Rhode, *Creating Abundance*. Extraordinary is the case where a textbook used a photograph showing the "effect of relative length of day and night on Douglas firs," which the credits acknowledge comes from R. J. Downs yet there is no mention of phytotrons even though such a picture must have come from the North Carolina State University phytotron. See credits and acknowledgments for image 27.27 in Starr, *Biology*, 454.

However, the *McGraw-Hill Encyclopedia of Science and Technology* does contain an entry on "phytotronics," which, the author Henry Hellmers stipulated, expanded the original definition of "research conducted in phytotronics," to "research using whole plants and conducted under controlled environmental conditions to determine responses to a single or known combination of environmental elements." Hellmers, "Phytotronics," 548. In addition, Salisbury and Ross, in *Plant Physiology* use once again the Douglas Fir image on p. 460, though without citation, and include a two-page autobiography of Frits Went on pp. 378–79. While most of Went's "reflections" concern his auxin discovery, he does outline that the contribution of the "last 30 years [of] phytotrons have helped to make ecology an experimental rather than a descriptive science" (p. 379). It is important that only Went and another guest plant physiologist, Richard Pharis, actually mention "phytotrons" in the course of their reflections. See p. 525 for Pharis, a doctoral student with Paul Kramer (director of the Duke phytotron), and a postdoc at Caltech with Anton Lang, Harry Highkin, and Jan Zeevaart in 1963–64, before the Caltech team was "dismantled." Salisbury himself graduated from Caltech before moving to Utah for the rest of his career, a career that included becoming president of the American Society for Plant Physiology, and rewriting the society's pamphlet *Careers in Plant Physiology* first developed by James Bonner in 1954. Hanson, *History of the American Society*, 188, 273.

72. Ceruzzi, *A History of Modern Computing*, 10.

73. Ceruzzi notes this. Nathan Ensmenger stresses that the people, the computer

programmers, have likewise disappeared from a history overwhelmingly concerned with physical computers as things. Ensmenger, *The Computer Boys Take Over.*

74. About explicit references to "phytotronist" as a new scientific identity, see Letter from Harry Highkin to R. N. Robertson, December 20, 1959, in R. N. Robertson Papers, MS117/1/2. Forty years later, recalling a long experience with phytotrons and their community, Lloyd Evans lamented that "many phytotronists" had "trapped themselves into a continuing preoccupation with improving the design of their units rather than using them for experiments." Evans, "Memoirs of a Meandering Biologist," 45.

75. Introductions to the array of smaller phytotrons can be gleaned from Roussel, "The Patscherkofel Phytotron"; von Wettstein, *The Phytotron in Stockholm*; Smeets, "IVT Phytotron; Harper and Roberts, *Phytotron Manual*; Gorissen et.al., "EPAS: An Advanced Phytotron"; Tischner, "Facts and Figures"; Prabhu and Chandra, *Phytotrons for Agricultural Research*; Iida, "Genetics and 'Breeding as a Science,'" 454. The French CNRS phytotronic group at Gif-sur-Yvette published an annual newsletter beginning in 1971, and in 1980 complied a useful list of controlled-environment facilities and phytotrons. They knew of some forty-seven operating facilities. See CNRS Secrétariat Phytotronique, *Phytotronic Newsletter* 21 (1980), 23–30.

76. Chouard, "Phytotronics," 3.

77. Edwards, *A Vast Machine*, xxiii.

78. Kevles, *The Physicists*, ch. 20.

79. For geneticists, see Campos, *Radium and the Science of Life*, 198, 203; Keller, *Refiguring Life*, 88. Molecular biologists too "projected themselves as strategic disciples . . . of atomic physics." Abir-Am, "The Molecular Transformation," 502, 505; Soraya de Chadarevian readily admitted that "historians have, in their own reckoning, been obsessed with the contribution of physicists in the origin of molecular biology" (de Chadarevian, *Designs for Life*, 51); Zallen, "Redrawing the Boundaries," 77. Evolutionary biologists sought a "biology to par with the physical sciences," Smocovitis, "Unifying Biology," 1, 3, 18, 20; Smocovitis, *Unifying Biology*, 192–93, as did ecologists: Kingsland, *The Evolution of American Ecology*, 3, 4, 179–80, 189–205; Kingsland, "An Elusive Science," 168–169.

80. Went, "The Earhart Plant Research Laboratory," 93; Chouard, "Phytotronics," 1.

81. Keller, *Refiguring Life*, 35. Keller was concerned with the molecular biologists who sought, she argued, the fundamental units of biological analysis, namely, "genes," which became understood as units of information transfer encoded into DNA.

82. Beginning with Clarke and Fujimura, *The Right Tools for the Job*, a number of excellent studies now work at this important intersection, including Benson, *Wired Wilderness*; Cullather, "Miracle of Modernization"; Curry, "Accelerating Evolution"; and Rasmussen, *Gene Jockeys*, 10.

83. Beginning with David E. Nye, who explains that "in contextualist arguments, social forces control the invention, adoption, and development of machines and systems" (Nye, *Consuming Power*, 3), and then Ruth Swartz Cowan, in *More Work for Mother*, where the section on the history of the gas- versus electric-powered refrigerator continues to take my breath away. The two seminal cases of the air pump in Shapin and Shaffer, *Leviathan and the Air Pump*, and the prison in Foucault, *Discipline and Punish* concern not only how technological devices, whether they are air pumps or prisons,

shape institutions, and vice versa, but also how knowledge of the subject, the "spring of the air" or the "inmate," is shaped by that interaction. For large national technologies comparable to phytotrons, namely, nuclear reactors, see Hecht, *The Radiance of France.*

84. From the "techno-centric" campaigns against infectious diseases throughout the 1950s and the Green Revolution in India. Amrith, *Decolonizing International Health,* 3; Abraham, *The Making of the Indian Atomic Bomb,* 49; Edwards, *A Vast Machine,* 8. Of course, one of the most lastingly significant modernist ideals was rural electrification, a widespread and expensive project begun in the 1930s in the United States and the 1950s in Australia, but the provision of this modern convenience was as deeply appreciated by the housewives of the heartland as by those in the outback. The famed Tennessee Valley Authority was matched by the Australian Snowy Mountain hydroelectric scheme. See Macintyre, *Australia's Boldest Experiment,* 172, 424.

85. At the Missouri Botanical Garden, Went erected the Climatron, the "first in a series of new greenhouses to be built to replace the obsolete and deteriorated old ones," he said. See "The Climatron of St. Louis," handwritten notes, n. d. (ca. 1960), in Frits Went papers, Record Group 3/2/6/1, box 1, folder 14. Quote from "The Climatron Opens to the Public," 131. Pound quote in Gay, *Modernism,* 4.

86. For a valuable effort to recover the "cell" as the focus of the science of life, see Stadler, 'Models, the Cell, and the Reformations"; and Creager, *The Life of a Virus.*

87. Zeevaart, "My Journey from Horticulture," 4; Kenneth Thimann expressed the same equation as early as 1957 as "Hereditary potentialities" joined with "Environmental Factors" to create the "Internal Physiological and Biochemical Processes and Conditions," which only then would become expressed as the plant's phenotype, its "Plant Growth and Development." Thimann argued that physiologists well knew that plants not only grew at radically different rates in various climates but that the internal processes of plants were often just as significantly affected. See "Thimann Report," attached to Thimann to the Secretary of the AIBS, March 13, 1957. In Phytotron Records, box 2, p. 4. In France, N. de Bilderling gave the expression as a multiplication rather than an addition: "Phenotype = genotype × environment." de Bilderling, "Phytotrons and Their Possible Use," 16.

88. Oosting, *The Study of Plant Communities,* 4; For ecological methods studies, see pp. 43–55, esp. 44.

89. Ballard, *Growing Plants Indoors,* 3–4. Ballard had clearly adopted many of the principles of controlled environments to make her indoor gardens flourish, noting that "the most specific effect of day length is on flowering," in addition to the "striking" effect of the diminished "intensity of the light" across the year where "in midsummer, the level of illumination may be as high as 10,000 foot-candles" whereas in December it is "rarely above 5,000."

90. Kramer and Boyer, *Water Relations in Plants,* 9–10. Kramer and Boyer also saw H. Lundegårdh as a pioneer in understanding how "the environment acts on plants through its effects on their physiological processes," 7.

91. Went, "Phytotronics," 155.

92. Galston, *The Life of the Green Plant,* 63. It was a message that continued into encyclopedias as recently as 2007: as Henry Hellmers said, "the growth and development

of plants is an expression of their genetic capability as affected by environmental factors" (Hellmers, "Phytotronics," 549).

93. Richardson, *Physics in Botany*, 99.

94. Pauly, *Controlling Life*, 54.

95. Arthur, Guthrie, and Newell, "Some Effects of Artificial Climates."

96. Levins and Lewontin, *The Dialectical Biologist*, 174.

97. Eckardt, "Research on the Structure and Function," 384; A comparison is readily made with the rise of ecosystem science under the International Biological program. According to David C. Coleman, the IBP made extensive use of designated biome regions for observations and great use of "conceptual and simulation models." Coleman, *Big Ecology*, 72–73, wherein there was also mention of "Theoretical Ecology."

98. Went, *The Experimental Control of Plant Growth*, 2.

99. Panel for the Plant Sciences, *The Plant Sciences Now*, 4.

100. Panel for the Plant Sciences, *The Plant Sciences Now*, iii. Quote from Harvey Brooks's cover letter submitting the report to Seitz at the NAS.

101. For hybrids and traditional breeding techniques, see Kingsbury, *Hybrid*. For mutation breeding via X-rays, chemicals, and radiation, see Curry, "Accelerating Evolution." There is a substantial literature on another path of plant scientists during mid-century toward the evolutionary synthesis. See Mayr and Provine, *The Evolutionary Synthesis*, and Smocovitis, *Unifying Biology*.

102. Rasmussen, "The Forgotten Promise of Thiamin." James Bonner's hormone research at Caltech, for example, sought those bullets for many years without success. See Kay, *The Molecular Vision of Life*, 185; Allen, *Life Science in the Twentieth Century*, 19; Kohler, *Lords of the Fly*.

103. Rheinberger, "Heredity in the Twentieth Century," 483. For the debate over reductionism in biology, Bruno Strasser's claim to molecular biology as "a major step in the reductionist agenda of the life sciences" contrasts with Michel Morange's challenge to claims of the reductionist intentions of many of the founders of molecular biologists. Morange, "The Death of Molecular Biology?" 33; Strasser, "A World in One Dimension," 492.

104. Creager, *Life Atomic*, 4, 5.

105. Sapp, *Genesis*, 136; Peter Bowler stresses this key assumption for work following on from the early Mendelians. See Bowler, "Variation from Darwin." Similarly, Angela Creager noted that early proponents of using radioisotopes as tracers often "did not reckon with the biological effects of the radiation they put into their systems," and even claimed that "low-level amounts of radiation did not disturb fundamental living processes." Creager, *Life Atomic*, 223.

106. Lamb, *Climate, History and the Modern World*, 4. Likewise, the environment declined in importance for eugenicists and medical geneticists (whether for debatably racist or public health interests) over the course of the first half of the twentieth century. As Nathaniel Comfort notes, in the late 1930s William Allan, a foundational member of the American medical genetics community, shifted "perhaps [to] over emphasize the genetic end and think too little of the lack of good environment" as he confronted "physical and mental disaster." Comfort, *The Science of Human Perfection*, 61, 103.

107. Coen, "Living Precisely," 498.

108. Personal correspondence from William Laing to the author, April 19, 2015.

109. By its nature molecular biology is "insulated from the heterogeneity that is so central to the organization of higher organisms, [namely] the problem of differentiation and development." Keller, *Refiguring Life*, 93. Also, Strasser, "A World in One Dimension." A nice summary of the defining characteristics of "molecular biology" is offered by Doris Zallen, who in two of four categories notes its emphasis on "the molecular level" and "simplest model systems." Zallen, "Redrawing the Boundaries,' 68.

110. Thone, "Plant Physiology at the Ithaca Congress," 297.

111. Kramer and Kozlowski, *Physiology of Trees*, viii. In their sixty-eight- page bibliography, the authors cite Went's *Experimental Control of Plant Growth* in the 1960 edition, but not in their 1979 revised edition.

112. For the fetishism of DNA, see Lewontin, *It Ain't necessarily So*, 143. Levins and Lewontin, *The Dialectical Biologist*, 52.

113. Kay, *The Molecular Vision of Life*, 263; This follows directly from Francis Crick's own claims: "Both of us [Jim Watson and Francis Crick] had decided, quite independently of each other, that the central problem in molecular biology was the chemical structure of the gene." Crick, *What Mad Pursuit*, 74.

114. Keller, *Refiguring Life*, 21–22. The overwhelming emphasis of the history of biology, Helen Curry said, was "the story of geneticists' great successes in the twentieth century." Curry, "Accelerating Evolution," 426. Nickelsen, *Explaining Photosynthesis*, 2.

115. Bowler, "Science and the Environment?" 18. Some of the best works of history serve to debunk long-cherished myths, notably the story of the Loyalists during the American Revolutionary War and even more so the story of slaves who fought for freedom on the British side. See Bailyn, "The Historiography of the Losers," and Schama, *Rough Crossings*. Likewise, the recovery of the stories of thriving homosexual communities in Berlin and New York from the nineteenth century until the 1930s and 1940s, until they were erased by Nazis and anti-vice crusaders, respectively. Chauncey, *Gay New York*, and Beachy, *Gay Berlin*.

116. With a deep debt to Thompson, *The Making of the English Working Class*.

117. Went and the Editors of Time-Life Books, *The Plants*, 122.

118. Jost, *Lectures on Plant Physiology*, 2. Originally published in German in 1903.

119. Vines and Rees, *Plant and Animal Biology*, 163–87.

120. Vines and Rees, *Plant and Animal Biology*, 146–47.

121. Billings, "The Environmental Complex," 251. The holocoenotic diagram is on p. 256. Billings cites Stebbins, Went, W. M. Hiesey, along with early work from Theodore Kozlowski, the future director of the Biotron at the University of Wisconsin-Madison.

122. Augier, *Phytotrons et phytotronique*, 3. A few years earlier, de Bilderling advocated considering the aerial part of the plant separate from the root system, and thus he generated two equations "Aerial environment = (T + H + L + C + V + Po + Pa + Pb) × (d + i + pcd + pcs + g + q + o)" and "Root environment = (Tr + Hr + Rr + Ar + Nr + Er) × (d′ + i′ + pcd′ + pcs′ + g′ + q′)" where T = temperature; H = humidity; L = light; C = atmospheric composition; V = wind; Po = atmospheric pollution; Pa = atmospheric precipitation; and Pb = atmospheric pressure, while d = duration; i = intensity; pcd = cyclical diurnal periodicity; pcs = seasonal periodicity; g=geometric gradient; q = quantity; and o = ori-

entation. For the root system, he defined Er as the "possibility of extension for roots." From de Bilderling, "Phytotrons and Their Possible Use," 17, 19.

123. Downs, "Phytotrons," 450, 483.

124. Hellmers, "Phytotronics," 549.

125. Stebbins, *Darwin to DNA*, 25. Stebbins appealed interestingly to future studies in a "particular, constant environment," 76. Commemorating the evolutionary synthesis, for instance, Ernst Mayr noted the "great plasticity of the phenotype" in plants, while Stebbins himself highlighted the struggle of earlier botanists who, comparing a dwarf and a large plant, could not say "from simple observations how much of their difference is based on heredity and how much on the environment." Mayr, *The Evolutionary Synthesis*, 137; Stebbins, "Botany and the Synthetic Theory," 140.

126. Allen, "Naturalists and Experimentalists," 199. Later Joel Hagen revised Allen's claims, suggesting that many biologists of the twentieth century readily occupied both camps, and that substantial interdisciplinary interests and work clouds any ready distinction. Hagen, "Experimentalists and Naturalists." See also Müller-Wille and Rheinberger, *A Cultural History of Heredity*, 127.

127. For de Bilderling, the complex combinations of environmental factors at work to result in any one phenotype "constitute the 'background noise' of nature" a term evidently pulled from radio astronomy. See de Bilderling, "Phytotrons and Their Possible Use," 17.

128. Went, "Gene Action," 840.

129. Frankel, "Concluding Remarks," 440.

130. Pickering, *The Cybernetic Brain*; McCray, *The Visioneers*, 28;

131. Hellmers, "Phytotronics," 548.

132. de Bilderling, "Phytotrons and Their Possible Use," 16.

133. Edwards, *A Vast Machine*, fig. 1.2, pp. 4–5.

134. Lange, Schulze, and Koch, "Evaluation of Photosynthesis Measurements," 340.

135. Letter from Yuill to T. Kozlowski, October 7, 1986, Biotron Papers, Series 06/80, Box 3, file "Biotron Committee Information, 1984–1989."

136. S. Rajki to Paul Kramer, February 6, 1976, Paul J. Kramer papers, box 11, file "Phytotrons, 1972–76."

137. Shapin and Shaffer, *Leviathan and the Air Pump*, 15. The shaping of knowledge systems to replicate systems of social power and legitimacy has been an important theme in the sociology of science, where scientists produce "object-oriented structures for social authority structures." Knorr-Cetina, *Epistemic Cultures*, 171–74.

CHAPTER 1. "THE AWE IN WHICH BIOLOGISTS HOLD PHYSICISTS"

Epigraph: James Bonner, interview with Graham Berry, Pasadena, California, March 13–14, 1980. Oral History Project, California Institute of Technology Archives. http://resolver.caltech.edu/CaltechOH:OH_Bonner_J, pp. 17–18 (accessed January 5, 2014).

1. Heilbron and Seidel, *Lawrence and His Laboratory*; for Calutrons, see Gregg Herkin, "The University of California," 120.

2. Davis, *City of Quartz*. Incomparable insight about Los Angeles is presented in film history: "Los Angeles is where the relation between reality and representation gets muddled." *Los Angeles Plays Itself*, dir. Thom Andersen.

3. Quite generally, the wish for a land of milk and honey (Exodus 3:7–8) sits beneath all manner of human stories. Food was the all-consuming obsession in early modern European folktales, Darnton, "Peasants Tell Tales"; similarly, Ginzburg, *The Cheese and the Worms*, 77; as well as nineteenth-century Australian folktales ("That's where the magic comes in," explained Bill the Koala, "The more you eats the more you gets.") Lindsay, *The Magic Pudding*, 23; and is a dominant motif of twentieth-century American mythic imagery such as Norman Rockwell's, "The Four Freedoms: *Freedom from Want*" (1943). For a general overview of the issues related to hunger as a historical force, see Vernon, *Hunger*. For a parallel case of technological optimism in the 1940s envisioning huge algae fields feeding the vast hungry masses in Asia and Africa, see Belasco, "Algae Burgers?"

4. At the Boyce Thompson Institute, William Crocker had noted that one of the limitations of its first constant-condition greenhouses was that the "lighting made it difficult to air-condition the room." Crocker, *Growth of Plants*, 286.

5. Went, *The Experimental Control of Plant Growth*, 2.

6. Kay, *The Molecular Vision of Life*, 185.

7. California Institute of Technology, *Catalogue 1950–51* (California Institute of Technology, 1950), 88.

8. Letter from Millikan to Harry Earhart, January 7, 1946. Robert A. Millikan Papers, file 24.7.

9. Kay, *The Molecular Vision of Life*. "In the standard history of the field . . . Caltech and Cold Spring Harbor were the 'Mecca and Medina' of the phage group." Quoted in de Chadarevian, *Designs for Life*, 161; Rasmussen, "The Forgotten Promise of Thiamin"; Berg and Singer, *George Beadle*.

10. Traweek, *Beamtimes and Lifetimes*, x.

11. Kingsland, "Frits Went's Atomic Age Greenhouse," 316, 292, 312.

12. Went, "The Earhart Plant Research Laboratory," 18.

13. Salisbury and Ross, *Plant Physiology*, 378–79.

14. Kargon, "Temple to Science"; Goodstein, *Millikan's School*, esp. ch. 4–6.

15. Copy of a letter from Millikan to Allan Balch, Cunard Line, July 5, 1927, in James Bonner papers, file 21.1. In a move toward fundamental science, which remained a hallmark of the institute's policy, the journal *Science* carried the announcement that Caltech would move its physiological laboratories away from their association with medical schools. "Biology at the California Institute of Technology," n. d. (ca. 1923). Copy of the original in James Bonner papers, file 21.1.

16. Kohler, *Lords of the Fly*, 124; Müller-Wille, and Rheinberger, *A Cultural History of Heredity*, 146–48.

17. Schürch, "Der Schlüssel zum Pflanzenwachstum, " 1–2.

18. Memo from Bonner to Sturtevant, ca. September 26, 1945. James Bonner papers, file 48.9. The experience of the biologists at Caltech paralleled that of the physicists; see Kevles, *The Physicists*, 156.

19. "Biology at Caltech," in Biology Division Papers, file 14.4.

20. Evans, "Conjectures, Refutations, and Extrapolations," 5.

21. Robert L. Sinsheimer, Oral History Interview with Shelley Erwin, 1990 and 1991. http://oralhistories.library.caltech.edu/33/0/OH_Sinsheimer.pdf.

22. Most undergraduate biology textbooks spend a page or two on Went's "conclusive demonstration" of auxin. In particular, see the most widely used textbook for biology, Campbell and Reece *Biology*, 806–7; as well as Taiz and Zeiger, *Plant Physiology*, 468–70; Vines and Rees, *Plant and Animal Biology*, 613–16; Gould and Keeton, *Biological Science*, 918–19; Purves et al., *Life*, 654; Starr, *Biology*, 450. None, however, references the phytotron. Where they tell history, textbooks always tell Whig history.

23. de Chadarevian, "Laboratory Science," 37, 18. From the beginnings of laboratories and the acceptance of laboratory-produced knowledge, they have confronted the issues of trust, reliability, and repeatability about universal claims in science. Scientists' experimental practices, social relations, model organisms, societies, and journals have all served to ensure trust by reliable repetition. Shapin and Shaffer, *Leviathan and the Air Pump*; Kohler, *Landscapes and Labscapes*.

24. Weevers, *Fifty Years of Plant Physiology*, 4; de Chadarevian, "Laboratory Science," 32.

25. Hanson, *History of the American Society*, 57.

26. Schürch, "Der Schlüssel zum Pflanzenwachstum," 105.

27. Went, "Auxin."

28. See Weevers, *Fifty Years of Plant Physiology*, 173–75; de Chadarevian, "Laboratory Science," 37.

29. Weevers, *Fifty Years of Plant Physiology*, 36. Went's preface is on page 1. National styles overlapped with disciplinary struggles: the Dutch celebrated their dominant position within plant physiology up to 1945. Went himself wrote the preface for Weevers's book *Fifty Years of Plant Physiology*, noting how the "Dutch can be proud of the quality and quantity of their research" in plant physiology.

30. Steere, "Introduction."

31. Thimann, "Foreword," vi.

32. Wildman, "The Auxin-A, B Enigma," 38.

33. Gaudillière, "Globalization and Regulation," 254.

34. Rasmussen, "Plant Hormones"; Rasmussen, "The Forgotten Promise of Thiamin." For the work of other British and American groups on herbicides of the same class, see Evans, *Feeding the Ten Billion*, 126–28.

35. Rasmussen, "Plant Hormones in War and Peace," 295; Rasmussen, "The Forgotten Promise of Thiamin," 247.

36. James Bonner, Sterling Emerson, Norman Horowitz and Donald Poulson, interview by Judith Goodstein, Harriett Lyle, and Mary Terrall, November 6, 1978. http://resolver.caltech.edu/CaltechOH:OH_Joint_Biology (accessed January 5, 2014), 22.

37. Letter from J. Fisher Stanfield to James Bonner, July 16, 1953. James Bonner papers, file 23.9. The Australian physiologist Rutherford Robertson explained in his introductory lectures, physiological "methods of investigation are essentially chemical and physical, but have regard to the complexity of living systems. Physiologists studied the function of organisms. Processes investigated are the integrated results of complex organization of cell, tissue, organ or organism. Physiology thus differs from biochemistry, though there is no sharp dividing line." "Plant physiology. Notes for lectures. Sydney," 1954. R. N. Robertson Papers, MS117/7/8. In their undergraduate textbook, Bonner and Galston made much the same claim, "the developments of modern biology have

tended more and more to obscure whatever dividing line may once have existed between physiology and biochemistry." Bonner and Galston, *Principles of Plant Physiology,* v. For the same comments about biochemistry, see Long, "William McElroy," 781.

38. Went, "Plant Physiology," 519.

39. Nickelsen, *Explaining Photosynthesis,* 215, 243–46. Creager, *Life Atomic,* 236–37. Cycles, Angela Creager noted, became "paradigmatic" after Hans Krebs's "Krebs cycle" for citric acid (225).

40. Hanson, *History of the American Society,* 179. Went served as the ASPP president in 1944–45, and Calvin became president in 1963–64 (193). Went never received a Nobel Prize, unlike Calvin.

41. Went, "Thermoperiodicity," 146–47.

42. Wildman, "The Auxin-A, B Enigma," 44. Auxin A and Auxin B were two much heralded types of auxin, the structures of which were identified in a Dutch laboratory in the 1930s, but which remained controversial for many years. For the point here, Wildman notes the overwhelming effort made to stabilize the environment of the experiments, including eventually having to account for "atmospheric electromagnetic radiation or cosmic rays as influencing the potency, [the Dutch] shielded the test plants by zinc and lead with bakelite as a control."

43. Kingsland, "Frits Went's Atomic Age Greenhouse," 294; Craig, *Centennial History of the Carnegie Institution,* 62–63.

44. Went and Eversole, "The Air-Conditioned Greenhouses at the California Institute of Technology," n. d. Historical Files—Biology Division, file A3.2, p. 4; Went's initial technical description of their air-conditioned greenhouses is in Went, "Plant Growth."

45. Went, "Orchids in My Life." Eversole may have obsessed over orchids, but for Went "orchids were only a sideline." Only one species *Cattleya trianae* underwent any serious research in the Earhart Laboratory, Went recalled (29).

46. Marshall et al., "The Technical Features, I," 151.

47. Went and Cosper, "Plant Growth." Tomato plants had previously been used by Daniel Arnon in the late 1930s to show the uptake of phosphorus via radioactive phosphorus-32. See Creager, *Life Atomic,* 38–39.

48. Went and Eversole, "The Air-Conditioned Greenhouses," 4.

49. Went and Eversole, "The Air-Conditioned Greenhouses," 4.

50. Brown, Pais, and Pippard, *Twentieth Century Physics,* 731.

51. Belasco, "Algae Burgers?"

52. Went and Eversole, "The Air-Conditioned Greenhouses," 4.

53. "Projects considered [by Caltech's Biology Division]," n. d. (ca. 1946). Handwritten notes in James Bonner papers, file 48.9.

54. Kay, *The Molecular Vision of Life,* 227.

55. Letter from Millikan to Beadle, November 6, 1945. Biology Division Papers, file 48.8.

56. Letter from Beadle to Millikan, December 3, 1945. Robert A. Millikan Papers, file 24.7.

57. Letter from Millikan to Beadle, November 6, 1945, Biology Division Papers, file 48.8.

58. "If we are going to conserve the finest elements in Anglo-Saxon civilization, we must conserve the method of the private initiative and not depend primarily upon government aid," Millikan said in 1919. Kargon, *The Rise of Robert Millikan,* 91.

59. See Robert A. Millikan Papers, file 39.1. Letter from Earhart to Millikan, November 5, 1952; Letter from Earhart to Ingebretsen, July 25, 1952.

60. Letter from Earhart to James Ingebretsen, Spiritual Mobilization, July 25, 1952. Copy in Robert A. Millikan Papers, file 39.1.

61. Letter from Harry Earhart to Millikan, December 7, 1945. Robert A. Millikan Papers, file 24.7.

62. Letter from Millikan to Harry Earhart, December 22, 1945. Robert A. Millikan Papers, file 24.7.

63. Letter from Millikan to Harry Earhart, December 22, 1945. Robert A. Millikan Papers, file 24.7.

64. Kargon, *The Rise of Robert Millikan*, 74, 148.

65. Curry, "Accelerating Evolution," ch. 6. Smith, *A Peril and a Hope*, 77.

66. Proposal "The Efficiency of Light Utilization by Plants," n. d. (ca. 1951). Biology Division Papers, file 28.3.

67. Letter from L. A. DuBridge to E. C. Herron, Lamp Division, G. E., April 14, 1952. Biology Division Papers, file 28.3.

68. Letter from Beadle to Mr. and Mrs. Richard Earhart, October 31, 1955. Biology Division Papers, file 28.1. Archives. California Institute of Technology.

69. Letter from Flora Rhind to L. A. DuBridge, October 26, 1956. Biology Division Papers, file 24.4.

70. See newspaper clippings in Division of Biology Historical Files, file A3.2.

71. Berg and Singer, *George Beadle*, 195.

72. Kay, *The Molecular Vision of Life*, 225; Westwick, *Into the Black*.

73. Letter from G. W. Beadle to Frits Went, July 20, 1948. Biology Division Papers, file 27.11.

74. Letter from Frits Went to Beadle, July 31, 1948. Biology Division Papers, file 27.11. Confirmed with DuBridge, August 4, 1948. Biology Division Papers, file 27.11.

75. Outline from the Harvard plant physiologist's Kenneth Thimann's dedication for the opening of Caltech's Earhart Plant Research Laboratory, July 1949. Historical Files—Biology Division. File A 3.2. Archives. California Institute of Technology.

76. James Bonner, interview with Graham Berry, 17–18. A decade earlier, in 1972, Beadle had also queried the origin of "phytotron." As Bonner told him: "the word phytotron was coined at a morning coffee session in the old Greasy Spoon at Caltech, and during a discussion between Samuel G. Wildman and myself. . . . Sam and I started out with the hypothesis that anything as fancy as the proposed Earhart Laboratory shouldn't be called an air-conditioned greenhouse or anything simple like that, but should have a more magnificent name. We ended up with 'thermophotophytotron' but quickly slimmed it down to 'phytotron.' Frits Went was of course very annoyed at the term and thought that fun was being poked at his brain child. In the meantime, however, Dr. Millikan has received word of the coinage of this new term. He then proceeded to give a public talk at which he said approximately that this new machine, the phytotron, would do for biology what cyclotrons had done for physics, and he then continuously used the term. Frits was impressed and started to use it himself." Letter from Bonner to Beadle, September 9, 1970. James Bonner Papers, File 20.1.

77. Went, "The Earhart Plant Research Laboratory," 93.

78. Augier, *Phytotrons et phytotronique*, 4.

79. Proctor, . "'-Logos,' 'Ismos,' and 'Ikos,'" 307.

80. Downs and Hellmers, *Environment*, 3.

81. Went, "The Phytotron," 3.

82. Went, "The Phytotron," 3.

83. Went and Eversole, "The Air-Conditioned Greenhouses," 6.

84. Quoted in Nickelsen, *Explaining Photosynthesis*, 90.

85. Weevers, *Fifty Years of Plant Physiology*, 4.

86. Hanson, *History of the American Society*, 45.

87. James Bonner, "The Role of Plant Physiology in Agricultural Practice," n. d (ca. 1952). Biology Division Papers, file 58.11.

88. Dennison, "Physics and the Department of Physics," 127.

89. Bush, *Pieces of the Action*, 138.

90. Bonner, interview with Graham Berry, 9–10.

91. Undated and untitled pages from the Biology Division's "Christmas Entertainments." Historical Records—Biology Division. File A 3.9. There is something thing beneath biology, of course: the history and philosophy of science (HPS)! When Andrew Warwick's son became an undergraduate at University College London the conversation went:

> Dad: "Ever considered doing a course in HPS?"
> Son: "Oh come on dad, HPS is even lower than biology."

92. *The Martian*, dir. Ridley Scott.

93. Rudolph, *Scientists in the Classroom*, 138–40.

94. Kohler, "The Management of Science," 287.

95. Frits Went Diary, September 1, 1962. Frits Went papers. Record Group 3/2/6, box 22.

96. Kingsland, *The Evolution of American Ecology*, 3, 4, 179–80, 189–205; Kingsland, "An Elusive Science," 168–69.

97. Kingsland, "Frits Went's Atomic Age Greenhouse."

98. Smocovitis, "Unifying Biology," 1, 3, 18, 20; Smocovitis, *Unifying Biology*, 192–93.

99. Beadle Memorial, "George Wells Beadle and the New Biology," October 10, 1980. James Bonner Papers, file 20.1.

100. Zallen, "Redrawing the Boundaries," 77. Kramer and Boyer give an illuminating example of the "Ohm's Law Analogy." Popular in the 1960s, the analogy, they said, described the water flow through the soil–plant–atmosphere continuum as like the "flow of electricity" in a circuit. By the 1980s it had fallen out of favor because it assumed steady-state flow and constant resistance, which while valid for circuits was rarely seen in biological systems. As late as 1995, though, Kramer and Boyer thought the analogy "too useful to be abandoned." Kramer and Boyer, *Water Relations in Plants and Soils*, 9. In short, plant physiologists remained committed to conceptions drawn from the physical sciences.

101. de Chadarevian, *Designs for Life*, 51. A nice explanation is in Nicolas Rasmussen, "The Mid-Century Biophysics Bubble," 246.

102. Leroi, *The Lagoon*, 8.

103. Oosting, *The Study of Plant Communities*, 7.

104. See letter from Bonner to Englesberg, May 3, 1967. James Bonner Papers. File 21.5.

105. Weevers, *Fifty Years of Plant Physiology*, 4.

106. Bünning, "Fifty Years of Research," 14, 17. The director or the Duke phytotron, Paul Kramer, used a similar expression, only about physical chemistry. "Acceptance of the water potential concept was slow because of the confusion regarding terminology, the lack of convenient methods for measuring it, and the inadequate training of plant physiologists in physical chemistry." Kramer and Boyer, *Water Relations in Plants*, 7. In other words, the plant physiologists were in awe of the physical sciences but also felt that their own training had been lacking in the physical sciences. Kramer also conveniently connects training to two other big themes of phytotrons, terminology and measurement; the obsession over both will be detailed through the example of light units in chapter 5.

107. Went, "Gene Action," 848. Went also noted that if his argument against quantum mechanical causes of variability was accepted, "the physiologists and embryologists are now in the uncomfortable position where they will have to explain why genes, which were just proved by the geneticist and biophysicists to be following quantum mechanics, do not impart the degree of variability upon growth and development which one would expect from quantum mechanics" (848). The solution Went offered was that the environment played a greater part than quantum mechanics. Went, Bünning, and Hendricks were part of an intimate network: Bünning was mentioned by Hendricks in his memoir, and shared the pages of a conference volume on vernalization with Went. See Hendricks, "The Passing Scene," 6, and Murneek and White, *Vernalization and Photoperiodism*. Went and Hendricks campaigned together in 1959 in the pages of no less than *Science* for a national controlled environment facility, that would provide an impetus for the Biotron. See Hendricks and Went, "Controlled-Climate Facilities."

108. Nickelsen, and Govindjee, *The Maximum Quantum Yield Controversy*, 37–38.

109. As the Duke University ecologist Henry Oosting noted, "as in other sciences, ecology has become more precise as it has developed and, with its concern for greater details, has demanded accurate measurement." Oosting, *Study of Plant Communities*, 31. See also Rudolph, *Scientists in the Classroom*, ch. 6. Another evocative example was the rise of the ecosystem concept in ecology. Arthur Tansley coined the term "ecosystem" in 1935 to represent, he said, "the whole system (in the sense of physics), including not only the organism-complex, but also the whole complex of physical factors forming what we call the environment of the biome." Quoted in Creager, *Life Atomic*, 352.

110. Hendricks, "The Passing Scene," 1. As Nickelsen, and Govindjee show, the experimental setup of Warburg's measurement of the quantum yield in photosynthesis involved an array of "sophisticated photophysical techniques" including using specific wavelengths of light. See Nickelsen and Govindjee, *The Maximum Quantum Yield Controversy*, 11–13, 60–63.

111. Kargon, "Temple to Science."

112. Bonner's notes on "Millikan's Report to the Trustees—1923," in James Bonner Papers, file 21.1.

113. Kay, *Molecular Vision of Life*, ch. 6–7. Also Berg and Singer, *George Beadle*, 3, 173.

114. Letter from Weaver to DuBridge, January 6, 1954. Lee A. DuBridge Papers, file 1.15.

115. Went and Eversole, "The Air-Conditioned Greenhouses," 6. Likewise, Doris Zallen highlighted Eugene Rabinowitch's comment from 1945 that plant physiologists and physicists were the target of his compendium *Photosynthesis and Related Processes*. Zallen, "Redrawing the Boundaries," 75.

116. Went, "The Effects of Climate," 63.

117. For biochemistry, see Long, "William McElroy," 801, where she argues that the biochemists certainly saw molecular biology as uniting biology, but in addition, as the molecular biologists moved to dominate the meaning and practice of biology, the biochemists reinvented their own role in molecular biology so as to appear more united. See also, de Chadarevian and Gaudillère, "The Tools of the Discipline." The whole of volume 29 of *Journal of the History of Biology* is devoted to these issues. See especially de Chadarevian, 'Sequences, Conformation, and Information." For molecular biology, see Kay, *The Molecular Vision of Life*, 7, who explicitly argued that biochemistry evolved into the molecular biology of "interdisciplinary cooperation." For evolutionary biology, see Farber, *Finding Order in Nature*, 108. In addition, recent work by Nathaniel Comfort on Barbara McClintock has revealed that she too, right around the early 1950s, hoped her theory would usher in a "phase of integration" among several biological specialties. Comfort, "'The Real Point Is Control,'" 140.

118. Smocovitis, "Unifying Biology," 43.

119. Went, "The Effects of Climate," 56.

120. Bonner, "The Chemical Cure," 28.

121. California Institute of Technology, *Catalogue 1950–51* (California Institute of Technology, 1950), 89.

122. Long, "William McElroy," 781. For James Franck, see Nickelsen, *Explaining Photosynthesis*, 98.

123. See the notes on "Biology at Caltech," in Biology Division Papers, file 14.4. It seems worth highlighting further the sheer breadth of topics within the category "biology" at Caltech, including botany, anatomy, zoology, ecology, biophysics, chemical biology, physical biology, biochemistry, marine biology, plant physiology, plant biochemistry, immunochemistry, protein chemistry, enzymology, molecular structure, cytogenetics, cytology, genetics, chemical genetics, embryology, animal physiology, bioorganic chemistry, immunogenetics, virology, electron microscopy, and psychobiology. Likewise, the instruments arrayed included the marine laboratory, air-conditioned greenhouse, farm, radioactivity, X-ray equipment, high-speed centrifuge, animal rooms, phage laboratory, molecular structure laboratory, autoclaves, and a fly room.

124. Gaddis, *The United States*, 237.

125. Went, "Plant Physiology," 525.

126. Went, "The Phytotron," 6; Went used this same line in, "The Earhart Plant Research Laboratory," 93.

127. See Mindell, "Automation's Finest Hour."

128. "Summary of Observations made by F. W. Went," Summer 1950. Biology Division Papers, file 75.16.

129. Went, "Fifty Years of Plant Physiology," 106. Pickering, *The Cybernetic Brain* is evocative on the idea of postwar science's fixation on computerization and control.

CHAPTER 2. AT WORK IN THE CALTECH PHYTOTRON

Epigraph: Went, "The Role of Environment," 383.

1. "The Division of Biology Presents Selections from Its Christmas Entertainments of 1949 and 1950." Historical Files—Biology Division. File A 3.9. I welcome suggestions as to the tune.

2. Letter from Beadle to Mr. and Mrs. Richard Earhart, October 31, 1955. Biology Division Papers, file 28.1.

3. Went, "The Response of Plants to Climate," 489.

4. Went, "The Role of Environment," 383.

5. Müller-Wille and Rheinberger, *A Cultural History of Heredity*, 110.

6. Mayr noted that "Natural selection is meaningless if there is no variation on which it can work." Mayr, "The Role of Systematics," 128.

7. Latour, "Give Me a Laboratory."

8. Went, "The Role of Environment," 378, 379, 383.

9. Handwritten notes, n. d. (ca. 1950–51?). Frits Went papers. Record Group 3/2/6/1, box 4, folder 10. A decade later, during the 1962 Plant Science Seminar hosted by Campbell Soup, Went nuanced his "graduation" thesis to more of a physics reading based on the influence of forces upon bodies: "after organisms had graduated from molecular forces to gravitational ones, from thermodynamics to classical mechanics." See Went, "Phytotronics," 151.

10. Handwritten notes, n. d. (ca. 1950–51?). Frits Went papers. Record Group 3/2/6/1, box 4, folder 10.

11. Handwritten notes, n. d. (ca. 1951). Frits Went papers. Record Group 3/2/6/1, box 11, folder 38.

12. Hiesey, "Growth and Development of Species," 205; see also Went's "Notes regarding work of William M. Hiesey," January 1–September 20, 1950. Biology Division Papers, file 37.34.

13. Müller-Wille and Rheinberger, *A Cultural History of Heredity*, 140.

14. Bowler, "Variation from Darwin," 22. Nils Roll-Hansen distinguished between "more sophisticated Mendelians like Johannsen [who] stressed that heredity and environment are equally essential in the formation of the organism" and more simplistic Mendelians who all-too-readily equated the genotype with the phenotype. Roll-Hansen, *The Lysenko Effect*, 26.

15. After discussing recent twin studies, Richard Lewontin offered an excellent analysis of the flaws in the pursuit of heritable traits as the key to life: "to say that genes are somehow influential is to say nothing, because genes are somehow influential in all traits of all organisms." Lewontin, *It Ain't Necessarily So*, 38.

16. Roll-Hansen, "Sources of Wilhelm Johannsen's Genotype," 477. Genetics was a branch of a larger physiological trunk, one where, I suggest, the problem was simplified by declaring the variable "environment" equal to approximately zero. As Müller-Wille and Rheinberger noted that when William Bateson's professorship of biology

was dedicated to "genetics," genetics itself was a "particular class of physiological problems." Müller-Wille and Rheinberger, *A Cultural History of Heredity*, 157.

17. Bonneuil, "Plant Breeding," esp. 298–302, 303. Since the last "phytotronics" conference was held in 1979, the timing of this narrative fits at least circumstantially. James Bonner, describing potentialities of "controlled environment facilities" before a meeting of the United States Agricultural Research Institute in 1959, thought that "one of the great uses of a phytotron in the future will be in plant breeding work." James Bonner, "Fundamental Plant Physiology Research," 72. Müller-Wille and Rheinberger, *A Cultural History of Heredity*, 158.

18. Newman, *Plant Breeding in Scandinavia*, 27–28. Newman's work was part of H. K. Hayes and C. R. Burnham, "Suggested Literature for Students in Plant Genetics," p. 4 for biology students at Caltech. See A. H. Sturtevant papers, file 22.13; East, "The Relation of Certain Biological Principles," 52–53; the historian Charles Rosenberg stressed the emerging role of the scientist-entrepreneur in agricultural experiment stations, which seems especially apt in light of Went's career. Rosenberg, "Science, Technology, and Economic Growth," 190; Müller-Wille and Rheinberger, *A Cultural History of Heredity*, 133–34.

19. Curry, "Accelerating Evolution," ch. 6. Smith, *A Peril and a Hope*, 77.

20. Bonner, "Fundamental Plant Physiology Research," 72–73.

21. Went, *The Experimental Control of Plant Growth*, 101, 115.

22. See California Institute of Technology, *Catalogue, 1955–56* (California Institute of Technology, 1955), 21–22.

23. Letter from Went to Highkin, April 28, 1952. Biology Division Papers, file 55.1.

24. "The 'Bean Test' an assay for traumatic acid," n. d. (ca. 1945–1950?). James Bonner Papers, file 46.13.

25. It is worth pausing here to highlight that in the early twenty-first century and an era of greater environmental awareness of the narrow range of temperatures that a difference of just 3°C in the daytime combined with 6°C at night caused the distinct shift from all female to all male flowers. Even more significant is the upper limit, only 30°C, beyond which no flowers developed at all; an average increase of only a few degrees over a few generations from global warming might push certain plants into a range where they would simply never flower again. Current optimistic plans from representatives at the Copenhagen convention in 2009, for instance, merely aim to keep global climate change to around 2°C, while it currently appears more likely that global climate change will be around 4°C. Charlton, "Man-Made World," 45. In answer to the question, "How do we expect climate to evolve in the future?" the Australian Academy of Science has proffered, "if greenhouse gas emissions continue to grow rapidly, it is expected that, by 2100, the global average air temperature over the Earth's surface will warm by around 4°C above mid-19th century temperatures." From https://www.science.org.au/publications/scienceofclimatechange-q-and-a-2015/summary (accessed February 16, 2015). The sensitive response of plants like the gherkin to temperature, namely, a 3°C difference in daytime temperature, would shift whole ecologies as plants consistently produce more male flowers than female. However, as of July 2014, oil companies agree and have informed their investors, as Shell Oil has,

that "We do not see governments taking the steps now that are consistent with the 2°C scenario." Indeed, according to the *Economist*, oil firms are betting heavily that increasing population over the next half century will keep oil dependence and keep oil prices high (above $80/barrel, the current break-even point of new development and exploration). See "The Elephant in the Atmosphere," *Economist* July 19, 2014, 59. However, as of November 2015, at the COP21 meeting in Paris more than 180 nations managed to sign an agreement to work toward only a 2°C increase.

26. Nitsch, "The Role of Plant Hormones," 8.

27. Mary Lou Whaling (nee Slichter) was the wife of Caltech physicist Ward Whaling. See Ward Whaling, interview with Shelley Erwin, Pasadena, California, April–May 1999. Oral History Project, California Institute of Technology Archives. http://resolver.caltech.edu/CaltechOH:OH_Whaling_W (accessed March 23, 2011).

28. Bonner used the standard Hoagland solution plus 390 (mg?) sucrose, 1 or 2x10–15M IAA, 0.0025M K-maleate, 100mg/l Mn, 100mg/l arg.

29. See Bonner and Whaling, Laboratory Notebook: "Avena Growth (at various temperatures) book iv, 1955–56." In James Bonner Papers, box 42. Nearly twenty years later, the experimental procedure remained the same. In an experiment to measure photosynthesis, William Laing and his colleagues grew soybeans in "growth chambers (30 C day/20 C night, 14 hr photoperiod 550 μeinsteins m-2 sec-1) subirrigated with modified Hoagland solution." Significantly, "only the youngest mature trifoliate leaf was used in all measurements," being "cut from the plant" and then "maintained at the desired temperature by immersion in a water bath" "illuminated with 650 μeinsteins m-2 sec-1" and air "circulated" at "7.0 liters min-1." From there the team could measure the change in rates of photosynthesis with different gas mixtures. See Laing, Ogren, and Hageman, "Regulation of Soybean," 679.

30. Rheinberger, *An Epistemology of the Concrete*.

31. Harman, "C. D. Darlington," 311. See also Levins and Lewontin, "The Problem of Lysenkoism"; Roll-Hansen, *The Lysenko Effect*; Kingsbury, *Hybrid*, ch. 9.

32. Went, *The Experimental Control of Plant Growth*, 195.

33. Letter from Bonner to Noel Kefford, CSIRO Division of Plant Industry, July 23, 1957. James Bonner Papers, file "Australia."

34. Schürch, "Der Schlüssel zum Pflanzenwachstum, " 9.

35. Letter from Went to Dr. C. Stacey French, CIW, November 21, 1951. Biology Division Papers, file 28.1.

36. Examples drawn from Went, *The Experimental Control of Plant Growth*, 111, 113–14, 124, 140.

37. Untitled background information for ACS application, n. d. (ca. 1956). Biology Division Papers, file 7.3.

38. Letter from F. W. Went to Harry Weaver, April 5, 1956. Biology Division Papers, file 7.3.

39. Letter from Went to Weaver, April 5, 1956.

40. Letter from Went to French, November 21, 1951.

41. Interestingly, they required 10–25 mm over 10–24 hours, while short intense storms produced far lower rates of germination. Went, "The Earhart Plant Research Laboratory," 18. In Historical Files—Biology Division, file A3.2.

42. Letter from Went to DuBridge, May 3, 1954. Biology Division Papers, file 27.12. In addition, although Went hoped for a desert laboratory, he realized that Caltech would probably "never have enough investigators and students interested in desert problems to make efficient use of a desert research center." Instead, true to a community style of science, Went advocated a "joint venture of U.C.L.A., Pomona College, CalTech, and probably U.S.C."

43. Went, *The Experimental Control of Plant Growth*, 15.

44. Letter from G. W. Green to G. P. Larson, July 20, 1949. Biology Division Papers, file 5.12.

45. Memo from G. Hall to Dr. Beadle, June 12, 1949. Biology Division Papers, file 5.12.

46. Bonner, "Arie Jan Haagen-Smit"; Pitts and Stephens, "Arie Jan Haagen-Smit."

47. Letter from G. W. Beadle to Stanley Scott, February 27, 1950. Biology Division Papers, file 5.12.

48. Haagen-Smit et al., "Investigation on Injury to Plants," 20–21.

49. Bonner, " rie Jan Haagen-Smit," 199.

50. Pitts and Stephens, "Arie Jan Haagen-Smit," 517.

51. Went, *The Experimental Control of Plant Growth*, 68.

52. Went, *The Experimental Control of Plant Growth*, 47.

53. "General Instructions and Guide for Working at Earhart Laboratory," n. d. Copy in Phytotron Records, box 4, file "Phytotron-notes on Duke's 1st proposal"; James Bonner's tribute to George P. Keyes, May 1963. Historical Files—Biology Division, file A3.2. For the rigors of the decontamination process, see Edward Hutchings, "How Does Your Garden Grow?" *Westways* (April 1963), 8–10, in Historical Files—Biology Division, file A3.2; and Went, *The Experimental Control of Plant Growth*, 47.

54. Went, *The Experimental Control of Plant Growth*, 47.

55. See Kohler, *Lords of the Fly*; Creager, *The Life of a Virus*, esp. ch. 8; Rader, *Making Mice*; Davies, *The Microbial Models*. Ankeny, "Wormy Logic." See also Müller-Wille and Rheinberger, *A Cultural History of Heredity* (ch. 6) for larger connections. For a handy general overview, see Endersby, *A Guinea Pig's History*.

56. Went, *The Experimental Control of Plant Growth*, 97; Went developed "a number of 'botanical Drosophilas,' small plants with a short life cycle" to "study natural evolution in the laboratory." See "Operation of the Earhart Plant Research Laboratory," April 1, 1952. Biology Division Papers, file 63.8. For Babcock's research program, see Smocovitis, "The 'Plant Drosophila'".

57. Went, *The Experimental Control of Plant Growth*, 96.

58. Leonelli, "*Arabidopsis*, the Botanical *Drosophila*." Lawton, "Ecological Experiments"; May, *Stability and Complexity*; Charles-Edwards, Doley, and Rimmington, *Modelling Plant Growth*. For significant insights into the styles of biology centered on both model organisms and models more generally, see Nickelsen, *Explaining Photosynthesis*.

59. Went, *The Experimental Control of Plant Growth*, 12.

60. Went, *The Experimental Control of Plant Growth*, 12.

61. Rabinowitch and Govindjee, *Photosynthesis*, 6–7.

62. Frits Went Diary, August 28, 1962. Frits Went Papers. Record Group 3/2/6, box 22.

63. Richardson, *Physics in Botany*, 45.

64. Dolin, *Leviathan*, 12.

65. McFarlane, “Light.”

66. Personal correspondence with William Laing, July 15, 2014.

67. Energy (E) = Planck’s constant (h) × the speed of light (c) / wavelenth (λ).

68. Personal correspondence with William Laing, July 15, 2014.

69. Adams, “An Air Pollution Phytotron,” 473.

70. Stephan Thiel et al., “A Phytotron for Plant Stress Research,” 458. See also https://www.helmholtz-muenchen.de/en/eus/facilities/phytotron/index.html (accessed November 1, 2016).

71. McFarlane, “Light,” 21. A “nano-Einstein: is $6.02\text{x}10^{4}$ photons of light.

72. Harper and Roberts, *Phytotron Manual*, appendix 14.

73. Thomas, Downs, and Saravitz, *Phytotron Procedural Manual*.

74. Adams, “An Air Pollution Phytotron,” 474. Shown in barometeric-like diagrams, the intensity differed in room 1 from 1,200 to 1700 ft-c, or nearly 70 percent.

75. Richardson, *Physics in Botany*, 96, 97.

76. Went, “Fifty Years of Plant Physiology,” 105.

77. Pickering, *The Cybernetic Brain*.

78. Richardson, *Physics in Botany*, 84.

79. Went, *The Experimental Control of Plant Growth*, 11.

80. Went, *The Experimental Control of Plant Growth*, 16.

81. Went, *The Experimental Control of Plant Growth*, 38.

82. Downs, “Phytotrons,” 474.

83. Nitsch, “Phytotrons,” 42.

84. Rees et al., “Preface,” ix.

85. Todes, “Pavlov’s Physiology Factory,” 214.

86. The questioner was probably James Shapiro. Highkin, “The Effect of Constant Temperature Environment,” 231, 238. At the International Biological Program (IBP) conference on photosynthesis in 1969, Theodore Alberta from the Dutch Institute for Biological and Chemical Research on Agricultural Crops in Wageningen, asked, “Should we have a so-celled ‘square climate’ in which factors like temperature and light intensity are changed from one level to another within minutes, or should we try and imitate the changes as they occur in nature? Here no definite choice can be made,” he concluded. Alberda, “On the Use of Plants,” 592.

87. Gleick, *Chaos*.

88. May, *Stability and Complexity*, 109, 109–10.

89. See, for instance, the memo from Beadle to Went, April 5, 1951. Biology Division Papers, file 75.16.

90. See Beadle to Went, March 1953; Went to Beadle, March 24, 1953; Beadle to Went, March 26, 1953. Biology Division Papers, file 75.16.

91. Frits Went Diary, August 19, 1958. Frits Went papers. Record Group 3/2/6, box 22.

92. Frits Went Diary, March 3, 1959. Frits Went papers. Record Group 3/2/6, box 22.

93. Frits Went Diary, August 19, 1958.

94. “Report of Committee for Screening of the Applications for Graduate Standing,” February 17, 1954. Biology Division Papers, file 14.10.

95. "Screening Committee Graduate Application" recommendations, February 13, 1956. Biology Division Papers, file 16.4.

96. National Science Foundation, "Financial Support Available for Graduate Students," August 1, 1953. Biology Division Papers, file 52.5. It was the same in astronomy. As the head of Caltech's graduate program, Jesse Greenstein lamented, astronomy's existence depended on attracting "brilliant young men interested in pure science" with offers that were competitive with the "fields of industry, engineering, and physics where the ultimate salaries are high and financial security greater." "Draft Proposal for Support of Astronomy by the National Science Foundation," n. d. (ca. 1951). Greenstein Papers, folder 27.2.

97. "Auxin a" was declared $C_{18}H_{32}O_5$, or trihydroxy-monocarboxylic acid.

98. Wildman, "The Auxin-A, B Enigma," 38. Sorting out the enigma has been a minor crusade for plant physiologists. See Karlson, "Ectohormones and Phytohormones," and Troyer, "Error or Fraud in Science."

99. Wildman, "The Auxin-A, B Enigma," 54. While Karlson was satisfied to lay the blame for the auxin-A, B fraud entirely at the door of Erxleben, Wildman could not concur. Wildman "had a feeling that Kögl and Haagen-Smit were not entirely innocent victims of Erxleben's machinations" (58).

100. See letter from Bonner to Karlson, November 4, 1982. James Bonner Papers, file "auxin a&b." For the period of excitement around the initial identifications, see Weevers, *Fifty Years of Plant Physiology*, 173–75.

101. Certainly Went still believed in auxin a and b in 1957. See Went, "Fifty Years of Plant Physiology,: 107; Letter from Bonner to Karlson, November 4, 1982. Went's own biographers for the fellows of the American National Academy of Science "found him fiercely loyal" as well as also a "vigorous opponent to his antagonists, holding firm convictions and sometimes championing highly unpopular ideas. For example, despite growing evidence that 3-indoleacetic acid is the major native auxin of plants, he continued to believe in the existence of auxin *a* until his death." See Galston and Sharkey, "Frits Warmolt Went," 353.

102. Letter from Bonner to Wildman, February 3, 1995. James Bonner Papers, file "auxin a&b."

103. Went, *The Experimental Control of Plant Growth*. Cf. pp. 97 and 96.

104. Letter to Beadle, April 14, 1952. Biology Division Papers, file 28.1.

105. Juhren, Hiesey, and Went, "Germination and Early Growth," 288.

106. The subject of orchids in the phytotron is very revealing of Went's conception of his new facility, particularly as he struggled to use popular mechanisms to support his work. Reading the accounts of the phytotron, rather than Went's glamorous picture in *The Experimental Control of Plant Growth*, or newspaper stories, or late-in-life reminiscences, we note how the interest in phytotrons among the cymbidium society's members declined steadily throughout the 1950s until eventually they had only one truck among six hundred in the facility shortly before Went moved to Saint Louis, Missouri, in 1957. While Went told his readers that the Earhart Laboratory received "financial support . . . from the American Orchid Society, the Cymbidium Society, and a number of individual orchid growers" (Went, *The Experimental Control of Plant Growth*, 148), actually in August 1955, Casamajor maintained 16/570 trucks for orchids in the phytotron, but by November 1955, it was only

4/595, and finally by May 1957, it was only 1/696. See 31.VIII.55; November 30, 1955; May 3, 1957. Biology Division Papers, file 27.10. In short, applied science paid for basic science.

107. "The Division of Biology Present Selections."

108. Letter from Bonner to George Irving, ARS, September 24, 1964. James Bonner Papers, file 22.3.

109. Memo from Lang to Bonner, October 13, 1964. James Bonner Papers, file 22.3.

110. Memo from Bonner to Hellmers/Brady, October 12, 1964. James Bonner Papers, file 22.3.

111. Letter from Bonner to Lang, September 25, 1964. James Bonner Papers, file 22.3.

112. Anton Lang, "Director's Report," in *Plant Research '66: Annual report of the MSU/AEC Plant Research Laboratory*. 1966. UA3.0, box 313, folder 36, p. 1. Michigan State University Archives.

113. Letter from Gene Wilson to Chuck Newtown, April 13, 1973. Historical Files—Biology Division, file A3.3.

114. Brief Biography of Sinsheimer attached to agenda for November 10, 1952. Biology Division Papers, file 14.10.

115. Rasmussen, *Gene Jockeys*, ch. 3.

116. Little though they needed it, the Lucy Mason Clark Fund enriched molecular biology's coffers, continuing to award fellowships even though a "specific program [of plant research] no longer exists in the division of Biology." Letter from Wilson to Newtown, April 13, 1973.

117. "Activities and Plans for the Biology Division," May 18, 1970. James Bonner Papers, file 20.20.

118. Letter from Bonner to Ben Burr, BNL, January 27, 1981. James Bonner Papers, file 21.5.

119. Delbrück, "A Physicist Looks at Biology."

120. Letter from Bock to Beadle, July 1970. Biotron Papers, Series 06/80, Box 2, file "Biotron dedication."

121. "Standardization of Controlled-Environment Research," n. d. (ca. September 1974). Biotron Papers, Series 06/80, Box 6, file "Project #73016."

122. Evans, "Memoirs of a Meandering Biologist," June 2005 (manuscript). NLA, MS9885, 45–46.

CHAPTER 3. THE CLIMATRON

Epigraph: Went, "The Climatron of St. Louis," handwritten notes, n. d. (ca. 1960). In Frits Went papers. Record Group 3/2/6/1, box 1, folder 14. Archives. Missouri Botanical Gardens.

1. Sandweiss, "From a Garden Looking Out."

2. Record Group 3/2/6, box 21, folder 7. Archives. Missouri Botanical Garden.

3. Frits Went Diary, October 10, 1958. Frits Went papers. Record Group 3/2/6, box 22. Archives. Missouri Botanical Garden.

4. Rader and Cain, *Life on Display*.

5. I am drawing ideas about scientific infrastructure from Edwards, *A Vast Machine*, 22. Chapters 1 and 8 are especially valuable.

6. Went to Brookings Smith, May 1, 1958. Frits Went Papers. Record Group 3/2/6/1, box 1, folder 18.

7. Layton, Layton, and Rohrbach, "Master Plan," 36. Record Group 3/2/6, box 20, folder 14. Archives. Missouri Botanical Garden.

8. Sheller, *Aluminum Dreams*; Cooper, *Air-Conditioning America*.

9. Frits Went Diary, April 21, 1958. Frits Went papers. Record Group 3/2/6, box 22. Went echoed the Modernist exclamation of Ezra Pound—"Make it New!"—when he called for old greenhouses to be swept aside and replaced with phytotrons. See, Gay, *Modernism*, 4.

10. Gordon, *Mapping Decline*; Sugrue, *The Origins of the Urban Crisis*, 164–65.

11. Mumford, *Climatron 50*, 38–43.

12. Misa, "The Compelling Tangle," 5.

13. *The Arithmetic Teacher* 9, no. 6 (1962), 329; *Art Education* 29, no. 3 (1976), 26; *Journal (American Water Works Association)* 73, no. 4 (1981), 18.

14. "Plants of the Bible," n. d. (ca. 1958). Record Group 3/2/6, box 16, folder 15. Archives. Missouri Botanical Garden.

15. Faherty, *Henry Shaw*, 78–89.

16. Frits Went Diary, April 21, 1958. Frits Went papers. Record Group 3/2/6, box 22.

17. Kleinman, "His Own Synthesis."

18. Frits Went Diary, April 19, 1958. Frits Went papers. Record Group 3/2/6, box 22. Archives. Missouri Botanical Garden.

19. Frits Went Diary, April 21, 1958.

20. Robert Brookings Smith was the most prominent member of the cultural elite of Saint Louis in the story of the creation of the Climatron. The meaning of cultural elite can be gained from his Memorial in the *Princeton Alumni Weekly* from March 2004. "Robert Brookings Smith '26. Born in St. Louis, he prepared at Lawrenceville and while at Princeton participated in football, crew, Triangle Club, and Cottage Club. He left Princeton in 1924 and joined the St. Louis brokerage firm of Smith, Moore & Co., where he became a limited partner. . . . Bob joined Mercantile Trust Co., later becoming a board member and vice chairman. Still later, he started two businesses: Cashex Inc., specializing in automatic check-authorization cards, and National Cache Card, a developer of smart cards for universities. He entered the Navy in 1942, serving as lieutenant commander and navigator aboard the aircraft carrier USS White Plains in WWII, and was decorated for heroism in the Battle of the Philippine Sea. Until his death, he was a managing partner of Common Bond Associates of St. Louis, a research and development partnership. He was a longtime trustee of Washington U. in St. Louis. The Class of 1926." To understand such a memorial means you are part of the cultural elite, whereas if you do not you are not.

21. Went to Brookings Smith, May 1, 1958. Frits Went papers. Record Group 3/2/6/1, box 1, folder 18.

22. Frits Went Diary, May 2, 1958. Frits Went papers. Record Group 3/2/6, box 22.

23. Memo from Went, May 6, 1959. Record Group 3/2/6, box 16, folder 25.

24. "General Outline of Proposed Program of Use of Missouri Botanical Arboretum," n. d. (ca. 1959). Record Group 3/2/6, box 16, folder 25.

25. Faherty, *A Gift to Glory In*, 171–200.

26. Layton, Layton, and Rohrbach, "Master Plan," 6. Record Group 3/2/6, box 20, folder 11. Archives. Missouri Botanical Garden.

27. Kleinman, "His Own Synthesis," 317.

28. John Holloway to the Director of the New Zealand Forest Service, July 29, 1955. Research Phytotron (Controlled Climate Facilities) 1955–1973, F 1 W3129 (Box 247) 41/7. Archives New Zealand.

29. Collection 1, RG 3/1 Series 2, box 1, file "News Releases 1958–59." Archives. Missouri Botanical Garden.

30. Went to Levin, February 25, 1959. Record Group 3/2/6, box 1, folder 2. Archives. Missouri Botanical Garden.

31. Letter from Went to Eversole and Clark, May 20, 1958. Frits Went papers, Record Group 3/2/6/1, box 1, folder 18.

32. "Employees of Missouri Botanical Garden." 1958. Record Group 3/2/6, box 16, folder 7. Archives. Missouri Botanical Garden.

33. Handwritten notes, n. d. (ca. 1958). Frits Went papers. Record Group 3/2/6/1, box 1, folder 18; Faherty, *A Gift to Glory In*, 178–79.

34. Report of Kenneth Smith to Went, October 28, 1958. Record Group 3/2/6, box 16, folder 11. Archives. Missouri Botanical Garden.

35. Minutes of Meeting between Went, Robert Smith, Hugh Cutler, Richman, Layton, and Rohrbach concerning the Palm House reconstruction, February 11, 1959. Archives. Missouri Botanical Garden. Collection 1, RG 3/2/6 Series 4, Box 19. File Misc Subjects: 1/14/59–2/11/59.

36. Minutes of meeting, November 27, 1957. Went to Levin, June 16, 1960. Record Group 3/2/6, box 1, folder 7.

37. Anderson and Cutler, "Development Program for Research at the Missouri Botanical Garden," n. d. (ca. 1958). Record Group 3/2/6, box 1, folder 7. Archives. Missouri Botanical Garden.

38. Anderson and Cutler, "Development Program for Research at the Missouri Botanical Garden."

39. Rohrback to Went, February 20, 1959. Record Group 3/2/6, box 20, folder 1. Archives. Missouri Botanical Garden.

40. Went to Layton, February 16, 1960. Record Group 3/2/6, box 20, folder 6. Archives. Missouri Botanical Garden.

41. Rohrback to Went, February 20, 1959

42. Rohrbach to Went, March 13, 1959. Record Group 3/2/6, box 20, folder 2. Archives. Missouri Botanical Garden.

43. Cutler to Alan Waterman, NSF, December 22, 1958. Record Group 3/2/6, box 1, folder 1, p. 3. Archives. Missouri Botanical Garden.

44. See Glaessner to the Business Committee, May 16, 1958. Frits Went papers. Record Group 3/2/6/1, box 1, folder 18.

45. Went to Levin, February 25, 1959.

46. Frits Went Diary, March 3, 1959. Frits Went papers. Record Group 3/2/6, box 22. These were serious names, as the future director of the garden, Peter Raven reminded

his readers years later, "Went had previously considered *plantosphere, sylvarium* and *floradome.*" "The Climatron at Twenty," 1.

47. "The Climatron of St. Louis," handwritten notes, n. d. (ca. 1960). Frits Went papers. Record Group 3/2/6/1, box 1, folder 14.

48. "The Climatron Opens to the Public," 131.

49. "The Climatron of St. Louis," handwritten notes.

50. "The Climatron Opens to the Public," 131.

51. Pantheon: Climatron. File "Reynolds Award." Record group 3/2/6, box 21, folder 4. Archives. Missouri Botanical Garden; File "Climatron-maps, diagrams, and index sheets." Archives. Missouri Botanical Garden.

52. Layton to Went, August 13, 1959. Record Group 3/2/6, box 20, folder 4. Archives. Missouri Botanical Garden.

53. Layton to Went, August 13, 1959.

54. Went to Layton, September 1, 1959. Record Group 3/2/6, box 20, folder 5. Archives. Missouri Botanical Garden.

55. "The Climatron of St. Louis," handwritten notes.

56. "The Climatron Opens to the Public," 131. Robert Moses wanted to sacrifice Washington Square Park to a new freeway running up Fifth Avenue, while Le Corbusier wanted to sweep clean most of the right bank of Paris to build new apartment buildings.

57. "Fact Sheet: The Climatron," n. d. (ca. 1960). Record Group 3/2/6. Box 21, folder 4. Archives. Missouri Botanical Garden.

58. "The Climatron Opens to the Public," 131.

59. All quotes from "Los Angeles Country Arboretum" and Climatron of St. Louis, in Frits Went papers. Record Group 3/2/6/1, box 1, folder 14.

60. "The Climatron of St. Louis," handwritten notes.

61. "Los Angeles Country Arboretum" and Climatron of St. Louis, 29.

62. "Los Angeles Country Arboretum" and Climatron of St. Louis, 15–20. As well as the finished product, "Teachers' and Group-Leaders' Guide to the Climatron." Record Group 3/2/6, box 16, folder 14. Archives. Missouri Botanical Garden.

63. "The Climatron Opens to the Public," 131.

64. See American Institute of Architects news release, April 2, 1961. Frits Went biographical file. Archives, Missouri Botanical Garden.

65. See American Institute of Architects news release, April 2, 1961.

66. "The Climatron as a Source of Revenue," n. d. (ca. 1961). Record Group 3/2/6, box 16, folder 13. Archives. Missouri Botanical Garden.

67. Handwritten notes, n. d. (ca. 1964). Missouri Botanical Garden Archives. Collection 1, RG 3/2/6/1, Box 1. Folder 8: Climatron 1963–64.

68. "The Climatron as a Source of Revenue."

69. "The Climatron as a Source of Revenue."

70. Frits Went Diary, March 3, 1959. Frits Went papers. Record Group 3/2/6, box 22.

71. Attached notes to letter from James van Sant to Went, February 8, 1961. Record Group 3/2/6, box 20, folder 8. Archives. Missouri Botanical Garden. There is a broader connection to the rise of experimental evolution as well. Sharon Kingsland has noted that the shape of ecology as a discipline in the twentieth century was a result of a larg-

er project to expand research in experimental evolution. Kingsland, *The Evolution of American Ecology*, 5.

72. Frits Went Diary, March 3, 1959.

73. Went to Levin, June 16, 1960.

74. "With the redevelopment program of the Garden, completely new sources of revenue can be envisaged, along a number of different lines:- rendering of services, providing of education and the creation of special experiences, in which the spectacular is wedded to entertainment and education. The Climatron belongs to the last-mentioned group." "The Climatron as a Source of Revenue."

75. Went to Levin, June 16, 1960.

76. Letter from Went to Hitchcock, November 21, 1962. Frits Went papers. Record Group 3/2/6/1, box 1, folder 21; see attached "Memorandum to the Board of Trustees of the Missouri Botanical Garden." Frits Went papers. Record Group 3/2/6/1, box 1, folder 21.

77. Handwritten notes, n. d. (ca. 1963). Frits Went papers. Record Group 3/2/6/1, box 1, folder 21.

78. Trustee Board minutes, April 30, 1963. Frits Went papers. Record Group 3/2/6/1, box 1, folder 21.

79. Notes attached to letter from James van Sant to Went, February 8, 1961.

80. Trustee Board minutes, April 30, 1963.

81. Letter from Went to Kramer, December 13, 1963. Phytotron Records, box 4, file "Phytotron correspondence thru Feb 1964." Duke University Archives.

82. Anderson, "The First Five Years," 1.

83. *Silent Running*, dir. Douglas Trumball; Mumford, *Climatron* 50, 91.

84. Keller, "Nature, Nurture," 290.

CODA I

Epigraph: Went, Handwritten notes, n. d. (ca. 1950–51?). Frits Went papers. Record Group 3/2/6/1, box 4, folder 10.

1. Billings, "The Environmental Complex," 256.

2. Harland-Jacobs, *Balancing Tradition and Innovation*, 28; Wilson, *Naturalist*, ch. 12.

3. Went, "Centenary of Schleiden's Textbook," 147.

4. Galston and Sharkey, "Frits Warmolt Went," 354, 359.

5. Letter from Arditti to Went, May 24, 1984. Frits Went papers. Record Group 3/2/6/1, box 12, folder 5.

6. Salisbury and Ross, *Plant Physiology*, 379.

7. Went, "Fifty Years of Plant Physiology," 105.

8. "Breve Conclusion," in Rapport d'Activite Scientifique di Phytotron, 1971–72. Versement 850101LABOS, article 10, dossier "phytotron," 178. Dépôt des Archives du CNRS (Paris).

9. Galston and Sharkey, "Frits Warmolt Went," 355.

10. Talk titled "Le Phytotron comme trait d'union entre l'interpretation et la réalité d'une plante," n. d. (ca. 1976). Frits Went papers. Record Group 3/2/6/1, box 5, folder 21.

CHAPTER 4. THE POSTCOLONIAL SCIENCE OF THE AUSTRALIAN PHYTOTRON

Epigraph: "Speeches at the Opening of the Phytotron—Prime Minister Robert Menzies," 1962. CSIRO Archives, Series 400.

1. "Speeches at the Opening of the Phytotron—Prime Minister Robert Menzies."

2. Lloyd Evans, "Memoirs of a Meandering Biologist," June 2005 (manuscript). NLA, MS9885, p. 38.

3. Evans, Wardlaw, and King, "Plants and Environment," 209.

4. Evans, Wardlaw, and King, "Plants and Environment," 216.

5. Evans, "CERES," 1142.

6. Radio astronomy, computing facilities, university nuclear reactors, and controlled environment laboratories no doubt addressed particular local issues, yet considered together the similarities are striking. Chronologically, by the time Sputnik launched, these four endeavors had become standard throughout the Western world. Australia started building its radio telescope at Parkes, had seen the Lucas Heights nuclear reactor become operational, and the Australian government had just approved millions in funding for a phytotron. In France, an early exploration adventure in radio astronomy had given way to a cooperative venture led by the Netherlands, but had established the GIF phytotron outside Paris, and its first centralized computer came online in 1958. In the United Kingdom too, the story runs eerily parallel. By 1957, the University of Manchester had built and was running one of the earliest electronic computers, the University of Manchester and Cambridge both had erected large radio telescopes, and Imperial College London had bid successfully to administer Britain's first "teaching reactor" at Silwood, which was destined to train the teams of nuclear engineers that any large-scale nuclear power industry would require. Among many more, for computing, see Beard and Pearcey, "The Genesis of an Early Stored-Program Computer"; Croarken, "The Beginnings of the Manchester Computer"; Akera, *Calculating a Natural World*. For atomic energy, see Cawte, *Atomic Australia*; Home, "The Rush to Accelerate"; Krige, "The Installation of High-Energy Accelerators"; Hecht, *The Radiance of France*; Gay, *The History of Imperial College*, 289, 659. And for radio astronomy, see Sullivan, "Early Years of Australian Radio"; Agar, *Science and Spectacle*; Pestré, "Studies of the Ionosphere."

7. "Speeches at the opening of the Phytotron—Prime Minister Robert Menzies."

8. When India conducted the second round of atomic tests in 1998, Itty Abraham relates the comments of an Indian political critic who "couldn't escape the current glee [at] the thought of what other nations would say—they wouldn't be able to kick us around as before." Abraham, *The Making of the Indian Atomic Bomb*, 1. See also Gabrielle Hecht, *Being Nuclear*, introduction.

9. Gunn, *High Encounters*, 109. The Australian government specifically reequipped QANTAS with the "Super Constellations," "as Australia's instrument in international aviation." Gunn, *High Encounters*, 1. When James Bonner flew the Pan Am clipper to Australia, the company opened its booklet with the claim: "you selected the most modern way to travel." See "Flying Clipper=Wise," James Bonner Papers, file "Australia." The moment is indicative of the resurgent modernism of the postwar, postcolonial period. Literary modernism, Michael North argued, "has been transformed in the general estimation from [a] brilliant young technocrat into a doddering old paranoiac

. . . a history that seems to match that of Howard Hughes better than that of [T. S.] Eliot or [James] Joyce." North, *Reading 1922*, 10. Likewise, architectural historians have long known that "ocean liners, grain elevators, and airplanes of the industrial West took center stage as the exclusive sources of twentieth-century modernism." Bozdogan, *Modernism and Nation Building*, 4. As Thomas Misa asked rhetorically, "Is there anything more assertively modern and more thoroughly technological than an airport?" Misa, "The Compelling Tangle," 1.

10. Quoted in Cullather, "Miracle of Modernization," 227. Adas, *Machines as the Measure*.

11. The debate rages still. Ward, *Australia and the British Embrace*. Contrast Carl Bridge, "Anglo-Australian Attitudes," 196 to Cottle, "A Bowyang Historian," 172. Similarly, the Australian minister for Immigration in the late 1960s who maintained that Australia "should have a monoculture . . . We don't want pluralism," while his own leader constantly returned to the theme of Australia as a "son grown to adulthood." Bolton, *The Oxford History of Australia*, 107, versus Watson, "Rabbit Syndrome," 37. David Lee illustrates this ambiguity nicely, noting that the Menzies government "endeavored to throw off Australia's dependence on the sterling area" but also advocated a "readiness to fit into Anglo-American strategic plans." Lee, *Search for Security*, 108. It is the parallel existence of both that presents, ironically, a uniquely Australian identity in this period. To quote Don Watson: "Australia is as much a lifestyle as it is a nation."

12. Oakman, *Facing Asia*, 86.

13. Oakman, *Facing Asia*, 86. It is also worth noting, though shameful to many Australians now, that the "White Australia" policy remained the official immigration policy until the early 1970s.

14. Ward, "Security," 251.

15. Oakman, *Facing Asia*, 88.

16. Westad, *The Global Cold War*.

17. Evans, Wardlaw, and King, "Plants and Environment," 208.

18. MacLeod, "On Visiting the 'Moving Metropolis,'" esp. 220–21.

19. Munns, *A Single Sky*, 41.

20. Schedvin, "Environment, Economy," 114.

21. Macintyre, *Australia's Boldest Experiment*, 166.

22. For budget and staff levels, see The Parliament of the Commonwealth of Australia, *Annual Report of the Council of Scientific and Industrial Research*, for the year ending 1946. Twentieth Annual Report (June 1945–46); The Parliament of the Commonwealth of Australia, *Annual Report of the Council of Scientific and Industrial Research*, for the year ending 1948. Twenty-second Annual Report (June 1947–48); The Parliament of the Commonwealth of Australia, *Annual Report of the Commonwealth Scientific and Industrial Research Organisation*, for the year ending 1952. Fourth Annual Report (June 1951–52). N.b., the value of the Australian pound circa 1950 was £1 = US$4.

23. Ian Clunies-Ross to D. G. Catcheside, March 30, 1951. Series A10651, file ICR 18/10. National Archives of Australia.

24. Schedvin, "The Culture of CSIRO," 85.

25. Schedvin, "The Culture of CSIRO," 86.

26. The working contemporary definition said that applied research had a "definite economic objective," whereas basic research had "no recognizable economic objective." These economically based definitions are unsurprising due to the larger aim of CSIRO to foster Australia's national development. See Gillespie, "Research Management," 14.

27. This brief biographical sketch is formed from Evans, "Sir Otto Frankel"; Gavin McCarthy Oral Interview with Otto Frankel, NLA 1985; M. Blythe, videotaped interview AAS, November 15, 1993.

28. O. H. Frankel, "Plant Industry—Reorganisation and Research Planning," March 24, 1952. NAA A9778, B1/5/62, p. 1.

29. Frankel, "Plant Industry," 2.

30. Evans, "Sir Otto Frankel," 12.

31. Letter from Frankel to Bonner, August 26, 1960. James Bonner Papers, box 12, folder "Australia 1958–59 & 60." For Bonner, see Letter from Bonner to P. Pearson, Ford Foundation, September 19, 1960. James Bonner Papers, box 12, folder "Australia 1950–59."

32. Letter from Russell to Went, March 5, 1953. Biology Division Papers, file 29.3.

33. Letter from Evans to Went, February 1, 1953. Biology Division Papers, file 29.3.

34. Evans, "Conjectures, Refutations, and Extrapolations," 5.

35. Letter from Went to Hammond, The Commonwealth Fund, January 21, 1955. Biology Division Papers, file 29.3.

36. Evans, "Memoirs of a Meandering Biologist," 47.

37. Evans, "Memoirs of a Meandering Biologist," 35.

38. Evans, "Memoirs of a Meandering Biologist," 38.

39. Evans, "Memoirs of a Meandering Biologist," 39.

40. Evans, "Memoirs of a Meandering Biologist," 47.

41. NAA. A4940, C2060. General Administrative Committee, Cabinet Minute—Submission No. 1106—Phytotron for CSIRO. April 29, 1958.

42. "Multi-science" quote in Frankel, "The Canberra Phytotron," 13. "Colombo" in *Rural Research in CSIRO* 41 (1962), Editorial.

43. Evans, "Memoirs of a Meandering Biologist," 42, 44.

44. Evans, "Memoirs of a Meandering Biologist," 44.

45. NAA, A463/61. 58/960. March 18, 1958. Prime Minister's Department Memo, RD to FM.

46. NAA. A4940, C2020. H. J. Goodes, First Assistant Secretary, to Treasurer, "Phytotron for CSIRO—Submission No. 1106. April 22, 1958.

47. Munns, *A Single Sky*, ch. 4.

48. Evans, "Memoirs of a Meandering Biologist," 40–41.

49. Harland, "Changing the Behaviour of Plants," 30.

50. File "Australian National University: Tandem Generator Proposal," June–July 1958. Australian Academy of Science. I very much thank Rosanne Walker at the Basser Library for locating this material.

51. Titterton's Exposition, December 4, 1957. File "Australian National University: Tandem Generator Proposal," (1037/1957) Australian Academy of Science.

52. NAA, A4940. C2060. March 28, 1958. Letter from Ian Clunies-Ross to R. G. Casey.

53. Frankel, "The Canberra Phytotron," 13.

54. NAA. A4940, C2060. March 28, 1958. "What Is Needed for an Australian Phytotron?" Attached to a letter from Ian Clunies-Ross to R. G. Casey.

55. NAA. A4940, C2060. March 28, 1958. "What Is Needed for an Australian Phytotron?"

56. Fred White CSIRO Archives, Series 400. 1962. "Speeches at the opening of the Phytotron." Fundamentally reductionist in nature, the same classic narrative of science was powerfully embraced by molecular biology in the same period. See Morange, *A History of Molecular Biology*, 5.

57. Traweek, *Beamtimes and Lifetimes*, 124.

58. NAA, A4940. C2060. March 28, 1958. Letter from Ian Clunies-Ross to R. G. Casey. The ability of the Phytotron to attract overseas scientists was lauded in Frankel, "The Canberra Phytotron," 12, where the captions to the photographs of the building said "[CERES] will be open to . . . overseas guest workers, particularly from the Colombo Plan area."

59. NAA. A4940, C2060. General Administrative Committee, Notes on Cabinet Submission No. 1106. April 28, 1958. This argument was also used with the Rockefeller Foundation two years later. See JJM Diary excerpt, July 5, 1960, folder 76, box, 7, Record Group, 410D, Rockefeller Foundation Archives, Rockefeller Archive Center, North Tarrytown, New York (hereafter RAC). JJM's statement is a compact version of the formal statement of Frankel in the letter requesting Rockefeller support. For the application, see O. H. Frankel to A. H. Moseman, November 4, 1960, folder 76, box, 7, Record Group, 410D, RAC.

60. NAA, A4940. C2060. March 28, 1958. "What Is Needed for an Australian Phytotron?" Attached to Letter from Ian Clunies-Ross to R. G. Casey.

61. Kaiser, "Cold War Requisitions"; Munns, "If We Build It."

62. Charrier, "Britain, India and the Genesis,"10.

63. Charrier, "Britain, India and the Genesis," 125.

64. Oakman, *Facing Asia*, 4.

65. Oakman, *Facing Asia*, 3.

66. Oakman, *Facing Asia*, 38.

67. Oakman, *Facing Asia*, 82.

68. Nanda, "Problems of Fundamental and Applied," 61.

69. Evans, "Sir Otto Frankel," 13.

70. Evans, "Memoirs of a Meandering Biologist," 42.

71. See https://www.science.org.au/about-us/shine-dome/architecture .

72. Cullather, "Miracle of Modernization," 233–34.

73. Frankel, "The IRRI Phytotron."

74. Menzies CSIRO Archives, Series 400. 1962. "Speeches at the Opening of the Phytotron"; Fred White. CSIRO Archives, Series 400. 1962. "Speeches at the Opening of the Phytotron"; Frankel CSIRO Archives, Series 400. 1962. "Speeches at the Opening of the Phytotron."

75. Frits Went Diary, August 24, 1962. Frits Went papers. Record Group 3/2/6, box 22.

76. Frankel, "The Canberra Phytotron," 13. Other examples of the popularization of the phytotron's engineering ingenuity are seen in "The CSIRO Phytotron."

77. Discussion of L. T. Evans's paper, "Extrapolation from Controlled Environments," in Evans, *Environmental Control of Plant Growth*, 435.

78. Went, "The Concept of a Phytotron," 3.

79. Went, 'The Concept of a Phytotron,' 3.

80. Frits Went Diary, August 25, 1962. Frits Went papers. Record Group 3/2/6, box 22.

81. Letter from Harry Highkin to R. N. Robertson, December 20, 1959. R. N. Robertson Papers. MS117/1/2.

82. Robertson, "Phytotrons, Spectrophotometers and Productivity," 314.

83. Letter from Highkin to Robertson, December 20, 1959.

84. Evans, "Memoirs of a Meandering Biologist," 45.

85. Letter from Highkin to Robertson, December 20, 1959.

86. Frankel, "The Social Responsibility," 302.

87. Anderson and Adams, "Pramoedya's Chickens," 183.

88. Abraham, *The Making of the Indian Atomic Bomb*, 20.

89. Abraham, *The Making of the Indian Atomic Bomb*, 10.

90. McCray, *The Visioneers*, 24–26.

91. Frankel, "Internationalism in Agricultural Science," 314.

92. Oakman, *Facing Asia*, 73.

93. Weinberg, *Reflections on Big Science*, 2–3.

94. Frankel, "The Social Responsibility," 301.

95. Childs, *Modernism and the Post-Colonial*, 6.

96. Young, *Postcolonialism*.

97. Chrisman and Williams, *Colonial Discourse*, 8. For a summary of most critiques of postcolonialism see, Hall, "When Was 'The Post-Colonial'"? esp. 242, 245, 249.

98. Bullard, "Imperial Networks," 198, 202 for valorization. The category "transcultural" itself is seemingly named for the journal established in 1956, *Transcultural Psychiatry*, which stresses its multicultural and interdisciplinary publications and readership, and seeks to relegitimatize the universal basis of science after the exposure of Imperial knowledge imposed upon the colonized.

99. Bullard, "Imperial Networks," 205–6. Likewise, mathematics possesses the cultural imprint of Western values revealed, tellingly, in the very act of imposing "universal" mathematics on nonwestern societies. As Alan Bishop outlined, the process of rationalism, objectism, and power and control through mathematics has been almost a secret weapon of cultural imperialism through a generally assumed universality and neutrality. See Bishop, "Western Mathematics."

100. Worboys, "Science and the Colonial Empire," 13–14.

101. North, *Reading 1922*, 14.

102. CSIRO Archives, Series 400. 1962. "Speeches at the Opening of the Phytotron."

103. Boehmer, *Colonial and Postcolonial Literature*, 237.

CHAPTER 5. THE TWIN PHYTOTRONS OF THE RESEARCH TRIANGLE BETWEEN DUKE AND NORTH CAROLINA STATE

Epigraph: Paul J. Kramer, "Botany in a Changing World," n. d. (ca. 1961). Duke University Archives. Biographical File—Kramer, Paul, 5.

1. Paul Kramer, Unpublished Autobiographical Notes, ca. 1988. Duke University Archives. Biographical Files—Paul J. Kramer, 14.

2. Paul J. Kramer, "Botany in a Changing World," n. d. (ca. 1961). Duke University Archives. Biographical File—Kramer, Paul, 5.

3. Paul Kramer, Unpublished Autobiographical Notes. ca. 1988. Duke University Archives. Biographical Files—Paul J. Kramer, 24.

4. Johnson to Hellmers, December 4, 1964. Phytotron Records, box 4, file "Hellmers." Duke University Archives.

5. Kramer to Hellmers, December 22, 1964. Phytotron Records, box 4, file "Hellmers." Duke University Archives.

6. Harland-Jacobs, *Balancing Tradition and Innovation*, 25.

7. At least, so says Stephen Hawking, "Foreword."

8. Kohler, *Lords of the Fly*, 12.

9. Johnson to Hellmers, December 4, 1964. Phytotron Records, box 4, file "Hellmers." Duke University Archives.

10. Kohler, "The Ph.D. Machine," 658.

11. Warwick, *Masters of Theory*. See also Kaiser and Warwick, *Pedagogy and the Practice of Science*; Jackson, "Visible Work."

12. Harland-Jacobs, *Balancing Tradition and Innovation*, 25.

13. The subheading of this section is the tagline used in R. J. Reynolds Tobacco Company's print advertisements for Camel cigarettes in the 1960s.

14. "A Preliminary Proposal for the Development of a Phytotron," June 1961. Records of the Associate Dean and Director, North Carolina Agricultural Research Center. UA 101.001, carton 42. North Carolina State University Archives.

15. Keller to Bonner, August 8, 1961. Records of the Associate Dean and Director, North Carolina Agricultural Research Center. UA 101.001, carton 42. North Carolina State University Archives. Went believed that the North Carolina group "based their proposal very much on my book." Frits Went Diary, March14, 1962. Frits Went papers. Record Group 3/2/6, box 22. Archives. Missouri Botanical Garden.

16. "A Phytotron for Tobacco Research," November 29, 1961; Letter from N.C. State President John T. Caldwell to William Friday, November 28, 1961. Papers of the Office of the Dean of the College of Agricultural and Life Sciences. UA 100.001, box 28, folder 2. North Carolina State University Archives.

17. "A Phytotron for Tobacco Research," November 29, 1961.

18. "A Phytotron for Tobacco Research," November 29, 1961.

19. "A Preliminary Proposal for the Development of a Phytotron," June 1961.

20. Letter from D. W. Colvard, President Mississippi State, to Bosman Gray, R. J. Reynolds, May 22, 1962. In Papers of the Office of the Dean of the College of Agricultural and Life Sciences. UA 100.001, box 28, folder 2. NC States University Archives.

21. Kramer to Cole, March 15, 1962. Phytotron Records, box 4, file "Phytotron, 1961–62." Duke University Archives.

22. "Report of the President to the Board of Trustees," March 8, 1961. Papers of the Office of the President, J. Deryl Hart, box 54, file 232 "President's Reports and Speeches." Duke University Archives. p. 4, 8.

23. Harland-Jacobs, *Balancing Tradition and Innovation*, 26.

24. Harland-Jacobs, *Balancing Tradition and Innovation*, 26; Wilson, *Naturalist.*

25. "Remarks to Graduating Classes," June 4, 1962. Papers of the Office of the President, J. Deryl Hart, box 54, file 232, "President's Reports and Speeches," 2. Duke University Archives.

26. Harland-Jacobs, *Balancing Tradition and Innovation*, 27.

27. Kramer(?), Memorandum, "Some Questions Likely to Be Asked by the Phytotron Panel," n. d. (ca. 1963). Papers of the Office of the President Douglas Knight, box 21, file, "Phytotron, 1963–69." Duke University Archives.

28. Kramer to Cole, March 15, 1962.

29. "Memorandum of agreement for the construction of a phytotron," July 23, 1962. Papers of the Office of the President, J. Deryl Hart, box 54, file 232 "President's Reports and Speeches." Duke University Archives.

30. Kramer to Cole, March 15, 1962.

31. Robinson to Levin, February 11, 1959. Phytotron Records, box 4, file "Phytotron-concerning Duke's 1st proposal-1958–60." Duke University Archives.

32. Robinson to Levin, February 11, 1959.

33. Harper to Levin, NSF, January 12, 1959. Phytotron Records, box 4, file "Phytotron-concerning Duke's 1st proposal-1958–60." Duke University Archives.

34. "Some Forest Research Problems . . ." n. d. (ca. 1959). Phytotron Records, box 4, file "Phytotron-concerning Duke's 1st proposal-1958–60." Duke University Archives.

35. Appel, *Shaping Biology*, Appendix B.

36. Kramer to Spencer, March 9, 1964. Papers of the Office of the President Douglas Knight, box 21, file, "Phytotron, 1963–69." Duke University Archives.

37. "A Proposal to the National Science Foundation Division of Biological and Medical Sciences for Construction of a Two-Unit Phytotron by Duke University and North Carolina State," 1963. Papers of the Office of the Dean of the College of Agricultural and Life Sciences. UA 100.001, box 28, folder 1. NC States University Archives.

38. "A Proposal to the National Science Foundation Division of Biological and Medical Sciences for Construction of a Two-Unit Phytotron by Duke University and North Carolina State," 1963.

39. Kramer to Spencer, March 9, 1964.

40. "Statement from John T. Caldwell, Chancellor of N.C. State. Attached to "A Proposal to the National Science Foundation Division of Biological and Medical Sciences for Construction of a Two-Unit Phytotron by Duke University and North Carolina State," 1963. Papers of the Office of the Dean of the College of Agricultural and Life Sciences. UA 100.001, box 28, folder 1. NC States University Archives.

41. Kramer to Hellmers, September 5, 1970. Paul Kramer papers, box 10, file "Phytotron correspondence, 1970–74." Duke University Archives.

42. "W. E. Splinter, "Remarks pertaining to the establishment of a phytotron at North Carolina State College," n. d. (ca. 1963). Phytotron Records, box 4, file "Notes for Phytotron Planning-1962–64." Duke University Archives.

43. Kramer's handwritten notes, n. d. (ca. 1958). Phytotron Records, box 4, file "Phytotron-notes on Duke's 1st proposal." Duke University Archives.

44. Kramer to Went, June 5, 1958. Phytotron Records, box 4, file "Phytotron-notes on Duke's 1st proposal." Duke University Archives.

45. Kramer's handwritten notes, n. d. (ca. 1958).

46. "W. E. Splinter, "Remarks pertaining to the establishment of a phytotron at North Carolina State College," n. d. (ca. 1963).

47. Letter from Keller to Lybrook, R. J. Reynolds, June 22, 1965. Papers of the Office of the Dean of the College of Agricultural and Life Sciences. UA 100.001, box 28, folder 2. NC States University Archives.

48. "Revised Schematic Stage Report," February 1, 1965. Phytotron Records, box 4. Duke University Archives.

49. Kramer's handwritten notes, n. d. (ca. 1958).

50. "A Proposal to the National Science Foundation Division of Biological and Medical Sciences for Construction of a Two-Unit Phytotron by Duke University and North Carolina State," 1963.

51. Billings, "The Environmental Complex," 251. The holocoenotic diagram is on p. 256. Billings cites Stebbins and Went, as well as Keck, Hiesey, and Clausen's 1948 C. I.W. paper, along with early work from Theodore Kozlowski, the future director of the Biotron at the University of Wisconsin-Madison.

52. Tesha, "Thermoperiod Effects," 3.

53. Kramer, Hellmers, and Downs, "SEPEL," 1203.

54. Tesha, "Thermoperiod Effects," 3.

55. Tesha, "Thermoperiod Effects," 13.

56. Tesha, "Thermoperiod Effects," 30–31.

57. R. J. Downs, "Progress Report for 1972," March 1, 1973. Records of the Assistant Director, North Carolina Agricultural Research Service. UA 101.005, carton 7. North Carolina State University Archives.

58. Downs, "Progress Report for 1972," March 1, 1973.

59. Downs, "Progress Report for 1972," March 1, 1973.

60. Handwritten notes included with the Minutes of the Phytotron Board meeting of November 19, 1971. Paul Kramer papers, Box 10, file "Phytotron Board, 1970–79." Duke University Archives.

61. Bartholomew to Downs, May 22, 1973. Paul Kramer papers, box 10, file "Phytotron correspondence, 1970–74." Duke University Archives.

62. Downs to Bartholomew, June 5, 1973. Paul Kramer papers, box 10, file "Phytotron correspondence, 1970–74." Duke University Archives.

63. Hellmers to Kramer, August 31, 1973. Paul Kramer papers, box 10, file "Phytotron correspondence, 1970–74." Duke University Archives.

64. Hellmers to Kramer, November 14, 1973. Paul Kramer papers, box 10, file "Phytotron correspondence, 1970–74." Duke University Archives.

65. Arthur Findeis, NSF American Embassy Tokyo, to Kramer, June 11, 1971. Records of the Associate Dean and Director, North Carolina Agricultural Research Center. UA 101.001, carton 42. North Carolina State University Archives.

66. Edwards, *A Vast Machine*, 372.

67. Hellmers to Kramer, November 14, 1973.

68. Judith Thomas to Hellmers, August 2, 1976. Paul Kramer papers, box 11, file "Phytotrons, 1972–76." Duke University Archives.

69. Compton and Benson, *Living and Working in Space,* ch. 7.

70. Marino and Odum, "Biosphere 2," notes for figure 3, p. 7. Sharon Kingsland has an excellent discussion of Odum's circuit metaphors. See Kingsland, *The Evolution of American Ecology,* 189–205.

71. Utilization of Plant Outputs," July 13, 1976, and "Utilization of Animal Outputs" attached to Judith Thomas to Hellmers, August 2, 1976.

72. Handwritten notes, n. d. attached to Judith Thomas to Hellmers, August 2, 1976.

73. Reider, *Dreaming the Biosphere,* 8. Gentry and Liptak, *The Glass Ark.*

74. Minutes of the SEPEL Board Meeting, August 6, 1974. Records of the Associate Dean and Director, North Carolina Agricultural Research Center. UA 101.001, carton 42. North Carolina State University Archives.

75. Appel, *Shaping Biology,* 259.

76. Kramer to Hellmers, December 4, 1973. Paul Kramer papers, box 10, file "Phytotron correspondence, 1970–74." Duke University Archives.

77. Hellmers to Kramer, November 14, 1973.

78. Wolfe, *Competing with the Soviets,* 103; Kaiser, "Cold War Requisitions." Nelkin, *The University and Military Research.*

79. Handwritten notes— "Discussion with Clark," June 6, 1975. Paul Kramer papers, box 10, file "Phytotron management problems-970s." Duke University Archives.

80. Handwritten notes—"Clark 2," n. d. (ca. June 9, 1975.) Paul Kramer papers, box 10, file "Phytotron management problems-1970s." Duke University Archives.

81. Handwritten notes— "Clark 2," n. d. (ca. June 9, 1975.)

82. Handwritten notes—"Clark 2," n. d. (ca. June 10, 1975.) Paul Kramer papers, box 10, file "Phytotron management problems -1970s." Duke University Archives.

83. Kramer to Knipling, December 9, 1977. Papers of the Office of President Sanford, box 107, file "Phytotron." Duke University Archives.

84. "Memorandum Concerning NSF and Phytotron Funding." n. d (ca. 1977). Papers of the Office of President Sanford, box 107, file "Phytotron." Duke University Archives.

85. Frederic Cleaveland to Sievers, October 11, 1978. Papers of the Office of President Sanford, box 107, file "Phytotron." Duke University Archives.

86. Appel, *Shaping Biology,* 260–61.

87. Boyd Strain to Kramer, October 4 1979. Paul Kramer papers, box 10, file "Phytotron Board, 1970–79." Duke University Archives.

88. "Table 1. Summary of Research Activities in the Duke University Phytotron," Strain(?) to Tyler, Biological Research Resources, NSF, August 12, 1982. Paul Kramer papers, box 10, folder "Phytotron-AAU visit 1970–84." Duke University Archives.

89. Strain to Kramer, October 2, 1979. Paul Kramer papers, box 10, file "Phytotron Board, 1970–79." Duke University Archives.

90. Strain to Cleaveland, September 6, 1978. Papers of the office of President Sanford, box 107, file "Phytotron." Duke University Archives.

91. Kramer, "Botany in a Changing World," n. d. (ca. 1961), 3.

CHAPTER 6. BIG BIOLOGY IN THE BIOTRON

Epigraph: Minutes of Ad Hoc Committee on Campus Plant Growth Facilities. October 1, 1992. Biotron Papers, Series 06/80, Box 4, file "UW Graduate School Plant Growth Facilities Committee, 1992–93." Archives, University of Wisconsin-Madison.

1. Edwards, *A Vast Machine*; Westwick, *Into the Black*; Gabrielle Hecht, *The Radiance of France*.

2. Committee of the Botanical Society of American, "Controlled-Climate Facilities," February 1960. In Biotron Papers, Series 06/80, Box 1, file "Botanical Society of American—1959" Archives, University of Wisconsin-Madison.

3. Memo, November 16, 1959. Biotron Papers, Series 06/80, Box 1, file "Biotron Concepts." Archives, University of Wisconsin-Madison.

4. Noted in Frits Went, Diary, May 6, 1956. Archives. Missouri Botanical Garden.

5. "Biotron Conference," December 10–12, 1959. Biotron Papers, Series 06/80, Box 1, file "Biotron Conference," 35. Archives, University of Wisconsin-Madison.

6. See the essays in Galison and Hevly, *Big Science*; Capshew and Rader, "Big Science." See also Kevles, "Big Science and Big Politics."

7. Weinberg, "Impact of Large-Scale Science"; see also his more complete treatment in *Reflections on Big Science*. In the words of the noted Cold War technocrat Lloyd Berkner, "we have reached the point where we are now prepared to spend literally millions of dollars on a single research tool" (Needell, "The Carnegie Institution," 62). Paul Forman characterized the increase in monetary outlay to American research after the Korean War, as opting for "guns *and* butter" (Forman, "Behind Quantum Electronics," 158).

8. Hendricks and Went, "Controlled-Climate Facilities," 510.

9. Westwick, *The National Labs*; Munns, "If We Build It"; Hoddeson, Kolb, and Westfall, *Fermilab*.

10. "Report on Meeting for Consideration of Controlled Environmental Facility," Oak Ridge Tennessee, July 22–23, 1956. In Phytotron Records, box 2, file "Controlled Environments-discussions/committees before 1962." Duke University Archives.

11. Contract, attached to Letter from Lilienthal to Graham, March 18, 1947. Folder "Oak Ridge Institute of Nuclear Studies," Box 10, entry group Subject Files of David Lilienthal, RG326, NARA (DC). Though an important institution operating for many years, Oak Ridge Institute of Nuclear Studies is still understudied. For the medical and radioisotope training programs, see Creager, *Life Atomic*, 317–18, and for its connection to reactor design and research technique training in the physical sciences and the Oak Ridge School of Reactor Technology, see Johnston, *The Neutron's Children*, 138–42.

12. Paul Kramer to Paul Pearson, July 17, 1956. Phytotron Records, box 2, file "Controlled Environments-discussions/committees before 1962." Duke University Archives.

13. Rasmussen, *Gene Jockeys*, 25. I consider the primary document on the continuous internecine debate over the definition of "basic research" to be Alan Waterman's "Basic Research in the United States." The debate grew out of Vannevar Bush's view in 1945 that if the United States was to promote new industries, new jobs, and new knowledge for defense then "this essential, new knowledge can only be obtained through basic scientific research" (Bush, *Science*, 1). I agree with Stephen Toulmin, who claimed that administratively labels such as "basic," "applied," "pure," "mission-orientated," and "fundamental"

reflected needs of the patrons and courtiers to conduct a relationship within policy guidelines of institutions. In the worlds of high-energy physics or the biomedical sciences the labels basic, applied, or mission-orientated science were imposed by a federal patron anxious to generate scientific research for broad "national goals" (Toulmin, "A Historical Reappraisal," 24). Scientists struggled to adopt and use these labels to secure funding or prestige, but in the rapidly shifting institutional and patronage landscape of the 1970s many scientists were caught, like the phytotronists, on the wrong label at the wrong time!

14. Siddiqi, "Fighting Each Other," 194–95.

15. Westwick, *The National Labs*, 6. Likewise, the Naval Research Laboratory's definition of basic research "rested explicitly upon institutional needs, internal management, independence from the Navy hierarchy, and recruitment of new scientific personnel" (Hevly, "Basic Research," 4–6).

16. Appel, *Shaping Biology*.

17. Panel for the Plant Sciences, *The Plant Sciences Now*, 4. Suggesting just how similar the realities of the American and Soviet Cold War were, both American technocrats and Soviet scientists embraced the rhetoric of support for science turning into technologies and industries. Cf. Berkner, *The Scientific Age* and Schmid, "Defining (Scientific) Direction."

18. "The New World of DNA," *Economist* 404, no. 8801 (September 8–14, 2012), 76–77.

19. Woese, "A New Biology," 175, 179.

20. A number of institutional histories can now be assembled to give a general portrait of the transformative effects of a military-industrial science complex on American universities, notably Leslie, *The Cold War and American Science*; Lowen, *Creating the Cold War University*; Goldman, "National Science in the Nation's Heartland"; Needell, "Preparing for the Space Age."

21. Memo from Van R. Potter, June 18, 1963, attached to letter from Willard to Alberty, June 27, 1963. Biotron Papers, Series 06/80, Box 1, file "UW Graduate School Correspondence." Archives, University of Wisconsin-Madison.

22. Letter from Senn to Willard, August 30, 1961. Biotron Papers, Series 06/80, Box 1, file "UW Graduate School Correspondence." Archives, University of Wisconsin-Madison. The biotron's proposed animal studies, however, looked more like NIH purview, it was noted.

23. Harraway, *Primate Visions*, ch. 9.

24. Gaudillière, "Paris-New York Roundtrip," 395.

25. Thimann and Skoog, "Studies on the Growth Hormone," 715; Skoog, "A Deseeded *Avena* Test."

26. Skoog, "The Effect of X-Rays."

27. Committee of the Botanical Society of American, "Controlled-Climate Facilities," February 1960, 20.

28. Committee of the Botanical Society of American, "Controlled-Climate Facilities," February 1960, 20.

29. Skoog(?), December 3, 1958. Biotron Papers, Series 06/80, Box 1, file "Biotron Concepts." Archives, University of Wisconsin-Madison.

30. Oosting, *The Study of Plant Communities*, 7.

31. Frits Went, Diary, May 6, 1956. Archives. Missouri Botanical Garden.

32. "Research proposal to the NSF," 1959. Biotron Papers, Series 06/80, Box 2, file "Nat Sci Found Building Proposal 1959," 7. Archives, University of Wisconsin-Madison.

33. "Criteria," n. d. (ca. 1957). Biotron Papers, Series 06/80, Box 1, file "Biotron Concepts." Archives, University of Wisconsin-Madison.

34. Frits Went, Diary, May 6, 1958; Paul Kramer's detailed notes also fill out the picture of the various voices weighing in on the BSA proposals. See Phytotron Records, box, file "Controlled Environments-discussions, committees before 1962." Duke University Archives. He also notes Schmidt-Nielsen's antipathy toward cooperative botanical and zoological study.

35. Committees spread like acronyms throughout the Cold War. Patrick McCray pinpointed that it was "the interlocking system of boards and committees that shaped postwar science" (McCray, "Project Vista," 343).

36. See Hamblin, *Poison in the Well*, 66, 71; Rudolph, *Scientists in the Classroom*, ch. 5 and 6; Wolfe, *Competing with the Soviets*, 50.

37. Committee of the Botanical Society of American, "Controlled-Climate Facilities," February 1960, 21.

38. Letter to Skoog from the Department of Veterinary Science. March 30, 1959. Biotron Papers, Series 06/80, Box 1, file "Potential Experiments, 1959–61." Archives, University of Wisconsin-Madison. In another experiment, the veterinary scientists reported finding that fluctuating temperatures "rather than a given temperature modified the course of infection."

39. H. R. Bird, "Environment and Respiratory Diseases of Turkeys." February 21, 1961. Biotron Papers, Series 06/80, Box 1, file "Potential Experiments, 1959–61." Archives, University of Wisconsin-Madison.

40. Skoog(?), December 3, 1958.

41. "Biotron Conference," December 10–12, 1959. Biotron Papers, Series 06/80, Box 1, file "Biotron Conference," 24. Archives, University of Wisconsin-Madison.

42. Committee of the Botanical Society of American, "Controlled-Climate Facilities," February 1960, 22

43. Creager, "Nuclear Energy."

44. "Criteria," n. d. (ca. 1957).

45. "Biotron Conference," December 10–12, 1959, 12.

46. "Biotron Conference," December 10–12, 1959, 30.

47. "Biotron Conference," December 10–12, 1959, 42.

48. "Biotron Conference," December 10–12, 1959, 42.

49. "Biotron Conference," December 10–12, 1959, 8. On Levin, see Appel, *Shaping Biology*, 45.

50. J. W., Memo, December 14 and 15, 1961. Biotron Papers, Series 06/80, Box 1, file "UW Graduate School Correspondence." Archives, University of Wisconsin-Madison.

51. J. W., Memo, December 14 and 15, 1961.

52. Letter from Harry Highkin to R. N. Robertson, December 20, 1959. R. N. Robertson Papers. Basser Library, AAS. MS117/1/2.

53. Letter from Willard to Senn, August 19, 1961. Biotron Papers, Series 06/80, Box 1, file "UW Graduate School Correspondence." Archives, University of Wisconsin-Madison.

54. Letter from Riker to Harrar, August 16, 1961. Biotron Papers, Series 06/80, Box 1, file "UW Graduate School Correspondence." Archives, University of Wisconsin-Madison.

55. Letter from Senn to Willard, August 24, 1961. Biotron Papers, Series 06/80, Box 1, file "UW Graduate School Correspondence." Archives, University of Wisconsin-Madison.

56. Letter from Senn to Willard, August 24, 1961.

57. Letter from Andres Karstens, Colonel, USAF to Senn, March 7, 1962. Biotron Papers, Series 06/80, Box 3, file "Senn Activities, 60–69." Archives, University of Wisconsin-Madison.

58. "Controlled Atmosphere Suite," n. d. (ca. 1960). Biotron Papers, Series 06/80, Box 1, file "Biotron Building Committee." Archives, University of Wisconsin-Madison.

59. Oswald, Golueke, and Horning, "Closed Ecological Systems," 45; Golueke, Oswald, and Gee, "A Study of Fundamental Factors," 1.

60. Lewontin, "The Cold War," 16.

61. Dubos, "Environmental Biology."

62. Minutes of Biotron Building Committee," July 3, 1961. Biotron Papers, Series 06/80, Box 1, file "Biotron Building Committee." Archives, University of Wisconsin-Madison.

63. J. W., Memo, December 14 and 15, 1961.

64. Letter from Senn to Kramer, October 15, 1962. Biotron Papers, Series 06/80, Box 3, file "Senn Activities, 1960–69." Archives, University of Wisconsin-Madison.

65. Senn to Jordan, July 31, 1970. Biotron Papers, Series 06/80, Box 3, file "NSF Grant Corres." Archives, University of Wisconsin-Madison.

66. Minutes of Biotron Committee, May 25, 1960. Biotron Papers, Series 06/80, Box 1, file "Concept Committee. . ." Archives, University of Wisconsin-Madison.

67. Senn to Bock, February 25, 1971. Biotron Papers, Series 06/80, Box 3, file "UW Graduate School Corres, 1970–72." Archives, University of Wisconsin-Madison.

68. Senn to Bock, June 22, 1971. Biotron Papers, Series 06/80, Box 3, file "Senn Activities, 1960–69." Archives, University of Wisconsin-Madison.

69. Senn to Bock, February 25, 1971.

70. Kohler, *Landscapes and Labscapes*, 3.

71. Committee of the Botanical Society of American, "Controlled-Climate Facilities," February 1960, 21.

72. Highkin, "The Effect of Constant."

73. Frits Went, Diary, May 7, 1958. Archives. Missouri Botanical Garden.

74. Letter from Pitt to Skoog, September 22, 1959. Biotron Papers, Series 06/80, Box 1, file "Director Search, 1959–60." Archives, University of Wisconsin-Madison.

75. Letter from Pitt to Skoog, September 22, 1959.

76. Letter from Senn to Lazenby, May 5, 1960. Biotron Papers, Series 06/80, Box 1, file "Design Proposals, 1964." Archives, University of Wisconsin-Madison.

77. Letter from Senn to Godschalx, May 26, 1961. Biotron Papers, Series 06/80, Box 1, file "Design Proposals, 1964." Archives, University of Wisconsin-Madison.

78. Letter from Senn to Godschalx, May 26, 1961. Biotron Papers, Series 06/80, Box 1, file "Design Proposals, 1964." Archives, University of Wisconsin-Madison.

79. Letter from Senn to Bock, August 5, 1970. Biotron Papers, Series 06/80, Box 3, file "UW Graduate School Corres, 1970–72." Archives, University of Wisconsin-Madison.

80. Handwritten notes included with the Minutes of the Phytotron Board meeting of November 19, 1971. Paul Kramer papers, Box 10, file "Phytotron Board, 1970–79." Duke University Archives.

81. November, *Biomedical Computing*, 241.

82. Letter from Bock to Sievers, July 2, 1976. Biotron Papers, Series 06/80, Box 3, file "NSF Grant Corres." Archives, University of Wisconsin-Madison.

83. Letter from Sievers to Bock, August 26, 1976. Biotron Papers, Series 06/80, Box 3, file "NSF Grant Corres." Archives, University of Wisconsin-Madison.

84. Letter from Sievers to Bock, August 26, 1976.

85. Budget summary attached to Senn to Berman, November 9, 1971. Biotron Papers, Series 06/80, Box 3, file "UW Graduate School Corres, 1970–72." Archives, University of Wisconsin-Madison.

86. "Biotron Conference," December 10–12, 1959, 4.

87. Biotron Papers, Series 06/80, Box 4, file "68002 D. Anderson & B. Easterday, Vet Sci Dept." Archives, University of Wisconsin-Madison.

88. Alford and Tibbitts, "Circadian Rhythm of L," 99.

89. Alford and Tibbitts, "Circadian Rhythm of leaves," 99, 102.

90. T. W. Tibbitts, "A proposal for the support of research on environmental factors regulating circadian rhythm in *Phaseolus* leaves," October 15, 1968. In Biotron papers, box 4, file "68003 T. Tibbitts, Horticulture." Archives. University of Wisconsin-Madison.

91. Tibbitts, "A Proposal for the Support of Research," October 15, 1968.

92. Senn to Tibbitts, February 21, 1966. In Biotron papers, box 4, file "68003 T. Tibbitts, Horticulture." Archives. University of Wisconsin-Madison.

93. Tibbitts to Alberty, March 2, 1966. In Biotron papers, box 4, file "68003 T. Tibbitts, Horticulture." Archives. University of Wisconsin-Madison.

94. Skoog to Senn, April 28, 1966. In Biotron papers, box 4, file "68003 T. Tibbitts, Horticulture." Archives. University of Wisconsin-Madison.

95. Skoog to Senn, April 28, 1966.

96. Appel, *Shaping Biology*, 178.

97. Handwritten notes, 1987. Biotron Papers, Series 06/80, Box 3, file "UW Grad School Corres, 1987–90." Archives, University of Wisconsin-Madison

98. Notes for John Wiley, July 25, 1989. Biotron Papers, Series 06/80, Box 3, file "UW Grad School Corres, 1987–90." Archives, University of Wisconsin-Madison.

99. "Conversation with Dean Wiley," July 25, 1989. Biotron Papers, Series 06/80, Box 3, file "UW Grad School Corres, 1987–90." Archives, University of Wisconsin-Madison.

100. "Environmental Requirements in Biotron, Year 1990–91." Biotron Papers, Series 06/80, Box 4, file "UW Graduate School—Biological Sciences Review, 1991." Archives, University of Wisconsin-Madison.

101. Tibbitts, Cao, and Wheeler, *Growth of Potatoes*, iii.

102. Roach, *Packing for Mars*, ch. 15.

103. Tibbitts, Cao, and Wheeler, *Growth of Potatoes*, vi–vii.

104. Minutes of Ad Hoc Committee on Campus Plant Growth Facilities. October

1, 1992. Biotron Papers, Series 06/80, Box 4, file "UW Graduate School Plant Growth Facilities Committee, 1992–93." Archives, University of Wisconsin-Madison.

CODA II: THE PASSING OF THE AGE OF BIOLOGY

Epigraph: Monteith, "Introduction," 108.

1. Letter from William Hiesey to David Keck, May 8, 1952. David Keck Papers, Box 4, folder "Correspondence–William Hiesey." Archives of the New York Botanical Garden.

2. Wisnioski, "'Liberal Education Has Failed.'" 757–58. See also Wisnioski, *Engineers for Change*, ch. 5 and 6.

3. For a view of technological optimism in France, see Bess, "Ecology and Artifice"; Hecht, *The Radiance of France*. For the crisis of confidence, see Wolfe, *Competing with the Soviets*, ch. 7.

4. Roland, *The Military-Industrial Complex*, 43.

5. Monteith, "Introduction," 107.

6. Monteith, "Introduction," 107.

7. Monteith, "Introduction," 108.

8. Monteith, "Introduction," 109.

9. Monteith, "Introduction," 108.

10. Monteith, "Introduction," 109.

11. Downs, "Phytotrons," 483. Downs did not provide any reference for the claim, but went on to emphasize that illuminance could be "duplicated" "when accompanied by a description of the light source."

12. Some numbers from Dissertation Abstracts give an indication of the magnitude of molecular biology's domination over biology.

CONCLUSION: THE NEW AGE OF CLIMATE

Epigraph: Klein, *This Changes Everything*, 15.

1. Kois was relieved that Ridley Scott included some wonder in the film version. http://www.slate.com/articles/arts/books/2015/10/book_review_the_martian_movie_gives_the_book_a_dose_of_wonder.2.html.

2. Huxley, *Brave New World*, 23.

3. Huxley, *Brave New World*, 18.

4. Heinlein, *Stranger in a Strange Land*, 20.

5. Ira Levin, *The Boys from Brazil*, 189. Levin extended and made explicit the earlier authors' concerns with similar growth and development by emphasizing the potential "psychological" similarities from certain events that were of "paramount importance in shaping [a] psyche," 191.

6. Teleplay by Adam Belanoff and Michael Piller, "The Masterpiece Society." *Star Trek: The Next Generation*, Season 5, episode 13. Aired February 10, 1992. Paramount Television.

7. Herbert, *Dune*, 137, 463; Appendix, 477–486.

8. See Went and Munz, "A Long-Term Test."

9. Sign on the exterior of the original CNRS phytotron at Gif-sur-Yvette.

10. Evans, *Feeding the Ten Billion*, 214.

11. Kolbert, *Field Notes from a Catastrophe*, preface and ch. 4. Oreskes and Conway, *Merchants of Doubt*, ch. 6.

12. Klein, *This Changes Everything*, 15.

13. Thompson, *The Making of the English Working Class*, 13.

14. Wilson, *Naturalist*, ch. 12.

15. Watson, "A Personal View of the Project," 164. After DNA, the molecular biologists were insufferable: Chargaff recalled "the tone" of the announcement of the structure of DNA: "somehow oracular and imperious, almost decalogous," which readers familiar with "Honest Jim" Watson's *Double Helix* will think actually understates the case. Chargaff, "Building the Tower of Babble," 778.

16. Smith, *The Genomic Age*, 2. "The "gene,"" opens Siddhartha Mukherjee's monumental book, "the basic unit of all biological information" (Mukherjee, *The Gene*, 9). Businesses have taken a leading role in boosting the role of genomics. For example, Smith's *The Genomic Age* is published by the AMA, standing for American Management Association, and not the American Medical Association in this case. Though the "genomic age" may be a speculation bubble, the major business promise is in health care, where its hyped promises play on fears of disease and death: "A cure for cancer would be virtually impossible if we did not understand the role of cancer genes" (Ridley, *Genome*, 2).

17. Sally Smith Hughes, *Genentech: The Beginnings of Biotech* (University of Chicago Press, 2011), 63.

18. While the quote is from the book of the series, Bronowski, *The Ascent of Man*, 317, like Carl Sagan's *Cosmos* over a decade later, the full impact is gained from watching the original broadcasts. Monod's declaration, Evelyn Fox Keller notes, was not true, not even for *E. coli* (Keller, *A Feeling for the Organism*, 6).

19. Gunther Stent also compared the era of molecular biology to the Renaissance, and the discovery of DNA to the end of the Middle Ages (Stent, "Introduction," xi).

20. Wilson, *Naturalist*, 223.

21. Bowler and Morus, *Making Modern Science*, 209.

22. Went, "Phytotronics," 151.

23. Chargaff, "Building the Tower of Babble," 779. Chargaff had revealed that in DNA the proportions of adenine and thiamine are the same (A = T), as are those of guanine and cytosine (G = C).

24. Nitsch, "The Role of Plant Hormones," 1.

25. Hanson, *History of the American Society*, 2.

26. Woese, "A New Biology," 175, 179.

27. Quote from Nelkin and Lindee, *The DNA Mystique*, 10.

28. Keller, *A Feeling for the Organism*, xi.

29. Galston, "Plant Photobiology," 433; Galston is credited with discovering defoliant chemicals later incorporated into products like Agent Orange, which Galston then passionately lobbied against the use of. Galston is a member of the phytotronists' community, working under Frits Went and James Bonner, and alongside Sam Wildman at Caltech from 1947 to 1955 before moving to Yale. His research on "flavin-based photoreception," though eventually disproved, operated in the wider phytotronist community of Folke Skoog (Biotron) and Kenneth Thimann (coauthor with Went). See Galston, "An Accidental Plant Biologist."

30. Evans, "Conjectures, Refutations, and Extrapolations," 14.

31. Kramer and Boyer, *Water Relations in Plants,* 12–13.

32. Personal correspondence from William Laing to the author, April 19, 2015.

33. Pickering, *The Cybernetic Brain,* 33.

34. Chargaff, "Building the Tower of Babble," 778. Kremer, "Physiology." Kremer's article barely mentions plant physiology, though the category is included in his lists of self-identified bioscientists. Plant physiology makes its appearance in 1921, reaches its height in 1949, and last appears in 1960. The category is not mentioned in either 1976 or 1989. See Kremer, 362–63. Kremer's article fills an important absence. Physiology, for example, was notably "not included" in one major collection, *The American Development of Biology,* 4, 9. Likewise, Gerald Geison noted in the first major historical treatment of physiology (which also did not include plant physiology) how the topic of physiology was absent in writings on American science up to the 1980s (Geison, "Toward a History of American Physiology," 9).

35. Quoted in Hanson, *History of the American Society,* 192, 237.

36. Keller, *A Feeling for the Organism,* xiv.

37. Gilbert, "Resurrecting the Body," 573.

38. Oreskes and Conway, *The Collapse of Western Civilization,* 14.

39. Klein, *This Changes Everything,* 13.

40. Evans, Wardlaw, and King, "Plants and Environment," 212.

41. Gorissen et al., "ESPAS"; Thiel et al., "A Phytotron for Plant Stress Research"; Lodovica Gullino et al., "New Phytotron for Studying the Effects."

42. Lawton et al., "The Ecotron," 181.

43. Lawton et al., "The Ecotron," 188–89.

44. L. Finér, et al, "The Joensuu Dasotrons," 137, 140, 142, 149.

45. Estabrook, *Tomatoland,* 141.

46. See http://www.nytimes.com/2013/06/24/booming/you-call-that-a-tomato.html (accessed July 2013).

47. Estabrook, *Tomatoland,* 15.

48. Estabrook, *Tomatoland,* 15. See also Kingsbury, *Hybrid,* 273–77.

49. Estabrook, *Tomatoland,* 143–44. A similar story was published in the United Kingdom, where scientists were quoted as saying that they could "actually target the genes that control flavour separately from those that control shelf life." Matt McGrath, "Flavour Changer: Genome Could Enhance Tomato Taste," www.bbc.co.uk/news/science-environment-18253577 (accessed July 10, 2012).

50. Wark, *Molecular Red,* xiv.

51. Wark, *Molecular Red,* xx.

52. "The Elephant in the Atmosphere," *Economist,* July 19, 2014, 59. Klein, *This Changes Everything,* 49. See also Edwards, *A Vast Machine.*

53. Oswald and Golueke, "Environmental Control Studies," 184.

BIBLIOGRAPHY

ARCHIVAL COLLECTIONS

Archives Nationales (Fontainebleau), France.

Archives New Zealand Te Rua te Mahara o te Kāwanatanga.

Australian Society of Plant Physiologists Papers. MS97. Basser Library, Australian Academy of Science.

Biology Division Papers. Archives. California Institute of Technology.

Biotron Papers. University of Wisconsin-Madison Archives.

James Bonner Papers. Archives. California Institute of Technology.

College of Agriculture and Life Sciences records. North Carolina State University Archives.

Commonwealth Scientific and Industrial Research Organisation Archives. Canberra.

Dépôt des Archives du CNRS (Gif-sur-Yvette).

Dépôt des Archives du CNRS (Paris).

Director's Papers, Frits Went. Record Group 3/2/6. Archives. Missouri Botanical Garden.

Lee A. DuBridge Papers. Archives. California Institute of Technology.

Duke University Archives.

Lloyd Evans Papers. MS213. Basser Library, Australian Academy of Science.

Lloyd Evans. "Memoirs of a Meandering Biologist." June 2005 (manuscript). MS9885. National Library of Australia.

Jesse Greenstein Papers. Archive. California Institute of Technology.

Historical Files—Biology Division. Archives. California Institute of Technology.

David Keck Papers. Archives of the New York Botanical Garden.

Paul J. Kramer Papers. Duke University Archives.

Michigan State University Archives.

Missouri Botanical Garden Archives.

National Archives of Australia. Canberra.

North Carolina State University Archives.

Papers of the Office of the President Douglas Knight. Duke University Archives.

Papers of the Office of President Sanford, Duke University Archives.

Phytotron Records. Duke University Archives.

"Phytotron vertical file." LuEster T. Mertz Library of the New York Botanical Garden, Bronx, New York.

Office of the Chancellor. North Carolina State University Archives.

Otto Frankel Papers. MS106. Basser Library, Australian Academy of Science.

Records of the Associate Dean and Director, North Carolina Agricultural Research Center. North Carolina State University Archives.

Robert A. Millikan Papers. Archives. California Institute of Technology.

R. N. Robertson Papers. MS117. Basser Library, Australian Academy of Science.

A. H. Sturtevant Papers. Archives. California Institute of Technology.

Merle Tuve Papers. Library of Congress.

Frits Went Papers. Record Group 3/2/6/1. Archives. Missouri Botanical Garden.

ORAL HISTORIES

James Bonner, Interview with Graham Berry, 1980. Oral History Project, Archives. California Institute of Technology.

James Bonner, Sterling Emerson, Norman Horowitz, and Donald Poulson. Interview by Judith Goodstein, Harriett Lyle, and Mary Terrall. November 6, 1978. Oral History Project, Archives, California Institute of Technology.Lincoln Constance. Interview with Ann Lage, 1986. University History Series, University of California at Berkeley.

Jesse Greenstein. Interview with Spencer Weart, April 7, 1977. *Sources for History of Modern Astrophysics*, American Institute of Physics.

Robert L. Sinsheimer. Oral History Interview with Shelley Erwin, 1990 and 1991. Archives, California Institute of Technology.

Folke Skoog. Interview with Barry Teicher, 1993. Archives, University of Wisconsin-Madison.

Ward Whaling. Interview by Shelley Erwin. April–May 1999. Oral History Project, Archives, California Institute of Technology.

PUBLISHED WORKS

Abir-Am, Pnina. "The Molecular Transformation of Twentieth-Century Biology." In *The Companion to Science in the Twentieth Century*, edited by John Krige and Dominique Pestre, 495–524. London: Routledge World Reference, 2003.

Abraham, Itty. *The Making of the Indian Atomic Bomb: Science, Secrecy and the Postcolonial State*. London: Zed Books, 1998.

Adams, Donald. "An Air Pollution Phytotron: A Controlled Environment Facility for Studies into the Effects of Air Pollutants on Vegetation." *Journal of the Air Pollution Control Association* 11 (1961), 470–76.

Adas, Michael. *Machines as the Measure of Men: Science, Technology, and Ideologies of Western Dominance*. Ithaca, NY: Cornell University Press, 1989.

Agar, Jon. *Science and Spectacle: The Work of Jodrell Bank in Post-war British Culture*. Amsterdam: Harwood Academic, 1998.

Akera, Atsushi. *Calculating a Natural World: Scientists, Engineers, and Computers during the Rise of U.S. Cold War Research*. Cambridge, MA: MIT Press, 2007.

Alberda, Th. "On the Use of Plants Grown in Controlled Environments for Analysing Their Photosynthetic Characteristics." In *Prediction and Measurement of Photosynthetic Productivity*, 591–93.

Alexander, Jennifer Karns. "Thinking Again about Science and Technology." *Isis* 103 (2012), 518–26.

Alford, D. K., and T. W. Tibbitts. "Circadian Rhythm of Leaves of *Phaseolus angularis* Plants Grown in a Controlled Carbon Dioxide and Humidity Environment." *Plant Physiology* 46 (1970), 99–102.

Allen, Garland E. *Life Science in the Twentieth Century*. New York: Wiley, 1975.

Allen, Garland E. "Naturalists and Experimentalists: The Genotype and the Phenotype." *Studies in the History of Biology* 3 (1979), 179–209.

American Greenhouse Manufacturing Company (AGMC). *American Greenhouses*. AGMC, 1928.

Amrith, Sunil S. *Decolonizing International Health: India and Southeast Asia, 1930–1965*. New York: Palgrave, 2006.

Anderson, Edgar. "The First Five Years of the Climatron." *Missouri Botanical Garden Bulletin* 53, no. 6 (1965), 1–28.

Anderson, Warwick, and V. Adams. "Pramoedya's Chickens: Postcolonial Studies of Technoscience." In *The Handbook of Science and Technology Studies*, 3rd ed., edited by Edward J. Hackett, Olga Amsterdamska, Michael Lynch, and Judy Wajcman, 181–204. Cambridge, MA: MIT Press, 2008.

Ankeny, Rachel. "Wormy Logic: Model Organisms as Case-Based Reasoning." In Creager, Lunbeck, and Wise, *Science without Laws*, 46–58.

Anker, Peder. *From Bauhaus to Ecohouse: A History of Ecological Design*. Baton Rouge: Louisiana State University Press, 2010.

Appel, Toby. *Shaping Biology: The National Science Foundation and American Biological Research, 1945–1975*. Baltimore: Johns Hopkins University Press, 2000.

Arthur, John, John Guthrie, and John Newell. "Some Effects of Artificial Climates on the Growth and Chemical Composition of Plants." *American Journal of Botany* 17 (1930), 416–82.

Augier, Henry. *Phytotrons et phytotronique: la bioclimatologie expérimentale*. Marseilles: Centre Regional de Documentation Pedagogique, 1972.

Bailyn, Bernard. "The Historiography of the Losers." In *The Ordeal of Thomas Hutchinson*, 397–408. Cambridge, MA: Belknap Press of Harvard University Press, 1968.

Ballard, Ernesta Drinker. *Growing Plants Indoors: A Garden in Your House*. New York: Barnes and Noble Books, 1971.

Bamford, James. *The Puzzle Palace: A Report on America's Most Secret Agency*. New York: Penguin, 1983.

Beachy, Robert. *Gay Berlin: Birthplace of a Modern Identity*. New York: Knopf, 2014.

Beard, M., and T. Pearcey. "The Genesis of an Early Stored-Program Computer: CSIRAC." *Annals of the History of Computing* 6 (1984), 106–15.

Belasco, Warren. "Algae Burgers for a Hungry World? The Rise and Fall of Chlorella Cuisine." *Technology and Culture* 38 (1997), 608–34.

Belasco, Warren. "Food and the Counterculture." In *Food and Global History*, edited by Raymond Grew, 273–92. Boulder, CO: Westview Press, 1999.

Benson, Etienne. *Wired Wilderness: Technologies of Tracking and the Making of Modern Wildlife*. Baltimore: Johns Hopkins University Press, 2010.

Berg, Paul, and Maxine Singer. *George Beadle: An Uncommon Farmer.* Cold Spring Harbor, NY: Cold Spring Harbor Laboratory Press, 2003.

Berkner, Lloyd V. *The Scientific Age: The Impact of Science on Society.* New Haven, CT: Yale University Press, 1964.

Bess, M. "Ecology and Artifice: Shifting Perceptions of Nature and High Technology in Postwar France." *Technology and Culture* 12 (1995), 830–62.

Biagioli, Mario. *Galileo Courtier: The Practice of Science in the Culture of Absolutism.* Chicago: University of Chicago Press, 1993.

Bijker, W. et al., eds. *The Social Construction of Technological Systems: New Directions in the Sociology of Technology.* Cambridge, MA: MIT Press, 1987.

Billings, W. D. "The Environmental Complex in Relation to Plant Growth and Distribution." *Quarterly Review of Biology* 27, no. 3 (1952), 251–65.

Bishop, Alan J. "Western Mathematics: The Secret Weapon of Cultural Imperialism." In *The Post-Colonial Studies Reader,* edited by Bill Ashcroft, Gareth Griffiths, and Helen Tiffin, 80–83. Oxford: Routledge, 2006.

Bobbit, Philip. *The Shield of Achilles: War, Peace, and the Course of History.* New York: Knopf, 2002.

Bodenheimer, F. S. *Studies in Biology and Its History.* Vol. 1. Biological Studies, 1957.

Boehmer, Elleke. *Colonial and Postcolonial Literature: Migrant Metaphors.* Oxford: Oxford University Press, 2005.

Bohr, Niels. "Unity in Knowledge." In *Atomic Physics and Human Knowledge,* 67–81. New York: Wiley, 1958.

Bolton, Geoffrey. *The Oxford History of Australia: The Middle Way, 1942–1995.* Oxford: Oxford University Press, 1996.

Bonner, James. "Arie Jan Haagen-Smit: 1900–1977."' *Biographical Memoirs of the National Academy of Sciences* (1989), 188–216.

Bonner, James. "The Chemical Cure for Climatic Lesions." *Engineering and Science* 20 (1957), 28–30.

Bonner, James. "Fundamental Plant Physiology Research in Controlled Environmental Facilities." *Research Outlook on Soil, Water, and Plant Nutrients: Proceedings of the Eighth Annual Meeting of the Agricultural Research Institute, October 1959.* Publication 785, National Academy of Sciences-National Research Council, 1960, 71–78.

Bonner, James. "Summary and Observations: Plant Science Symposium." *Proceedings: Plant Science Symposium 1962,* 213–23. Camden, NJ: Campbell Soup Company, 1962.

Bonner, James, and Galston, Arthur. *Principles of Plant Physiology.* San Francisco: Freeman, 1952.

Bonneuil, Christophe. "Plant Breeding and Experimental Cultures: Agriculture and the Development of Genetics in France." *Journal of the History of Biology* 39 (2006), 281–309.

Bowler, Peter. "Science and the Environment: New Agendas for the History of Science?" In *Science and Nature: Essays in the History of the Environmental Sciences,* edited by M. Shortland, 1–21. London: British Society for the History of Science, 1993.

Bowler, Peter J. "Variation from Darwin to the Modern Synthesis." In *Variation,* edited by Benedikt Hallgrímsson and Brian K. Hall, 9–27. Elsevier, 2005.

Bowler, Peter J., and Iwan Rhys Morus. *Making Modern Science: A Historical Survey*. Chicago: University of Chicago Press, 2005.

Bowler, Peter J., and John V. Pickstone, eds. *Cambridge History of Science: The Modern Biological and Earth Sciences*. Cambridge: Cambridge University Press, 2009.

Bozdogan, Sibel. *Modernism and Nation Building: Turkish Architectural Culture in the Early Republic*. Seattle: University of Washington Press, 2001.

Brady, N. C. "Foreword." In *Proceedings of the Symposium on Climate and Rice*, 1. International Rice Research Institute, 1976.

Bridge, Carl. "Anglo-Australian Attitudes: Remembering and Re-reading Russel Ward." *Journal of Australian Colonial History* 10, no. 2 (2008), 187–200.

Broecker, Wallace S., and Robert Kunzig. *Fixing Climate: What Past Climate Changes Reveal about the Current Threat—and How to Counter It*. New York: Hill and Wang, 2009.

Bronowski, J. *The Ascent of Man*. British Broadcasting Corporation, 1973.

Brown, L., A. Pais, and Sir B. Pippard, eds. *Twentieth Century Physics. Vol. 2*. London: Institute of Physics Publishing and American Institute of Physics, 1995.

Buderi, Robert. *The Invention that Changed the World: How a Small Group of Radar Pioneers Won the Second World War and Launched a Technological Revolution*. New York: Simon and Schuster, 1996.

Bullard, Alice. "Imperial Networks and Postcolonial Independence: The Transition from Colonial to Transcultural Psychiatry." In *Psychiatry and Empire*, edited by Sloan Mahone and Megan Vaughan, 197–219. New York: Palgrave, 2007.

Bünning, Erwin. "Fifty Years of Research in the Wake of Wilhelm Pfeffer." *Annual Review of Plant Physiology* 28 (1977), 1–22.

Bush, Vannevar. *Pieces of the Action*. New York: Morrow, 1970.

Bush, Vannevar. *Science: The Endless Frontier, A Report to the President July 1945*. Washington, DC: Government Printing Office, 1945.

Bush, Vannevar. "Science in Medicine and Related Fields." *Medical Annals of the District of Columbia* 22, no. 1 (1953), 1–7.

Cairns, J., Gunther S. Stent, and J. D. Watson, eds. *Phage and the Origins of Molecular Biology*. Cold Spring Harbor, NY: Cold Spring Harbor Laboratory Press, 1966.

Campbell, Neil A., and Jane B. Reece. *Biology*. 6th ed. San Francisco: Benjamin Cummings, 2002.

Campos, Luis A. *Radium and the Science of Life*. Chicago: University of Chicago Press, 2015.

Capshew, James, and Karen Rader. "Big Science: Price to the Present." *Osiris 2nd Series* 7 (1992), 3–25.

Carson, Rachel. *Silent Spring*. Boston: Houghton Mifflin, 1962.

Cawte, Alice. *Atomic Australia, 1944–1990*. University of New South Wales Press, 1992.

Ceruzzi, Paul E. *A History of Modern Computing*. Cambridge, MA: MIT Press, 1998.

Champagnat, P. "Pierre Chouard, 1903–1983: un marriage réussi entre science et foi." In *Biologistes et Naturalistes français du XXe siècle*, 61–102. Paris: Hermann, 2012.

Chargaff, Erwin. "Building the Tower of Babble." *Nature* 248 (1974), 776–79.

Charles-Edwards, David A., David Doley, and Glyn M. Rimmington. *Modelling Plant Growth and Development*. New York: Academic Press, 1986.

Charlton, Andrew. "Man-Made World: Choosing between Progress and Planet." *Quarterly Essay* 44 (2011), 1–76.

Charrier, Philip "Britain, India and the Genesis of the Colombo Plan, 1945–1951." PhD diss., Cambridge University, 1995.

Chauncey, George. *Gay New York: Gender, Urban Culture, and the Makings of the Gay Male World, 1890–1940*. New York: Basic Books, 1994.

Childs, Peter. *Modernism and the Post-Colonial: Literature and Empire, 1885–1930*. London: Continuum, 2007.

Chouard, P. "Introduction." In Chouard and de Bilderling, *Phytotronique*, 1–3.

Chouard, P. "Phytotronics, a Technique and a New Strategy of Research with Reference to Scientific Horticulture." In Chouard and de Bilderling, *Phytotronics*, 1–5.

Chouard, P., and N. de Bilderling, eds. *Phytotronics in Agricultural and Horticultural Research*. Paris: Gauthier-Villars, 1974.

Chouard, Pierre, and Jean-Paul Nitsch, eds. *The World of Plants. Encyclopedia of the Life Sciences*, Vol. 3. Garden City, NY: Doubleday, 1965.

Chrisman, Laura, and Patrick Williams. *Colonial Discourse and Post-Colonial Theory: A Reader*. New York: Harvester, 1993.

Clarke, Adele, and Joan H. Fujimura, eds. *The Right Tools for the Job: At Work in Twentieth-Century Life Sciences*. Princeton, NJ: Princeton University Press, 1992.

Clausen, Jens, and William M. Hiesey. *Experimental Studies on the Nature of Species: IV Genetic Structure of Ecological Races*. Washington, DC: Carnegie Institution of Washington, 1958.

"The Climatron at Twenty." *Missouri Botanical Garden Bulletin* 69, no. 1 (1981), 1, 4.

"The Climatron Opens to the Public." *Missouri Botanical Garden Bulletin* 48, no. 8 (1960), 131–32.

Coen, Deborah R. "Living Precisely in Fin-de-Siècle Vienna." *Journal of the History of Biology* 39 (2006), 493–523.

Cohen-Cole, Jamie. "Instituting the Science of Mind: Intellectual Economics and Disciplinary Exchange at Harvard's Center for Cognitive Studies." *British Journal for the History of Science* 40 (2007), 567–97.

Coleman, David C. *Big Ecology: The Emergence of Ecosystem Science*. Berkeley: University of California Press, 2010.

Comfort, Nathaniel. "'The Real Point Is Control': The Reception of Barbara McClintock's Controlling Elements." *Journal of the History of Biology* 32 (1999), 133–62.

Comfort, Nathaniel. *The Science of Human Perfection: How Genes Became the Heart of American Medicine*. New Haven, CT: Yale University Press, 2012.

Compton, W. David, and Charles D. Benson. *Living and Working in Space: A History of Skylab*. NASA, 1983.

Conway, Gordon. *The Doubly Green Revolution: Food for All in the Twenty-First Century*. Ithaca, NY: Cornell University Press, 1997.

Cooke, G. Dennis. "Ecology of Space Travel." In Odum, *Fundamentals of Ecology*, 498–509.

Cooper, Gail. *Air-Conditioning America: Engineers and the Controlled Environment, 1900–1960*. Baltimore: Johns Hopkins University Press, 2002.

Cottle, Drew. "A Bowyang Historian in the Cold War Antipodes: Russel Ward and the Making of *The Australian Legend*." *Journal of Australian Colonial History* 10, no. 2 (2008), 171–86.

Cowan, Ruth Swartz. *More Work for Mother: The Ironies of Household Technology from the Open Hearth to the Microwave*. New York: Basic Books, 1983.

Craig, Patricia. *Centennial History of the Carnegie Institution of Washington: Volume 4. The Department of Plant Biology*. Cambridge: Cambridge University Press, 2005.

Creager, Angela N. H. *Life Atomic: A History of Radioisotopes in Science and Medicine*. Chicago: University of Chicago Press, 2013.

Creager, Angela N. H. *The Life of a Virus: Tobacco Mosaic Virus as an Experimental Model, 1930–1965*. Chicago: University of Chicago Press, 2002.

Creager, Angela N. H. "Nuclear Energy in the Service of Biomedicine: The U.S. Atomic Energy Commission's Radioisotope Program, 1945–1955." *Journal of the History of Biology* 39 (2006), 649–84.

Creager, Angela N. H. "Tracing the Politics of Changing Postwar Research Practices: The Export of 'American' Radioisotopes to European Biologists." *Studies in the History and Philosophy of Biological and Biomedical Sciences* 33 (2002), 367–88.

Creager, Angela N. H., and Hannah Landecker. "Technical Matters: Method, Knowledge and Infrastructure in Twentieth-Century Life Science." *Nature Methods* 6, no. 10 (2009), 701–5.

Creager, Angela N. H., Elizabeth Lunbeck, and M. Norton Wise, eds. *Science without Laws: Model Systems, Cases, Exemplary Narratives*. Durham, NC: Duke University Press, 2007.

Crick, Francis. *What Mad Pursuit: A Personal View of Scientific Discovery*. New York: Basic Books, 1988.

Croarken, Mary. "The Beginnings of the Manchester Computer Phenomenon: People and Influences." *IEEE Annals of the History of Computing* 15, no. 3 (1993), 9–16.

Crocker, William. *Growth of Plants: Twenty Years' Research at Boyce Thompson Institute*. New York: Reinhold, 1948.

"The CSIRO Phytotron." *Refrigeration Journal, Incorporating Air Conditioning and Heating* 14, no. 11 (1961), 34–35.

Cullather, Nick. *The Hungry World: America's Cold War Battle against Poverty in Asia*. Cambridge, MA: Harvard University Press, 2010.

Cullather, Nick. "Miracle of Modernization: The Green Revolution and the Apotheosis of Technology." *Diplomatic History* 28, no. 2 (2004), 227–54.

Curry, Helen Anne. "Accelerating Evolution, Engineering Life: American Agriculture and Technologies of Genetic Modification, 1925–1960." PhD diss., Yale University. 2012.

Darnton, Robert. "Peasants Tell Tales." In *The Great Cat Massacre and Other Adventures in French Cultural History*, ch. 1. New York: Basic Books, 1984.

Darwin, C. G. "Use of the Termination –tron in Physics." *Nature* 143 (1939), 602.

Davies, Rowland H. *The Microbial Models of Molecular Biology: From Genes to Genomes*. Oxford: Oxford University Press, 2003.

Davis, Mike. *City of Quartz: Excavating the Future in Los Angeles*. New York: Vintage, 1992.

de Bilderling, N. "Phytotrons and Their Possible Use in Horticulture." In Chouard and de Bilderling, *Phytotronics*, 14–26.

de Chadarevian, Soraya. *Designs for Life: Molecular Biology after World War Two*. Cambridge: Cambridge University Press, 2002.

de Chadarevian, Soraya. "Laboratory Science versus Country-House Experiments: The Controversy between Julius Sachs and Charles Darwin." *British Journal for the History of Science* 29 (1996), 17–41.

de Chadarevian, Soraya. "Sequences, Conformation, and Information: Biochemists and Molecular Biologists." *Journal of the History of Biology* 29 (1996), 361–86.

de Chadarevian, Soraya, and Jean-Paul Gaudillère. "The Tools of the Discipline: Biochemists and Molecular Biologists." *Journal of the History of Biology* 29 (1996), 327–30.

Delbrück, Max. "A Physicist Looks at Biology." In Cairns, Stent, and Watson, *Phage and the Origins of Molecular Biology*, 9–22.

Deleuze, Gilles. "Postscript on the Societies of Control." *October* 59 (1992), 3–7.

Dennis, Michael Aaron. "Our Monsters, Ourselves: Reimagining the Problem of Knowledge in Cold War America." In *Dreamscapes of Modernity: Sociotechnical Imaginaries and the Fabrication of Power*, edited by Sheila Jasanoff and Sang-Hyun Kim, 56–78. Chicago: University of Chicago Press, 2015.

Dennison, David M. "Physics and the Department of Physics since 1900." In *Research: Definitions and Reflections*, edited by Donald E. Thackrey, 120–136. Ann Arbor: University of Michigan Press, 1967.

Dolin, Eric Jay. *Leviathan: The History of Whaling in America*. New York: Norton, 2008.

Downs, R. J. "Phytotrons." *Botanical Review* 46 (1980), 447–89.

Downs, R. J., and H. Hellmers. *Environment and the Experimental Control of Plant Growth*. New York: Academic Press, 1975.

Downs, Robert J. *Controlled Environments for Plant Research*. New York: Columbia University Press, 1975.

Dubos, René. "Environmental Biology." *BioScience* 14, no. 1 (1964), 11–14.

DuBridge, Lee A. "Basic Research in the Private University." In *Symposium on Basic Research*, 108. Washington, DC: American Association for the Advancement of Science, 1959.

Dyer A. J., and F. J. Maher. "Automatic Eddy-Flux Measurement with the Evapotron." *Journal of Applied Meteorology* 4 (1965), 622–25.

East, E. M. "The Relation of Certain Biological Principles to Plant Breeding." Connecticut Agricultural Experiment Station *Bulletin* 158 (November 1907).

Eckardt, F. E. "Research on the Structure and Function of the Photosynthetic Apparatus under Field Conditions: Aims and Methods." In *Prediction and Measurement of Photosynthetic Productivity*, 379–84.

Eco, Umberto. *The Name of the Rose*. New York: Harcourt Brace Jovanovich, 1983.

Edwards, Paul N. *The Closed World: Computers and the Politics of Discourse in Cold War America*. Cambridge, MA: MIT Press, 1997.

Edwards, Paul N. *A Vast Machine: Computer Models. Climate Data, and the Politics of Global Warming*. Cambridge, MA: MIT Press, 2013.

"The Eggatron." *Rural Research in CSIRO* 41 (1962), 36.

Ellyard, David, and S. Cockburn. *Oliphant: The Life and Times of Sir Mark Oliphant*. Axiom Books, 1992.

Endersby, Jim. *A Guinea Pig's History of Biology*. Cambridge, MA: Harvard University Press, 2007.

Endersby, Jim. *Imperial Nature: Joseph Hooker and the Practices of Victorian Science*. Chicago: University of Chicago Press, 2010.

Ensmenger, Nathan. *The Computer Boys Take Over: Computers, Programmers, and the Politics of Technical Expertise*. Cambridge, MA: MIT Press, 2010.

Estabrook, Barry. *Tomatoland: How Modern Industrial Agricultural Destroyed Our Most Alluring Fruit*. Kansas City, MO: Andrews McMeel, 2011.

Evans, L. T. "CERES: An Australian Phytotron." *Nature* 195 (1962), 1142.

Evans, L. T. *Environmental Control of Plant Growth: Proceedings of a Symposium Held at Canberra, Australia, August, 1962*. New York: Academic Press, 1963.

Evans, L. T. *Feeding the Ten Billion: Plants and Population Growth*. Cambridge: Cambridge University Press, 1998.

Evans, L. T. "Sir Otto Frankel, 1900–1998." *Historical Records of Australian Science* 12, no. 4 (1999), 1–30.

Evans, L. T., I. F. Wardlaw, and R. W. King. "Plants and Environment: Two Decades of Research at the Canberra Phytotron." *Botanical Review* 51, no. 2 (1985), 203–72.

Evans, Lloyd. "Conjectures, Refutations, and Extrapolations." *Annual Review of Plant Biology* 54 (2003), 1–21.

Faherty, S. J., and William Barnaby. *A Gift to Glory In: The First Hundred Years of the Missouri Botanical Garden (1859–1959)*. Ocean Park, WA: Harris and Friedrich, 1989.

Faherty, S. J., and William Barnaby. *Henry Shaw: His Life and Legacies*. Columbia: University of Missouri Press, 1987.

Farber, Paul Lawrence. *Finding Order in Nature: The Naturalist Tradition from Linnaeus to E. O. Wilson*. Baltimore: Johns Hopkins University Press, 2000.

Feyerabend, Paul. *Against Method: Outline of an Anarchistic Theory of Knowledge*. London: Verso, 1993.

Finér, L. et al. "The Joensuu Dasotrons: A New Facility for Studying Shoot, Root, and Soil Processes." *Plant and Soil* 231 (2001), 137–49.

Forman, Paul. "Behind Quantum Electronics: National Security as a Basis for Physical Research in the United States, 1940–1960." *Historical Studies in the Physical and Biological Sciences* 18 (1987), 149–229.

Forman, Paul. "Inventing the Maser in Postwar America." *Osiris* 7 (1992), 103–34.

Foucault, Michel. *Discipline and Punish: The Birth of the Prison*. Penguin, 1991.

Fox, J. Eugene, and Mark Jacobs. "Preface." In Fox and Jacobs, *Molecular Biology of Plant Growth Control*.

Fox, J. Eugene, and Mark Jacobs, eds. *Molecular Biology of Plant Growth Control*. New York: Alan R. Liss, 1987.

Frankel, O. H. "The Canberra Phytotron." *Rural Research in CSIRO* 41 (1962), 12–14.

Frankel, O. H. "Concluding Remarks: The Next Decade." In *Environmental Control of Plant Growth: Proceedings of a Symposium Held at Canberra, Australia, August 1962*, edited by L. T. Evans, 435–45. New York: Academic Press, 1963.

Frankel, O. H. "Internationalism in Agricultural Science." *Australian Journal of Science* 28, no. 8 (1965), 314.

Frankel, O. H. "The IRRI Phytotron: Science in the Service of Human Welfare." In *Proceedings of the Symposium on Climate and Rice*, 3–10. International Rice Research Institute, 1976.

Frankel, O. H. "The Social Responsibility of Agricultural Science." *Australian Journal of Science* 25, no. 7 (1963), 301–7.

Gaddis, John Lewis. *The United States and the Origins of the Cold War*. New York: Columbia University Press, 1972.

Galison, Peter. *Image and Logic: A Material Culture of Microphysics*. Chicago: University of Chicago Press, 1997.

Galison, Peter. "Ten Problems in the History and Philosophy of Science." *Isis* 99, no. 1 (2008), 111–24.

Galison, Peter, and Bruce Hevly, eds. *Big Science: The Growth of Large-Scale Research*. Stanford, CA: Stanford University Press, 1992.

Galison, Peter, Bruce Hevly, and Rebecca Lowen. "Controlling the Monster: Stanford and the Growth of Physics Research, 1935–1962." In Galison and Hevly, *Big Science*, 46–77.

Galston, Arthur W. "An Accidental Plant Biologist." *Plant Physiology* 128, no. 3 (2002), 786–87.

Galston, Arthur W. *The Life of the Green Plant*. Englewood Cliffs, NJ: Prentice Hall, 1961.

Galston, Arthur W. "Plant Photobiology in the Last Half Century." *Plant Physiology* 54 (1974), 427–36.

Galston, Arthur W., and Thomas D. Sharkey. "Frits Warmolt Went." *Biographical Memoirs of the National Academy of Sciences* 74 (1998), 349–64.

Garner, W. W., and H. A. Allard. "Effect of Relative Length of Day and Night and Other Factors of the Environment on Growth and Reproduction in Plants." *Journal of Agricultural Research* 18 (1920), 553–606.

Gaudillière, Jean-Paul. "Globalization and Regulation in the Biotech World: The Transatlantic Debates over Cancer Genes and Genetically Modified Crops." *Osiris* 21 (2006), 251–72.

Gaudillière, Jean-Paul. "Paris–New York Roundtrip: Transatlantic Crossings and the Reconstruction of the Biological Sciences in Post-War France." *Studies in the History and Philosophy of Biological and Biomedical Sciences* 33, no. 3 (2002), 389–417.

Gay, Hannah. *The History of Imperial College 1907–2007: Higher Education and Research in Science, Technology, and Medicine*. London: Imperial College Press, 2007.

Gay, Peter. *Modernism: The Lure of Heresy from Baudelaire to Beckett and Beyond*. New York: Norton 2008.

Geison, Gerald. "Toward a History of American Physiology." In *Physiology in the American Context, 1850–1940*, edited by Gerald Geison, 1–15. New York: American Physiological Society, 1987.

Gentry, Linnea, and Karen Liptak. *The Glass Ark: The Story of Biosphere 2*. Puffin, 1991.

Gilbert, Scott F. "Resurrecting the Body: Has Postmodernism Had Any Effect on Biology?" *Science in Context* 8, no. 4 (1995), 563–77.

Gilbert, Walter. "A Vision of the Grail." In Kevles and Hood, *Code of Codes*, 83–97.

Gillespie, D. T. C. "Research Management in the Commonwealth Scientific and Industrial Research Organisation, Australia." *Public Administration (London)* 46, no. 1 (1964), 11–31.

Ginzburg, Carlo. *The Cheese and the Worms*. Penguin, 1992.

Ginzburg, Lev, and Mark Colyvan. *Ecological Orbits: How Planets Move and Populations Grow*. Oxford: Oxford University Press, 2004.

Gleick, James. *Chaos: The Making of a New Science*. New York: Viking, 1987.

Goldman, Joanne. "National Science in the Nation's Heartland: The Ames Laboratory and Iowa State University, 1942–1962." *Technology and Culture* 41 (2000), 432–59.

Golinski, Jan. *Making Natural Knowledge: Constructivism and the History of Science*. Cambridge University Press, 1998.

Golueke, C. G., W. J. Oswald, and H. K. Gee. "A Study of Fundamental Factors Pertinent to Microbiological Waste Conversion in Control of Isolated Environments." March 31, 1965. Project no. 8659 under Contract no. AF 19(628)-2462 for the Air Force Cambridge Research Laboratories, USAF.

Golueke, Clarence G., and William J. Oswald. "The Algatron: A Novel Microbial Culture System." *Sun at Work* 11, no. 1 (1966), 3–9.

Goodstein, Judith R. *Millikan's School: A History of the California Institute of Technology*. New York: Norton, 1991.

Gordon, Colin. *Mapping Decline: St. Louis and the Fate of the American City*. Philadelphia: University of Pennsylvania Press, 2009.

Gorissen. A., P. Kuikman, J. Ginkel, H. Beek, and A. Jansen. "EPAS: An advanced Phytotron for Measuring Carbon Dynamics in a Whole Plant-Soil System." *Plant and Soil* 179 (1996), 81–87.

Gould, James L., and William T. Keeton. *Biological Science*. New York: Norton, 1996. Originally published in 1967.

Guerlac, Henry. *RADAR in World War II*. Melville, NY: American Institute of Physics, 1987.

Guha, Ramachandra. *How Much Should a Person Consume? Environmentalism in India and the United States*. Berkeley: University of California Press, 2006.

Guha, Ramachandra. *The Unquiet Woods: Ecological Change and Peasant Resistance in the Himalaya*. Oxford: Oxford University Press, 1989.

Gullino, Maria Lodovica, Massimo Pugliese, Alessandro Paravicini, Enzo Casulli, Andrea Rettori, Mattia Sanna, and Angelo Garibaldi. "New Phytotron for Studying the Effects of Climate Change on Plant Pathogens." *Journal of Agricultural Engineering* 42, no. 1 (2012), 1–12.

Gunn, John. *High Encounters, 1954–1970*. Brisbane: University of Queensland Press, 1988.

Haagen-Smit, A. J., Ellis F. Darley, Milton Zaitlin, Herbert Hull, and Wilfred Noble. "Investigation on Injury to Plants from Air Pollution in the Los Angeles Area." *Plant Physiology* 27, no. 18 (1952), 18–34.

Habermas, Jürgen. "Modernity: An Incomplete Project." In *The Anti-Aesthetic: Essays on Postmodern Culture*, edited by Hal Foster, 3–15. Port Townsend, WA: Bay Press, 1983.

Hagen, Joel B. *An Entangled Bank: The Origins of Ecosystem Ecology*. New Brunswick, NJ: Rutgers University Press, 1992.

Hagen, Joel B. "Experimentalists and Naturalists in Twentieth-Century Botany: Experimental Taxonomy, 1920–1950." *Journal of the History of Biology* 17 (1984), 249–70.

Hagen, Joel B. "Teaching Ecology during the Environmental Age, 1965–1980." *Environmental History* 13 (2008), 704–23.

Hall, Stuart. "When Was 'The Post-Colonial'? Thinking at the Limit." In *The Post-Colonial Question: Common Skies, Divided Horizons*, edited by Iain Chambers and Lidia Curti, 242–60. New York: Routledge, 1996.

Hamblin, Jacob Darwin. "Let There Be Light . . . and Bread: The United Nations, the Developing World, and Atomic Energy's Green Revolution." *History and Technology* 25, no. 1 (2009), 25–48.

Hamblin, Jacob Darwin. *Poison in the Well: Radioactive Waste in the Oceans at the Dawn of the Nuclear Age*. New Brunswick, NJ: Rutgers University Press, 2009.

Hannaway, Owen. "Laboratory Design and the Aim of Science: Andreas Libavius versus Tycho Brahe." *Isis* 77 (1986), 585–610.

Hanson, J. B. *History of the American Society of Plant Physiologists*. American Society of Plant Physiologists, 1989.

Harland, S. C. "Changing the Behaviour of Plants." *New Scientist*, January 16, 1958, 30.

Harland-Jacobs, Jessica. *Balancing Tradition and Innovation: The History of Botany at Duke University, 1849–1996*. Durham, NC: Duke University Press, 1996.

Harmen, Oren, and Michael R. Dietrich, eds. *Rebels, Mavericks, and Heretics in Biology*. New Haven, CT: Yale University Press, 2008.

Harman, Oren Solomon. "C. D. Darlington and the British and American Reaction to Lysenko and the Soviet Conception of Science." *Journal of the History of Biology* 36 (2003), 309–52.

Harper, A. M., and D. W. A. Roberts. *Phytotron Manual*. Research Branch Agriculture Canada, 1988.

Harraway, Donna. *Primate Visions: Gender, Race, and Nature in the World of Modern Science*. New York: Routledge, 1989.

Harrison, George. "Life and Quantum Theory." In *Proceedings: Plant Science Symposium 1962*, 109–21. Camden, NJ: Campbell Soup Company, 1962.

Harvey, David. *The Condition of Postmodernity: An Enquiry into the Origins of Cultural Change*. Blackwell, 1990.

Harwood, Jonathan. "Comments on Experimentation in Twentieth-Century Agricultural Science." *History and Philosophy of the Life Sciences* 37 (2015), 326–30.

Harwood, Jonathan. "Did Mendelism Transform Plant Breeding? Genetic Theory and Breeding Practice, 1900–1945." In *New Perspectives on the History of Life Sciences and Agriculture*, edited by Denise Phillips and Sharon Kingsland, 345–70. Springer, 2015.

Hawking, Stephen. "Foreword." In *From Newton to Hawking: A History of Cambridge University's Lucasian Professors of Mathematics*, edited by Kevin C. Knox and Richard Noakes, xvii–xix. Cambridge: Cambridge University Press, 2003.

Hecht, Gabrielle. *Being Nuclear: Africans and the Global Uranium Trade*. MIT Press, 2014.

Hecht, Gabrielle. *The Radiance of France: Nuclear Power and National Identity after World War II*. Cambridge, MA: MIT Press, 2009.

Heilbron, John, and Robert Seidel. *Lawrence and His Laboratory: A History of the Lawrence Berkeley Laboratory*. Berkeley: University of California Press, 1989.

Heims, Steve Joshua. *Constructing a Social Science for Postwar American: The Cybernetics Group, 1946–1953*. MIT Press, 1993.

Heinlein, Robert A. *Stranger in a Strange Land*. Ace Trade, 1991.

Hellmers, Henry. "Phytotronics." In *McGraw-Hill Encyclopedia of Science and Technology*, 10th ed. (Par-Plan), 13:548–50. New York: McGraw-Hill, 2007.

Hendricks, S. B., and F. W. Went. "Controlled-Climate Facilities for Biologists." *Science* 128 (1958), 510–12.

Hendricks, Sterling B. "The Passing Scene." *Annual Review of Plant Physiology* 21 (1970), 1–11.

Herbert, Frank. *Dune*. New York: Ace Books, 1998.

Herkin, Gregg. "The University of California, the Federal Weapons Labs, and the Founding of the Atomic West." In *The Atomic West*, edited by Bruce Hevly and John M. Findlay, 119–35. Seattle: University of Washington Press, 1998.

Hevly, Bruce W. "Basic Research within a Military Context: The Naval Research Laboratory and the Foundations of Extreme Ultraviolet and X-Ray Astronomy." Ph.D. diss., Johns Hopkins University, 1987.

Hiesey, William M. "Growth and Development of Species and Hybrids of *Poa* under Controlled Temperatures." *American Journal of Botany* 40 (1953), 205–21.

Highkin, H. R. "The Effect of Constant Temperature Environment and of Continuous Light on the Growth and Development of Pea Plants." *Cold Spring Harbor Symposia on Quantitative Biology* 25 (1960), 231–38.

Hoddeson, Lillian, Adrienne W. Kolb, and Catherine Westfall. *Fermilab: Physics, the Frontier, and Megascience*. Chicago: University of Chicago Press, 2008.

Home, R. W. "The Rush to Accelerate: Early Stages in Nuclear Physics Research in Australia." *Historical Studies in the Physical and Biological Sciences* 36, no. 2 (2006), 213–41.

Hood, Leroy. "Biology and Medicine in the Twenty-First Century." In Kevles and Hook, *Code of Codes*, 136–63.

Hounshell, David. *From the American System to Mass Production, 1800–1932: The Development of Manufacturing Technology in the United States*. Baltimore: Johns Hopkins University Press, 1985.

Hudson, J. P., ed. *Control of the Plant Environment*. Butterworth Scientific Publications, 1957.

Hughes, Agatha C., and Thomas P. Hughes, eds. *Systems, Experts, and Computers: The Systems Approach in Management and Engineering, World War II and After*. Cambridge, MA: MIT Press, 2000.

Hughes, Sally Smith. *Genentech: The Beginnings of Biotech*. University of Chicago Press, 2011.

Hughes, Thomas P. *American Genesis: A Century of Invention and Technological Enthusiasm*. New York: Penguin, 1989.

Hughes, Thomas P. *Human-Built World: How to Think about Technology and Culture*. Chicago: University of Chicago Press, 2005.

Hughes, Thomas P. *Rescuing Prometheus: Four Monumental Project that Changed the Modern World*. New York: Vintage, 2000.

Hughes, Thomas P., and Agatha C. Hughes. "Introduction." In Hughes and Hughes, *Systems, Experts, and Computers*, 1–26.

Huxley, Aldous. *Brave New World*. New York: Penguin, n.d.

Iida, Kaori. "Genetics and 'Breeding as a Science': Kihara Hitoshi and the Development of Genetics in Japan in the First Half of the Twentieth Century." In *New Perspectives on the History of Life Sciences and Agriculture*, edited by Denise Phillips and Sharon Kingsland, 439–58. Springer, 2015.

Iliffe, Rob. *Newton: A Very Short Introduction*. Oxford: Oxford University Press, 2007.

Inoue, E., Z. Uchijima, T. Saito, S. Isobe, and K. Uemura. "The 'Assimitron': A Newly Devised Instrument for Measuring CO2 Flux in the Surface Air Layer." *Journal of Agricultural Meteorology, Tokyo* 25 (1969), 165–71.

Jackson, Catherine. "Visible Work: The Role of Students in the Creation of Liebig's Giesson Research School." *Notes and Records of the Royal Society* 62 (2008), 31–49.

Jensen, Merle H. "Food Production in Greenhouses." In *Plant Production in Closed Ecosystems: The International Symposium on Plant Production in Closed Ecosystems Held in Narita, Japan, August 26–29, 1996*, edited by E. Goto, K. Kurata, M. Hayashi, and S. Sase, 1–14. Kluwer, 1997.

John, Caroline Carlton. "A Phytotron Is Good for Plants and Other Living Things." *Duke Alumni Register* 56, no. 2 (1970), 3–5.

Johnson, Ann. "Modeling Molecules: Computational Nanotechnology as a Knowledge Community." *Perspectives on Science* 17 (2009), 144–73.

Johnston, Sean F. *The Neutron's Children: Nuclear Engineers and the Shaping of Identity*. Oxford: Oxford University Press, 2012.

Jost, Ludwig. *Lectures on Plant Physiology*. Translated by R. J. H. G. [1907]. BiblioBazaar, 2009.

Judson, Horace Freeland. *The Eighth Day of Creation*. New York: Simon and Schuster, 1979.

Judson, Horace Freeland. "A History of the Science and Technology behind Gene Mapping and Sequencing." In Kevles and Hood, *Code of Codes*, 37–80.

Judt, Tony. *Postwar: A History of Europe since 1945*. William Heinemann, 2005.

Juhren, Marcella, W. M. Hiesey, and F. W. Went. "Germination and Early Growth of Grasses in Controlled Conditions." *Ecology* 34 (1953), 288–300.

Kaiser, David. "Cold War Requisitions, Scientific Manpower, and the Production of American Physicists after World War II." *Historical Studies in the Physical and Biological Sciences* 33, no. 1 (2002), 131–59.

Kaiser, David. "Moving Pedagogy from the Periphery to the Center." In Kaiser and Warwick, *Pedagogy and the Practice of Science*, 1–8.

Kaiser, David. "The Postwar Suburbanization of American Physics." *American Quarterly* 56 (2004), 851–88.

Kaiser, David. "'A Ψ Is Just a Ψ?' Pedagogy, Practice, and the Reconstitution of General Relativity, 1942–1975." *Studies in the History and Philosophy of Modern Physics* 29, no. 3 (1998), 321–38.

Kaiser, David, and Andrew Warwick, eds. *Pedagogy and the Practice of Science: Historical and Contemporary Perspectives*. Cambridge: Cambridge University Press, 2005.

Kargon, Robert H. *The Rise of Robert Millikan: Portrait of a Life in American Science*. Ithaca, NY: Cornell University Press, 1982.

Kargon, Robert H. "Temple to Science: Cooperative Research and the Birth of the California Institute of Technology." *Historical Studies in the Physical Sciences* 8 (1977), 3–31.

Karlson, Peter. "Ectohormones and Phytohormones." *TIBS* (October 1982), 382–83.

Kay, Lily E. *The Molecular Vision of Life: Caltech, the Rockefeller Foundation and the Rise of the New Biology*. Oxford: Oxford University Press, 1993.

Kay, Lily E. "Who Wrote the Book of Life? Information and the Transformation of Molecular Biology, 1945–55." *Science in Context* 8, no. 4 (1995), 609–34.

Keller, Evelyn Fox. *A Feeling for the Organism: The Life and Work of Barbara McClintock*. San Francisco: Freeman, 1983.

Keller, Evelyn Fox. "Nature, Nurture, and the Human Genome." In Kevles and Hood, *Code of Codes*, 281–99.

Keller, Evelyn Fox. *Refiguring Life: Metaphors of Twentieth-Century Biology*. New York: Columbia University Press, 1995.

Kevles, Daniel J. "Big Science and Big Politics in the United States: Reflection on the Death of the SSC and the Life of the Human Genome Project." *Historical Studies in the Physical and Biological Sciences* 27, no. 2 (1997), 269–98.

Kevles, Daniel J. *The Physicists: The History of a Scientific Community in Modern America*. Cambridge, MA: Harvard University Press, 1995.

Kevles, Daniel J. "Principles and Politics in Federal R&D Policy, 1945–90: An Appreciation of the Bush Report." Preface in *Science: The Endless Frontier*, ix–xxxiii. Arlington, VA: National Science Foundation, 1990.

Kevles, Daniel, and Gerald L. Geison. "The Experimental Life Sciences in the Twentieth Century." *Osiris* 10 (1995), 97–121.

Kevles, Daniel J., and Leroy Hood, eds. *The Code of Codes: Scientific and Social Issues in the Human Genome Project*. Cambridge, MA: Harvard University Press, 1992.

Kingsbury, Noel. *Hybrid: The History and Science of Plant Breeding*. Chicago: University of Chicago Press, 2009.

Kingsland, Sharon E. "Defining Ecology as Science." In *Foundations of Ecology: Classic Papers with Commentaries*, edited by L. A. Real and J. H. Brown, 1–13. Chicago: University of Chicago Press, 1991.

Kingsland, Sharon E. "An Elusive Science: Ecological Enterprise in the Southwestern United States." In *Science and Nature: Essays in the History of the Environmental Sciences*, edited by Michael Shortland, 151–79. London: British Society for the History of Science, 1993.

Kingsland, Sharon E. *The Evolution of American Ecology*. Baltimore: Johns Hopkins University Press, 2005.

Kingsland, Sharon E. "Frits Went's Atomic Age Greenhouse: The Changing Landscape on the Lab–Field Border." *Journal of the History of Biology* 42 (2009), 289–324.

Kirsch, Scott. *Proving Grounds: Project Plowshare and the Unrealized Dream of Nuclear Earthmoving*. New Brunswick, NJ: Rutgers University Press, 2005.

Klein, Naomi. *This Changes Everything: Capitalism vs. the Climate*. New York: Simon and Schuster, 2014.

Kleinman, Daniel. *Politics on the Endless Frontier: Postwar Research Policy in the United States*. Durham, NC: Duke University Press, 1995.

Kleinman, Kim. "His Own Synthesis: Corn, Edgar Anderson, and Evolutionary Theory in the 1940s." *Journal of the History of Biology* 32 (1999), 293–320.

Kleinman, Kim. "The Museum in the Garden: Research, Display, and Education at the Missouri Botanical Garden since 1859." PhD diss., Union Institute College of Graduate Studies, 1997.

Knowles, Scott G., and Stuart W. Leslie. "'Industrial Versailles': Eero Saarinen's Corporate Campuses for GM, IBM, and AT&T." *Isis* 92 (2001), 1–33.

Knorr-Cetina, Karin. *Epistemic Cultures: How the Sciences Make Knowledge.* Cambridge, MA: Harvard University Press, 1999.

Kohler, Robert. *Landscapes and Labscapes: Exploring the Lab–Field Border in Biology.* Chicago: University of Chicago Press, 2002.

Kohler, Robert E. *Lords of the Fly:* Drosophila *Genetics and the Experimental Life.* Chicago: University of Chicago Press, 1994.

Kohler, Robert E. "The Management of Science: The Experience of Warren Weaver and the Rockefeller Foundation Programme in Molecular Biology." *Minerva* 14 (1976), 279–306.

Kohler, Robert E. *Partners in Science: Foundations and Natural Scientists, 1900–1945.* Chicago: University of Chicago Press, 1991.

Kohler, Robert E. "The Ph.D. Machine: Building on the Collegiate Base." *Isis* 81 (1990), 639–62.

Kohler, Robert E. "Plants and Pigeonholes: Classification as a Practice in American Ecology." *Historical Studies in the Natural Sciences* 38, no. 1 (2008), 77–108.

Kohlstedt, Sally G. *The Formation of the American Scientific Community: The American Association for the Advancement of Science, 1848–60.* Urbana: University of Illinois Press, 1976.

Kolbert, Elizabeth. *Field Notes from a Catastrophe: Man, Nature, and Climate Change.* Bloomsbury, 2006.

Kramer, P. J., H. Hellmers, and R. J. Downs. "SEPEL: New Phytotrons for Environmental Research." *Bioscience* 20, no. 22 (1970), 1201–8.

Kramer, Paul J., and Theodore T. Kozlowski. *Physiology of Trees.* McGraw-Hill, 1960.

Kramer, Paul J., and Theodore T. Kozlowski. *Physiology of Woody Plants.* New York: Academic Press, 1979.

Kramer, Paul Jackson, and John Strickland Boyer. *Water Relations in Plants and Soils.* New York: Academic Press, 1995.

Kremer, Richard. "Physiology." In Bowler and Pickstone, *Cambridge History of Science,* 358–66.

Krige, John. *American Hegemony and the Postwar Reconstruction of Science in Europe.* Cambridge, MA: MIT Press, 2006.

Krige, John. "The Installation of High-Energy Accelerators in Britain after the War: Big Equipment but Not Big Science." In *The Restructuring of Physical Sciences in Europe and the United States, 1945–60,* edited by Michelangelo De Maria, Mario Grilli, and Fabio Sebastiani, 488–501. Singapore: World Scientific, 1989.

Krige, John, and Kai-Henrik Barth. "Introduction: Science, Technology, and International Affairs." *Osiris* 21 (2006), 1–21.

Krizek, D. T. "Maximizing Plant Growth in Controlled Environments." In Chouard and de Bilderling, *Phytotronics*, 6–12.

Kuhn, Thomas S. *The Structure of Scientific Revolutions*. Chicago: University of Chicago Press, 1973.

Laing, William A., William L. Ogren, and Richard H. Hageman. "Regulation of Soybean New Photosynthetic CO_2 Fixation by the Interaction of CO_2, O_2, and Ribulose 1, 5-Diphosphate Carboxylase." *Plant Physiology* 54 (1974), 678–85.

Lamb, H. H. *Climate, History and the Modern World*. Methuen, 1982.

Lange, O. L., E. D. Schulze, and W. Koch. "Evaluation of Photosynthesis Measurements Taken in the Field." In *Prediction and Measurement of Photosynthetic Productivity*, 339–52.

Latour, Bruno. "Give Me a Laboratory and I Will Raise the World." In *Science Observed: Perspective on the Social Study of Science*, edited by Karin D. Knorr-Cetina and Michael Mulkay, 141–70. Sage, 1983.

Latour, Bruno, and Steve Woolgar. *Laboratory Life: The Construction of Scientific Facts* Princeton, NJ: Princeton University Press, 1986.

Lawrence, W. J .C. "The Glasshouse as an Environment for Plant Experiments." In *Control of the Plant Environment*, edited by J. P. Hudson, 129–38. London: Butterworth Scientific Publications, 1957.

Lawton, J. H. et al. "The Ecotron: A Controlled Environmental Facility for the Investigation of Population and Ecosystem Processes." *Philosophical Transactions: Biological Sciences* 341, no. 1296 (1993), 181–94.

Lawton, John H. "Are There General Laws in Ecology?" *OIKOS* 84 (1999), 177–92.

Lawton, John H. "Ecological Experiments with Model Systems: The Ecotron Facility in Context." In *Experimental Ecology: Issues and Perspectives*, edited by William J. Resetarits Jr. and Joseph Bernardo, 170–81. Oxford: Oxford University Press, 1998.

Lawton, John H. "The Ecotron Facility at Silwood Park: The Value of 'Big Bottle' Experiments." *Ecology* 77, no. 3 (1996), 665–69.

Le Corbusier. *Towards a New Architecture*. Translated by Frederick Etchells. Mineola, NY: Dover, 1986.

Lee, David. *Search for Security: The Political Economy of Australia's Postwar Foreign and Defence Policy*. Allen and Unwin with RSPAS, ANU, 1995.

Leonelli, Sabina. "*Arabidopsis*, the Botanical *Drosophila*: From Mouse Cress to Model Organism." *Endeavour* 31 (2007), 34–38.

Leroi, Armand Marie. *The Lagoon: How Aristotle Invented Science*. New York: Penguin, 2015.

Leslie, Stuart W. *The Cold War and American Science: The Military-Industrial-Academic Complex at MIT and Stanford*. New York: Columbia University Press, 1993.

Leslie, Stuart W. "Playing the Education Game to Win: The Military and Interdisciplinary Research at Stanford." *Historical Studies in the Physical and Biological Sciences* 18 (1987), 55–88.

Levin, Ira. *The Boys from Brazil*. New York: Random House, 1976.

Levins, Richard, and Richard Lewontin. *The Dialectical Biologist*. Cambridge, MA: Harvard University Press, 1985.

Levins, Richard, and Richard Lewontin. "The Problem of Lysenkoism." In Levins and Lewontin, *The Dialectical Biologist*, 163–96.

Lewontin, R. C. "The Cold War and the Transformation of the Academy." In *The Cold War and the University: Towards an Intellectual History of the Postwar Years*, 1–34. New York: New Press, 1997.

Lewontin, Richard. *Biology as Ideology*. HarperPerennial, 1991.

Lewontin, Richard. *It Ain't necessarily So: The Dream of the Human Genome and Other Illusions*. New York: New York Review Books, 2001.

Lindsay, Norman. *The Magic Pudding*. Angus and Robinson, 1976.

Long, S. P., and F. I. Woodward, eds. *Plants and Temperature*. Symposia of the Society for Experimental Biology Society for Experimental Biology, 1988.

Long, Tully. "William McElroy, the McCollum-Pratt Institute, and the Transformation of Biology at Johns Hopkins, 1945–1960." *Journal of the History of Biology* 42 (2009), 765–809.

Los Angeles Plays Itself. Directed by Thom Andersen. Written by Thom Andersen. Thom Andersen Productions, 2003.

Lowen, Rebecca. *Creating the Cold War University: The Transformation of Stanford*. University of California Press, 1997.

Macintyre, Stuart. *Australia's Boldest Experiment: War and Reconstruction in the 1940s*. Sydney: NewSouth, 2015.

MacKenzie, Donald. *Knowing Machines: Essays on Technical Change*. Cambridge, MA: MIT Press, 1998.

MacLeod, Roy. "On Visiting the 'Moving Metropolis': Reflections on the Architecture of Imperial Science." In Reingold and Rothenberg, *Scientific Colonialism*, 217–49.

Marino, Bruno D. V., and H. T. Odum. "Biosphere 2: Introduction and Research Progress." *Ecological Engineering* 13 (1999), 3–14.

Marshall, L. C., H. O. Eversole, A. J. Hess, and F. W. Went. "The Technical Features of Greenhouse Air Conditioning." *Refrigerating Engineering* 55, no. 2 (February 1948), 151–54.

Marshall, L. C., H. O. Eversole, A. J. Hess, and F. W. Went. "The Technical Features of Greenhouse Air Conditioning, II." *Refrigerating Engineering* 55, no. 3 (March 1948), 269–74.

The Martian. Directed by Ridley Scott. Written by Drew Goddard. Twentieth Century Fox, 2015.

May, Robert M. *Stability and Complexity in Model Ecosystems*. Princeton, NJ: Princeton University Press, 1974.

Mayr, Ernst. *The Growth of Biological Thought: Diversity, Evolution, and Inheritance*. Cambridge, MA: Belknap Press of Harvard University Press, 1982.

Mayr, Ernst. "The Role of Systematics in the Evolutionary Synthesis." In Mayr and Provine, *The Evolutionary Synthesis*, 123–136.

Mayr, Ernst, and William B. Provine, eds. *The Evolutionary Synthesis: Perspectives on the Unification of Biology*. Cambridge, MA: Harvard University Press, 1980.

McCray, W. Patrick. "Project Vista, Caltech, and the Dilemma of Lee DuBridge." *Historical Studies in the Physical and Biological Sciences* 34 (2004), 339–70.

McCray, W. Patrick. *The Visioneers: How a Group of Elite Scientists Pursued Space Colonies, Nanotechnologies, and a Limitless Future*. Princeton, NJ: Princeton University Press, 2013.

McElheny, Victor K. *Watson and DNA: Making a Scientific Revolution*. New York: Perseus, 2003.

McFarlane, J. Craig. "Light." In *A Growth Chamber Manual: Environmental Control for Plants*, edited by Robert W. Langhans, 15–44. Comstock, 1978.

Milburn, Colin. "Nanotechnology in the Age of Posthuman Engineering: Science Fiction as Science." In *Nanoculture: Implications of the New Technoscience*, edited by N. Katherine Hayles, 109–29. Bristol: Intellect Books, 2004.

Milburn, Colin. "Nanowarriors: Soldier Nanotechnology and Comic Books." *Intertexts* 9 (2005), 77–103.

Mindell, David. "Automation's Finest Hour: RADAR and System Integration in World War II." In Hughes and Hughes *Systems, Experts, and Computers*, 27–56.

Mindell, David. *Digital Apollo: Human and Machine in Spaceflight*. Cambridge, MA: MIT Press, 2011.

Misa, Thomas J. "The Compelling Tangle of Modernity and Technology." In *Modernity and Technology*, edited by Thomas J. Misa, Philip Brey, and Andrew Feenberg, 1–30. Cambridge, MA: MIT Press, 2003.

Mooney, Harold A. "Paul Jackson Kramer." *Proceedings of the American Philosophical Society* 143, no. 2 (1999), 341–43.

Monteith, J. L. "Introduction: Some Partisan Remarks on the Virtues of Field Experiments and Physical Analogues." In Rees et al., *Crop Processes in Controlled Environments*, 107–10.

Morange, Michel. "The Death of Molecular Biology?" *History and Philosophy of the Life Sciences* 30 (2008), 31–42.

Morange, Michel. *A History of Molecular Biology*. Cambridge, MA: Harvard University Press, 1998.

Mukherjee, Siddhartha. *The Gene: An Intimate History*. New York: Scribner, 2016.

Müller-Wille, Staffan, and Hans-Jörg Rheinberger. *A Cultural History of Heredity*. Chicago: University of Chicago Press, 2012.

Mumford, Eric. *Climatron 50: A Celebration of 50 Years*. St. Louis: Missouri Botanical Garden, 2010.

Munns, David P. D. "'The Awe in Which Biologists Hold Physicists': Frits Went's First Phytotron at Caltech, and an Experimental Definition of the Biological Environment." *History and Philosophy of the Life Sciences* 36, no. 2 (2014), 209–31.

Munns, David P. D. "Controlling the Environment: The Australian Phytotron and Postcolonial Science." *British Scholar* 2, no. 2 (2010), 197–226.

Munns, David P. D. "If We Build It, Who Will Come? Training Radio Astronomers and the Limitations of 'National' Laboratories in Cold War America." *Historical Studies in the Physical and Biological Sciences* 34, no. 1 (2003), 93–117.

Munns, David P. D. *A Single Sky: How an International Community Forged the Science of Radio Astronomy*. Cambridge, MA: MIT Press, 2013.

Murneek, A. E., and R. O. Whyte. *Vernalization and Photoperiodism: A Symposium*. Chronica Botanica, 1948.

Murphy, Michelle. *Sick Building Syndrome and the Problem of Uncertainty: Technoscience, Environmental Politics and Women Workers*. Durham, NC: Duke University Press, 2006.

Naeem, Shahid. "Experimental Validity and Ecological Scale in Evaluating Research Programs." In *Scaling Relations in Experimental Ecology*, edited by Robert H. Gardner, W. Michael Kemp, Victor S. Kennedy, and John E. Petersen, 223–50. New York: Columbia University Press, 2001.

Nanda, K. K. "Problems of Fundamental and Applied Plant Physiology Which Require Controlled Environments." In Chouard and de Bilderling, *Phytotronique*, 61–67.

National Council for Research and Development (NCRD). *A Phytotron in Israel: Controlled Environment Facilities for Biological & Agricultural Research* NCRD Report no. 12/61/M, 1961.

Needell, Allan A. "The Carnegie Institution of Washington and Radio Astronomy: Prelude to an American National Observatory." *Journal for the History of Astronomy* 22 (1991), 55–67.

Needell, Allan A. "Preparing for the Space Age: University-based Research, 1946–57." *Historical Studies in the Physical and Biological Sciences* 18 (1987), 89–109.

Nelkin, Dorothy. *Nuclear Power and Its Critics: The Cayuga Lake Controversy*. Ithaca, NY: Cornell University Press, 1971.

Nelkin, Dorothy. *The University and Military Research: Moral Politics at M.I.T.* Ithaca, NY: Cornell University Press, 1972.

Nelkin, Dorothy, and M. Susan Lindee. *The DNA Mystique: The Gene as a Cultural Icon*. San Francisco: Freeman, 1995.

Newman, L. H. *Plant Breeding in Scandinavia*. Ottawa: The Canadian Seed Growers Association, Canadian Building, n. d.

Nickelsen, Kärin. *Explaining Photosynthesis: Models of Biochemical Mechanisms, 1840–1960*. Springer, 2015.

Nickelsen, Kärin, and Govindjee. *The Maximum Quantum Yield Controversy: Otto Warburg and the "Midwest Gang."* Bern Studies in the History and Philosophy of Science, 2011.

Nitsch, J. P. "A Critical Comparison of Various Types of Phytotrons." In Chouard and de Bilderling, *Phytotronique*, 17–23.

Nitsch, J. P. "Phytotrons: Past Achievements and Future Needs." In Rees et al., *Crop Processes in Controlled Environments*, 33–56.

Nitsch, J. P. "The Role of Plant Hormones in Fruit Development." PhD diss., California Institute of Technology, 1951.

North, Michael. *Reading 1922: A Return to the Scene of the Modern*. Oxford: Oxford University Press, 1999.

November, Joseph. *Biomedical Computing: Digitizing Life in the United States*. Baltimore: Johns Hopkins University Press, 2012.

Nye, David E. *Consuming Power: A Social History of American Energies*. Cambridge, MA: MIT Press, 1998.

Nye, Mary Jo. *Before Big Science: The Pursuit of Modern Chemistry and Physics, 1800–1940*. London: Prentice Hall International, 1996.

Oakman, Daniel. *Facing Asia: A History of the Colombo Plan*. Canberra: Pandanus, 2004.

Odum, Eugene. "The Emergence of Ecology as a New Integrative Discipline." *Science* 195, no. 4284 (1977), 1289–93.

Odum, Eugene. *Fundamentals of Ecology*. Saunders, 1971.

Odum, Eugene P. *Ecology: A Bridge Between Science and Society*. Sunderland, MA: Sinauer, 1997.

Olesko, Kathryn M. *Physics as a Calling: Discipline and Practice in the Königsberg Seminar for Physics*. Ithaca, NY: Cornell University Press, 1991.

Olmstead, Alan L., and Paul W. Rhode. *Creating Abundance: Biological Innovation and American Agricultural Development*. Cambridge: Cambridge University Press, 2008.

Oosting, Henry J. *The Study of Plant Communities: An Introduction to Plant Ecology*. San Francisco: Freeman, 1956.

Oreskes, Naomi. "Introduction." In Oreskes and Krige, *Science and Technology in the Global Cold War*, 1–9.

Oreskes, Naomi, and Erik M. Conway. *The Collapse of Western Civilization*. New York: Columbia University Press, 2014 [2393].

Oreskes, Naomi, and Erik M. Conway. *Merchants of Doubt: How a Handful of Scientists Obscured the Truth on Issues from Tobacco Smoke to Global Warming*. Bloomsbury, 2011.

Oreskes, Naomi, and John Krige, eds. *Science and Technology in the Global Cold War*. Cambridge, MA: MIT Press, 2014.

Oswald, W. J., and C. G. Golueke. "Environmental Control Studies with a Closed Ecological System." *Proceedings of the 8th Annual Meeting of the Institute of Environmental Sciences* 8 (1962), 183–91.

Oswald, William J., Clarence G. Golueke, and Donald O. Horning. "Closed Ecological Systems." *Journal of the Sanitary Engineering Division: Proceedings of the American Society of Civil Engineers* 91, no. SA4 (1965), 23–45.

Panel for the Plant Sciences. *The Plant Sciences Now and in the Coming Decade: A Report on the Status, Trends, and Requirements of Plant Sciences in the United States*. Washington, DC: National Academy of Sciences, 1966.

Parolini, Giuditta. "Charting the History of Agriculture Experiments." *History and Philosophy of the Life Sciences* 37 (2015), 231–41.

Pauly, Philip J. *Biologists and the Promise of American Life*. Princeton, NJ: Princeton University Press, 2002.

Pauly, Philip J. *Controlling Life: Jacques Loeb and the Engineering Ideal in Biology*. Oxford University Press, 1987.

Pennazio, Sergio. "Mineral Nutrition of Plants: A Short History of Plant Physiology." *Biology Forum* 98 (2005), 251–36.

Perkins, John H. *Geopolitics and the Green Revolution: Wheat, Genes, and the Cold War*. New York: Oxford University Press, 1997.

Pestré, Dominique. "Studies of the Ionosphere and Forecasts for Radiocommunications: Physicists and Engineers, the Military and National Laboratories in France (and Germany) after 1945." *History and Technology* 13 (1997), 183–205.

Phillips, Denise, and Sharon Kingsland, eds. *New Perspectives on the History of Life Sciences and Agriculture*. Springer, 2015.

Picard, Jean-François. *La République des Savants: La recherche française et le C.N.R.S.* Flammarion, 1990.

Pickering, Andrew. *The Cybernetic Brain: Sketches of Another Future*. Chicago: University of Chicago Press, 2010.

Pickering, Andrew. *The Mangle of Practice: Time, Agency, and Science.* Chicago: University of Chicago Press, 1995.

Pitts, James N. Jr., and Edgar R. Stephens. "Arie Jan Haagen-Smit: 1900–1977." *Journal of the Air Pollution Control Association* 28 (1978), 516–17.

Prabhu, K. V., and P. Chandra. *Phytotrons for Agricultural Research.* Indian Council of Agricultural Research, 2000.

Pratchett, Terry, and Neil Gaiman. *Good Omens: The Nice and Accurate Prophecies of Agnes Nutter, Witch.* London: Corgi, 2006.

Prediction and Measurement of Photosynthetic Productivity: Proceedings of the IBP/PP Technical Meeting, Trebon, 14–21 September 1969. Wageningen: Centre for Agricultural Publishing and Documentation, 1970.

Proctor, Robert N. "'-Logos,' 'Ismos,' and 'Ikos': The Political Iconicity of Denominative Suffixes (or, Phonesthemic Tints and Taints in the Coining of Science Domain Names)." *Isis* 98 (2007), 290–309.

Purves, William K. et al. *Life: The Science of Biology.* 6th ed. San Francisco: Freeman, 2001.

Rabinowitch, Eugene, and Govindjee. *Photosynthesis.* Wiley, 1969.

Rader, Karen A. "Alexander Hollaender's Postwar Vision for Biology: Oak Ridge and Beyond." *Journal of the History of Biology* 39 (2006), 685–706.

Rader, Karen A. *Making Mice: Standardizing Animals for American Biomedical Research, 1900–1955.* Princeton, NJ: Princeton University Press, 2004.

Rader Karen A., and Victoria E. M. Cain. *Life on Display: Revolutionizing U.S. Museums of Science and Natural History in the Twentieth Century.* Chicago: University of Chicago Press, 2014.

Raper, David. "A Cost-Benefit Evaluation of Controlled-Environment Research versus Field Research." *Agronomy Journal* 65 (1973), 701–5.

Raper, C. David, and R. J. Downs. "Field Phenotype in Phytotron Culture: A Case Study for Tobacco." *Botanical Review* 42 (1976), 317–43.

Rasmussen, Nicolas. "The Forgotten Promise of Thiamin: Merck, Caltech Biologists, and Plant Hormones in a 1930s Biotechnology Project." *Journal of the History of Biology* 32 (1999), 245–61.

Rasmussen, Nicolas. *Gene Jockeys: Life Science and the Rise of Biotech Enterprise.* Baltimore: Johns Hopkins University Press, 2014.

Rasmussen, Nicolas. "The Mid-Century Biophysics Bubble: Hiroshima and the Biological Revolution in America, Revisited." *History of Science* 35 (1997), 245–93.

Rasmussen, Nicolas. *Picture Control: The Electron Microscope and the Transformation of Biology in America, 1940–60.* Stanford, CA: Stanford University Press, 1997.

Rasmussen, Nicolas. "Plant Hormones in War and Peace: Science, Industry, and Government in the Development of Herbicides in 1940s America." *Isis* 92 (2001), 291–316.

Raven, Peter H., and George B. Johnson. *Biology.* 6th ed. New York: McGraw-Hill, 2002.

Redfield, Peter. *Space in the Tropics: From Convicts to Rockets in French Guiana.* Berkeley: University of California Press, 2000.

Rees, A. R., K. E. Cockshull, D. W. Hand, and R. G. Hurd, eds. *Crop Processes in Controlled Environments.* New York: Academic Press, 1972.

Rees, A. R., K. E. Cockshull, D. W. Hand, and R. G. Hurd. "Preface." In Rees et al., *Crop Processes in Controlled Environments*, viii–ix.

Reider, Rebecca. *Dreaming the Biosphere: The Theater of All Possibilities*. Albuquerque: New Mexico University Press, 2009.

Reingold, Nathan, and Marc Rothenberg, eds. *Scientific Colonialism: A Cross-Cultural Comparison*. Washington, DC: Smithsonian Institution Press, 1987.

Repp, Gertraud. "New Climatic Measuring Chamber for Plant Physiological Research." In Chouard and de Biulderling, *Phytotronique*, 11.

Rheinberger, Hans-Jörg. *An Epistemology of the Concrete: Twentieth-Century Histories of Life*. Durham, NC: Duke University Press, 2010.

Rheinberger, Hans-Jörg. "Heredity in the Twentieth Century: Some Epistemological Considerations." *Public Culture* 25, no. 3 (2013), 477–93.

Rheinberger, Hans-Jörg. "Internationalism and the History of Molecular Biology." *Annals of the History and Philosophy of Biology* 11 (2006), 249–54.

Rhodes, Richard. *The Making of the Atomic Bomb*. New York: Penguin, 1988.

Richardson, J. A. *Physics in Botany*. London: Pitman, 1964.

Ridley, Matt. *Genome: The Autobiography of a Species in 23 Chapters*. New York: Perennial, 2000.

Roach, Mary. *Packing for Mars: The Curious Science of Life in the Void*. New York: Norton, 2010.

Robertson, R. N. "Phytotrons, Spectrophotometers and Productivity." *Australian Journal of Science* 26, no. 10 (1964), 313–17.

Roland, Alex. *The Military-Industrial Complex*. Washington, DC: Society for the History of Technology, 2001.

Roll-Hansen, Nils. *The Lysenko Effect: The Politics of Science*. Humanity Books, 2005.

Roll-Hansen, Nils. "Sources of Wilhelm Johannsen's Genotype Theory." *Journal of the History of Biology* 42 (2009), 457–93.

Rosenberg, Charles E. "Science, Technology, and Economic Growth: The Case of the Agricultural Experiment Station Scientist, 1875–1914." In *Nineteenth Century American Science*, edited by George H. Daniels, 181–209. Evanston, IL: Northwestern University Press, 1972.

Rossi, Bruno B. "The Decay of 'Mesotrons' (1939–1943): Experimental Particle Physics in the Age of Innocence." In *The Birth of Particle Physics*, edited by Laurie M. Brown and Lillian Hoddeson, 183–205. Cambridge: Cambridge University Press, 1986.

Rothenberg, Marc. "Organization and Control: Professionals and Amateurs in American Astronomy, 1899–1918." *Social Studies of Science* 11 (1981), 305–25.

Roussel, L. "The Patscherkofel Phytotron." *Translations of the New Zealand Forest Service* 2 (1963), n. p.

Rudd-Jones, D. "Controlled Environment." In Rees et al., *Crop Processes in Controlled Environments*, 1–6.

Rudolph, John. *Scientists in the Classroom: The Cold War Reconstruction of American Science Education*. New York: Palgrave, 2002.

Ruhland, W., ed. *Encyclopedia of Plant Physiology*. Springer-Verlag, 1961.

Sagan, Carl. *Billions and Billions: Thoughts on Life and Death at the Brink of the Millennium*. New York: Ballantine Books, 1998.

Sage, Linda C. *Pigment of the Imagination: A History of Phytochrome Research*. New York: Academic Press, 1992.

Salisbury, Frank B., Josef I. Gitelson, and Genry M. Lisovsky. "BIOS-3: Siberian Experiments in Bioregenerative Life Support." *BioScience* 47, no. 9 (1997), 575–85.

Salisbury, Frank B., and Cleon W. Ross. *Plant Physiology*. 4th ed. Belmont, CA: Wadsworth, 1992.

Sandweiss, Eric. "From a Garden Looking Out: Public Culture in Henry Shaw's St. Louis." In *St. Louis in the Century of Henry Shaw: A View beyond the Garden Wall*, edited by Eric Sandweiss, 1–16. Columbia: University of Missouri Press, 2003.

Sapolsky, Harvey M. *Science and the Navy: The History of the Office of Naval Research*. Princeton, NJ: Princeton University Press, 1990.

Sapp, Jan. *Genesis: The Evolution of Biology*. Oxford: Oxford University Press, 2003.

Schama, Simon. *Rough Crossings: Britain, the Slaves, and the American Revolution*. New York: HarperCollins, 2006.

Schedvin, C. B. "The Culture of CSIRO." *Australian Cultural History* 2 (1982/83), 76–89.

Schedvin, C. B. "Environment, Economy, and Australian Biology, 1890–1939." In Reingold and Rothenberg, *Scientific Colonialism*, 101–26.

Schedvin, C. Boris. *Shaping Science and Industry: A History of Australia's Council for Scientific and Industrial Research, 1926–49*. Allen and Unwin, 1987.

Schmid, Sonja D. "Defining (Scientific) Direction: Soviet Nuclear Physics and Reactor Engineering during the Cold War." In Oreskes and Krige, *Science and Technology in the Global Cold War*, 317–42.

Schürch, Caterina. "Der Schlüssel zum Pflanzenwachstum Interdisziplinäre Phytohormonforschung in Utrecht und Pasadena 1928–1938." Masters thesis, Ludwig Maximilian University of Munich, 2016.

Seidel, Robert. "Accelerating Science: the postwar transformation of the Lawrence Radiation Laboratory." *Historical Studies in the Physical Sciences* 13 (1983), 375–400.

Shapin, Steven, and Simon Shaffer. *Leviathan and the Air Pump: Hobbes, Boyle and the Experimental Life*. Princeton, NJ: Princeton University Press, 1989.

Sheller, Mimi. *Aluminum Dreams: The Making of Light Modernity*. Cambridge, MA: MIT Press, 2014.

Shortland, M., ed. *Science and Nature: Essays in the History of the Environmental Sciences*. London: British Society for the History of Science, 1993.

Siddiqi, Asif. "Fighting Each Other: The N-1, Soviet Big Science, and the Cold War at Home." In Oreskes and Krige, *Science and Technology in the Global Cold War*, 189–225.

Silent Running. Directed by Douglas Trumball. Written by Deric Washburn, Michael Cimino, and Steven Bochco. Universal Pictures, 1972.

Skoog, Folke. "The Effect of X-Rays on Growth Substance and Plant Growth." *Science* 79, no. 2046 (1934), 256.

Skoog, F. "A Deseeded *Avena* Test Method for Small Amounts of Auxin and Auxin Precursors." *Journal of General Physiology* 20 (1937), 331–34.

Sleeper. Directed by Woody Allen. Written by Woody Allen and Marshall Brickman. Rollins-Joffe Productions, 1973.

Smeets, L. "IVT Phytotron: 1953–1978: Plant Breeding Research under Controlled Conditions in Horticultural Crops." *Netherlands Journal of Agricultural Science* 26 (1978), 2–7, 8–12.

Smith, Alice Kimball. *A Peril and a Hope: The Scientists' Movement in America 1945–47.* Cambridge, MA: MIT Press, 1971.

Smith, Gina. *The Genomic Age: How DNA Technology is Transforming the Way We Live and Who We Are.* New York: American Management Association, 2005.

Smith, Robert W. "Engines of Discovery: Scientific Instruments and the History of Astronomy and Planetary Science in the United States in the Twentieth Century." *Journal for the History of Astronomy* 28 (1997), 49–77.

Smith, Robert W. *The Space Telescope: A Study of NASA, Science, Technology, and Politics.* Cambridge: Cambridge University Press, 1993.

Smith, Merritt Roe, and Leo Marx, eds. *Does Technology Drive History? The Dilemma of Technological Determinism.* Cambridge, MA: MIT Press, 2001.

Smocovitis, V. B. "Disciplining Botany: A Taxonomic Problem." *Taxon* 41 (1992), 459–70.

Smocovitis, V. B. "The 'Plant *Drosophila*': E. B. Babcock, the Genus *Crepis* and the Evolution of a Genetics Research Program at Berkeley, 1912–1947." *Historical Studies of the Natural Sciences* 39 (2009), 300–355.

Smocovitis, V. B. "Unifying Biology: The Evolutionary Synthesis and Evolutionary Biology." *Journal of the History of Biology* 25 (1992), 1–65.

Smocovitis, Vassiliki Betty. "Botany and the Evolutionary Synthesis: The Life and Work of G. Ledyard Stebbins, Jr." PhD diss., Cornell University, 1988.

Smocovitis, Vassiliki Betty. *Unifying Biology: The Evolutionary Synthesis and Evolutionary Biology.* Princeton, NJ: Princeton University Press, 1996.

Somerville, Chris. "The Twentieth Century Trajectory of Plant Biology." *Cell* 100 (2000), 13–25.

Stadler, Max. "Models, the Cell, and the Reformations of Biological Science, 1920–1960." PhD diss., Imperial College London, 2010.

Starr, Cecie. *Biology: Concepts and Applications.* Brooks/Cole, 1998.

Stebbins, G. Ledyard. "Botany and the Synthetic Theory of Evolution." In Mayr and Provine, *The Evolutionary Synthesis*, 139–52.

Stebbins, G. Ledyard, *Darwin to DNA, Molecules to Humanity*, San Francisco: Freeman, 1982.

Steere, William Campbell. "Introduction." In Went and the Editors of Time-Life Books, *The Plants*.

Stent, Gunther S. "Introduction." In Watson, *The Double Helix*, xi–xxii,

Strasser, Bruno J. "A World in One Dimension: Linus Pauling, Francis Crick and the Central Dogma of Molecular Biology." *History and Philosophy of the Life Sciences* 28 (2006), 491–512.

Sugrue, Thomas. *The Origins of the Urban Crisis: Race and Inequality in Postwar Detroit.* Princeton, NJ: Princeton University Press, 2005.

Sullivan, W. T. III. "Early Years of Australian Radio Astronomy." In *Australian Science in the Making*, edited by R. Home, 308–44. Cambridge: Cambridge University Press, 1988.

Sumner, James, and Graeme J. N. Gooday. "Introduction." *History of Technology* 28 (2008), 1–13.

Taiz, Lincoln, and Eduardo Zeiger. *Plant Physiology*. 4th ed. Sunderland, MA: Sinauer, 2006.

Taylor, R. J., and A. J. Dyer. "An Instrument for Measuring Evaporation from Natural Surfaces." *Nature* 181 (February 8, 1958), 408–9.

Terkel, Studs. *"The Good War": An Oral History of World War II*. New Press, 1997.

Tesha, Patrick V. "Thermoperiod Effects on the Seedling Growth of Three Tropical Pines." MSc diss., Duke University, 1970.

Thiel, Stephan, Thorsten Döhring, Matthias Köfferlein, Andre Kosak, Peter Martin, and Harald K. Seidlitz. "A Phytotron for Plant Stress Research: How Far Can Artificial Lighting Compare to Natural Sunlight?" *Journal of Plant Physiology* 148 (1996), 456–63.

Thimann, Kenneth V. "Foreword." In Murneek and Whyte. *Vernalization and Photoperiodism*, vi–viii.

Thimann, Kenneth V., and Folke Skoog. "Studies on the Growth Hormone of Plants III: The Inhibiting Action of the Growth Substance on Bud Development." *Proceedings of the National Academy of Sciences* 19, no. 7 (1933), 714–16.

Thomas, Judith, Robert J. Downs, and Carole H. Saravitz. *Phytotron Procedural Manual*. Technical Bulletin 244 North Carolina State University, North Carolina Agricultural Research Service, 1983.

Thompson, E. P. *The Making of the English Working Class*. New York: Penguin, 1978.

Thone, Frank. "Plant Physiology at the Ithaca Congress." *Plant Physiology* 1, no. 4 (1926), 293–305.

Tibbitts, T. W., W. Cao, and R. M. Wheeler. *Growth of Potatoes for CELSS*. NASA Contractor Report 177646. August 1994. Available at ntrs.nasa.gov.

Tibbitts, T. W., and D. L. Henninger. "Food Production in Space: Challenges and Perspectives." In *Plant Production in Closed Ecosystems: The International Symposium on Plant Production in Closed Ecosystems Held in Narita, Japan, August 26–29, 1996*, edited by E. Goto, K. Kurata, M. Hayashi, and S. Sase, 189–203. Kluwer, 1997.

Tischner, T. "Facts and Figures on the Martonvásár Phytotron." *Acta Agronomica Hungarica* 45 (1997), 353–59.

Todes, Daniel P. "Pavlov's Physiology Factory." *Isis* 88 (1997), 205–46.

Toulmin, Stephen. "A Historical Reappraisal." In *The Classification of Research*, edited by Oliver D. Hensley, 21–29. Lubbock: Texas Tech University Press, 1988.

Traweek, Sharon. *Beamtimes and Lifetimes: The World of High Energy Physics*. Cambridge, MA: Harvard University Press, 1988.

Tron. Directed by Steven Lisberger. Written by Steven Linsberger. Disney, 1982.

Tron: Legacy. Directed by Joseph Kosinski. Written by Edward Kitsis et al. Disney, 2010.

Troyer, James R. "Error or Fraud in Science: Auxins A and B and Animal Tumor Proteins." *Journal of the North Carolina Academy of Science* 124, no. 1 (2008), 1–5.

Uchijima, Z. "Carbon Dioxide Environment and Flux within a Corn Crop Canopy." In *Prediction and Measurement of Photosynthetic Productivity*, 179–96.

Vernon, James. *Hunger: A Modern History*. Cambridge, MA: Belknap Press of Harvard University Press, 2007.

Vidal, Gore. "The State of the Union." In *Matters of Fact and Fiction: Essays, 1973–1976*, 265–85. Heinemann, 1977.

Vines, A. E., and N. Rees. *Plant and Animal Biology*. Vol. 2, 4th ed. London: Pitman, 1974. First published in 1958.

von Wettstein, Diter. *The Phytotron in Stockholm (Skogshögskolans fytotronanläggning)*. Studia Forestalia Suecica no. 44. Royal College of Forestry Stockholm, 1967.

Waggoner, Paul E. "On Dynamics of Development of Photosynthetic Systems." In *Prediction and Measurement of Photosynthetic Productivity*, 583–87.

Waisel, Yoav, Amram Eschel, and Uzi Kafkafi, eds. *Plant Roots: The Hidden Half.* New York: Marcel Decker, 2002.

Walker, Charles T., and Glen A. Slack. "Who Named the –ON's?" *American Journal of Physics* 38, no. 12 (1970): 1380–89.

Wang, Jessica. *American Science in an Age of Anxiety: Scientists, Anticommunism, and the Cold War.* Chapel Hill: University of North Carolina Press, 1999.

Ward, Stuart. *Australia and the British Embrace: The Demise of the Imperial Ideal.* Victoria: Melbourne University Press, 2001.

Ward, Stuart. "Security: Defending Australia's Empire." In *Australia's Empire*, edited by Deryck M. Schreuder and Stuart Ward, 232–58. Oxford: Oxford University Press, 2008.

Wark, McKenzie. *Molecular Red: Theory for the Anthropocene*. London: Verso, 2015.

Warwick, Andrew. *Masters of Theory: Cambridge and the Rise of Mathematical Physics*. Chicago: University of Chicago Press, 2003.

Waterman, Alan. "Basic Research in the United States." In *Symposium on Basic Research*, edited by Dael Wolfle, 17–40. Washington, DC: American Association for the Advancement of Science, 1959.

Watson, Don. "Rabbit Syndrome: Australia and America." *Quarterly Essay* 4 (2001).

Watson, James D. *The Double Helix: A Personal Account of the Discovery of the Structure of DNA*. London: Weidenfeld and Nicolson, 1981.

Watson, James D. "A Personal View of the Project." In Kevles and Hood, *Code of Codes*, 164–73.

Weevers, Theodorus. *Fifty Years of Plant Physiology*. Amsterdam: Scheltema & Holkema, 1949.

Weinberg, Alvin. "Impact of Large-Scale Science on the United States." *Science* 134, no. 3473 (1961), 161–64.

Weinberg, Alvin M. *Reflections on Big Science*. Cambridge, MA: MIT Press, 1967.

Weinberger, Sharon. *Imaginary Weapons: A Journey through the Pentagon's Scientific Underworld*. New York: Nation Books, 2006.

Weir, Andy. *The Martian*. New York: Broadway Books, 2014.

Weisz, George. *Divide and Conquer: A Comparative History of Medical Specialization*. Oxford: Oxford University Press, 2006.

Went, F. W. "Auxin, the Plant Growth-Hormone." *Botanical Review* 1, no. 5 (1935), 162–82.

Went, F. W. "The Concept of a Phytotron." In Evans, *Environmental Control of Plant Growth*, 3–10.

Went, F. W. "The Earhart Plant Research Laboratory." *Chronica Botanica* 12, no. 2 (1950), 93–107.

Went, F. W. "The Effects of Climate on Plant Growth and Distribution." In *Proceedings of the Seventh International Botanical Congress, Stockholm 1950*, 56–63. Almqvist & Wiskell, 1950.

Went, F. W. "Fifty Years of Plant Physiology in the U.S.A." *American Journal of Botany* 44, no. 1 (1957), 105–10.

Went, F. W. "Gene Action in Relation to Growth and Development." *Proceedings of the National Academy of Sciences of the United States of America* 39 (1953), 838–48.

Went, F. W. "Phytotronics." *Proceedings: Plant Science Symposium 1962*, 149–61. Camden, NJ: Campbell Soup Company, 1962.

Went, F. W. "Plant Growth under Controlled Conditions—I: The Air-Conditioned Greenhouses at the California Institute of Technology." *American Journal of Botany* 30 (1943), 157–63.

Went, F. W. "Plant Physiology." In *Vistas in Botany: A Volume in Honor of the Bicentenary of the Royal Botanic Gardens, Kew*, edited by W. B. Turrill, 519–31. London: Pergamon, 1959.

Went, F. W. "The Response of Plants to Climate." *Science* 112, no. 2913 (1950), 489–94.

Went, F. W. "The Role of Environment in Plant Growth." *American Scientist* 44, no. 4 (1956), 378–98.

Went, F.W. "Thermoperiodicity." in Murneek and Whyte, *Vernalization and Photoperiodism*, 145–57.

Went, F. W. "Why Does a Biologist Become a Biologist?" *American Biology Teacher* 24, no. 6 (October 1962), 405–25.

Went, F. W., and Lloyd Cosper. "Plant Growth under Controlled Conditions —VI: Comparison between Field and Air-Conditioned Greenhouse Culture of Tomatoes." *American Journal of Botany* 32 (1945), 643–54.

Went, F. W., and Munz, P. A. "A Long-Term Test for Seed Longevity." *Al Aliso* 2 (1949), 63–75.

Went, Frits. "The Earhart Plant Research Laboratory." *Engineering and Science* 15, no. 6 (1952), 14–18.

Went, Frits. "The Phytotron." *Engineering and Science* 12, no. 9 (1949), 3–6.

Went, Frits W. "Centenary of Schleiden's Textbook of Botany." *Chronica Botanica* 7 (1942), 145–48.

Went, Frits W. *The Experimental Control of Plant Growth: With Reference to the Earhart Plant Research Laboratory at the California Institute of Technology. Chronica Botanica*, no. 17. New York: Ronald Press Company, 1957.

Went, Frits W. "Orchids in My Life." In *Orchid Biology, Reviews and Perspective— V*, edited by Joseph Ardetti, 21–36. Portland, OR: Timber Press, 1990.

Went, Frits W., and the Editors of Time-Life Books. *The Plants*. Time-Life Books, 1963.

Werner, Earl E. "Ecological Experiments and a Research Program in Community Ecology." In *Experimental Ecology: Issues and Perspectives*, edited by William J. Resetarits Jr. and Joseph Bernardo, 3–26. Oxford: Oxford University Press, 1998.

Westad, Odd Arne. *The Global Cold War: Third World Interventions and the Making of Our Times*. Cambridge University Press, 2007.

Westwick, Peter. *Into the Black: JPL and the American Space Program, 1976–2004*. New Haven, CT: Yale University Press, 2007.

Westwick, Peter. *The National Labs: Science in an American System*. Cambridge, MA: Harvard University Press, 2003.

Wildman, S. G. "The Auxin-A, B Enigma: Scientific Fraud or Scientific Ineptitude?" *Plant Growth Regulation* 22 (1997), 37–68.

Wilson, Edmund. *To the Finland Station*. New York: New York Review Books, 2003.

Wilson, Edward O. *Naturalist*. Island Press, 1994.

Wisnioski, Matthew H. *Engineers for Change: Competing Vision of Technology in 1960s America*. Cambridge, MA: MIT Press, 2012.

Wisnioski, Matthew H. "'Liberal Education Has Failed': Reading Like an Engineer in 1960s America." *Technology and Culture* 50, no. 4 (2009), 753–82.

Woese, Carl R. "A New Biology for a New Century." *Microbiology and Molecular Biology Reviews* 68, no. 2 (2004), 173–86.

Wolf, Friedrich. "Technical Description." In Chouard and de Bilderling, *Phytotronique*, 12–16.

Wolfe, Audra J. *Competing with the Soviets: Science, Technology, and the State in Cold War America*. Baltimore: Johns Hopkins University Press, 2014.

Worboys, Michael. "Science and the Colonial Empire, 1895–1940." In *Science and Empire: Essays in Indian Context*, edited by Deepak Kumar, 13–27. Anamika Praashan, 1991.

Young, Robert J. C. *Postcolonialism: A Very Short Introduction*. Oxford: Oxford University Press, 2003.

Zachary, G. Pascal. *Endless Frontier: Vannevar Bush Engineer of the American Century*. New York: Free Press, 1997.

Zallen, Doris T. "Redrawing the Boundaries of Molecular Biology: The Case of Photosynthesis." *Journal of the History of Biology* 26 (1993), 65–87.

Zeevaart, Jan A. D. "My Journey from Horticulture to Plant Biology." *Annual Review of Plant Biology* 60 (2009), 1–19.

INDEX